ENVIRONMENT AND RESETTLEMENT POLITICS IN CHINA

Environment and Resettlement Politics in China

The Three Gorges Project

GØRILD HEGGELUND
The Fridtjof Nansen Institute, Norway

Routledge
Taylor & Francis Group
LONDON AND NEW YORK

First published 2004 by Ashgate Publishing

Published 2017 by Routledge
2 Park Square, Milton Park, Abingdon, Oxon OX14 4RN
711 Third Avenue, New York, NY 10017, USA

First issued in paperback 2017

Routledge is an imprint of the Taylor & Francis Group, an informa business

British Library Cataloguing in Publication Data
Heggelund, Gørild
Environment and resettlement politics in China : the Three Gorges Project. - (King's SOAS studies in development geography)
1.Three Gorges Water Control Project 2.Dams - Environmental aspects - China - Yangtze River Gorges 3.Dams - Political aspects - China - Yangtze River Gorges 4.Dams - Social aspects - China - Yangtze River Gorges 5.Water resources development - Environmental aspects - China - Yangtze River Valley 6.Water resources development - Political aspects - China - Yangtze River Valley 7. Water resources development - Social aspects - China - Yangtze River Valley 8.Environmental policy - China 9.Forced migration - China - Yangtze River Valley
I.Title II.King's College, London III.University of London. School of Oriental and African Studies
333.9'16215'09512

Library of Congress Cataloging-in-Publication Data
Heggelund, Gørild, 1961-
Environment and resettlement politics in China : the Three Gorges Project / Gørild Merethe Heggelund.
p. cm. -- (King's SOAS studies in development geography)
Based on the author's doctoral thesis.
Includes bibliographical references and index.
ISBN 0-7546-3859-6
1. Environmental policy--China. 2. China--Economic policy. 3. Migration, Internal--China. 4. San Xia shui li shu niu (China) I. Title. II. Series.

HC430.E5H44 2003
333.91'6215'0951212--dc22

2003062723

ISBN 13: 978-1-138-26647-6 (pbk)
ISBN 13: 978-0-7546-3859-9 (hbk)

Contents

List of Maps, Figures and Tables

Maps

Figures

Tables

Acknowledgements

The book is based upon a PhD dissertation that was completed in March 2002. I wish to acknowledge the Norwegian Research Council and the Fridtjof Nansen Institute for their financial support. For their assistance during the write-up of this book, I would like to sincerely thank the following individuals: Dr. Eduard Vermeer, Head of the Documentation and Research Centre for Contemporary China, Sinological Institute, Leiden University, the Netherlands. His valuable comments have been important for the book, as well as his insight into the environmental, rural and political affairs in China. Professor Steinar Andresen, Senior Research Fellow at FNI and Professor of Political Science, Institute of Political Science, University of Oslo, has provided invaluable guidance. His sound advice regarding the book structure and contents has been crucial. Dr. Michael M. Cernea, the World Bank, for taking the time to read parts of the dissertation, for giving valuable advice regarding the Impoverishment Risks and Reconstruction model and for interesting e-mail conversations.

Most important of all were my Chinese friends and colleagues who have assisted me in different ways during this study either by arranging meetings, providing material, or discussing the project with me. In particular I would like to mention Dr. Chen Guojie, Chengdu, CAS, and Dai Qing for enlightening me about the issues of the resettlement and the environment decision-making process. I also wish to acknowledge Dr. Pan Jiahua, CASS for his valuable assistance and good advice during these years of writing the dissertation, and Dr. Hu Tao, SEPA, for valuable advice regarding China's environment. Furthermore, I am indebted to each of my interviewees for taking the time to answer my numerous questions and for giving me a deeper understanding of the issues in question. I would have liked to acknowledge each of you here, but some persons have asked for anonymity due to the controversy surrounding this project. I also wish to express my deep gratitude to Maryanne Rygg for enthusiastic help with the editing, language checks and formatting of the document. I am grateful to Zhu Rongfa, an external member of the FNI staff, who provided valuable assistance in the writing process by regularly sending newspaper articles about China's environmental developments.

In closing, I wish to thank my parents for their never-ending encouragement and care, and for giving their support twenty years ago when I told them that I wanted to go to China to study Chinese. Thanks to my husband Quan Gang who has given me moral support during the period of writing, and for our discussions about Chinese politics! Last but not least, this work is dedicated to my dear children Gard, Einar and Sofia, who have had to endure a tired, short-tempered and sometimes exhausted mother. You are the sunshine of my life.

Lysaker,
Gørild M. Heggelund

List of Abbreviations

CAS	Chinese Academy of Sciences
CASS	Chinese Academy of Social Sciences
CC	Central Committee
CCP	Chinese Communist Party
CCTV	China Central Television
CPPCC	Chinese People's Political Consultative Conference
CTGPN	China Three Gorges Project News
CWRC (YRWRC)	Changjiang Water Resources Commission (or Yangtze River Water Resources Commission)
CYTGPDC	China Yangtze Three Gorges Project Development Corporation
EIA	Environmental Impact Assessment
EPB	Environmental Protection Bureau
IRR	Impoverishment Risks and Reconstruction
MOFTEC	Ministry of Foreign Trade and Economic Co-operation
MOST	Ministry of Science and Technology
MWR	Ministry of Water Resources
MWREP	Ministry of Water Resources and Electric Power
NEPA	National Environmental Protection Agency
NEPB	National Environmental Protection Bureau
NPC	National People's Congress
PRC	People's Republic of China
SC	State Council
SDPC	State Development Planning Commission
SDRC	State Development and Reform Commission
SEPA	State Environmental Protection Administration
SETC	State Economic and Trade Commission
SOEs	State-Owned Enterprises
SPC	State Planning Commission (now SDPC)
SSTC	State Science and Technology Commission (now MOST)

TGPCC	Three Gorges Project Construction Committee
TGPDC	Three Gorges Project Development Corporation
TVEs	Township and Village Enterprises
USBR	United States Bureau of Reclamation
WCD	World Commission on Dams
YRWRC	Yangtze River Water Resources Commission
YVPO	Yangtze Valley Planning Office
YVWRPB	Yangtze Valley Water Resources Protection Bureau (Changjiang liuyu shuiziyuan baohuju)

Currency equivalents:

Date: 14.03.02
US$ 100 = RMB (Renminbi) 828

Measurements:

1 mu = 0.165 acre
1 *mu* = 0.0667 ha
1 *mu* = 667 m2
1 *jin* = 1/2 kilogram

Glossary of Chinese Terms

Romanization	English equivalent	Chinese characters
Anzhi hao yimin	Settle the displaced people well	安置好移民
Ban de chu, wen de zhu, zhubu neng zhifu	Move out, settle down to a stable life, and gradually become rich	搬得出，稳得住，逐步能致富
Banfa	Measures	办法
Baobanshi yimin anzhi fangzhen	An 'all-arrangement' plan	包办式移民安置方针
Bingku	Reservoirs that do not function according to standard	病库
Chaicao shan	Firewood hills	柴草山
Changjiang ji Sanxia shengtai huanjing jiance wang	Yangtze and the Three Gorges Ecology and Environment Monitoring Network	长江暨三峡生态环境监测网
Changjiang liuyu shuiziyuan baohuju	Yangtze Valley Water Resources Protection Bureau	长江流域水资源保护局
Chengjiao nongcun	Outskirts of towns	城郊农村
Da lian gangtie	To smelt iron and steel	大炼钢铁
Di er ci tudi chengbao	The second round of contract (for land)	第二次土地承包
Diaomin	Unruly people	刁民
Difang	Locale	地方
Dixia jingji	Underground economy	地下经济
Dongqian renkou	People to be resettled	动迁人口
Dongtai (shuzi)	Dynamic figure (includes inflation)	动态数字

Romanization	English equivalent	Chinese characters
Doufu zha gongcheng	*doufu* 'scum' projects (bad quality)	豆腐渣工程
Duikou zhiyuan	Counterpart support	对口支援
Erci yimin	'Secondary migrants/relocatees'	二次移民
Fagui	Regulations	法规
Faze	Regulations	法则
Fazhi	Rule by law	法治
Feinongye hukou	Non-agricultural household	非农业人口
Fei zhengfu zuzhi	Non-governmental organisation	非政府组织
Fen	Unit of area equal to 0.1 mu	分
Fensan anzhi	Scattered resettlement	分散安置
Fenzhi chi fan	Split-the-food (projects)	分着吃饭
Ganju	Citrus; tangerines and oranges	柑橘
Ganjudi	Citrus land	柑橘地
Gengdi	Cultivated land	耕地
Gonggong wushui gou	Public sewage channel	公共污水沟
Gongmin shehui	Civil society	公民社会
Guanxi	Relations	关系
Guihua	Set long term goals	规划
Guizhang	Decrees	规章
Guojia jingji tizhi gaige bangongshi	The Office for Restructuring the Economy	国家经济体制改革办公室
Handi	Non-irrigated land	旱地
Heise shouru	Black income	黑色收入
Hetandi	Flood land/riverside land	河滩地

Romanization	English equivalent	Chinese characters
Houshe	Mouthpiece	喉舌
Hukou	Permanent residence	户口
Huanbao qunzhong tuanti	Environmental mass organisation	环保群众团体
Huanjing guanli	Environmental management	环境管理
Jianding xinxin, kaituo jinqu, wei shixian "jiuwu" huanjing baohu mubiao er fendou	Strengthen confidence, and continue to forge ahead in order to struggle to realise the environmental protection goals of the Ninth Five-Year-plan	坚定信心 开拓进取 为实现 九五 环境保护目标而奋斗
Jiangli yu chufa	Award and penalty	奖励与处罚
Jianjie yanmo	Indirect inundation	间接淹没
Jihua	Concrete project planning	计划
Jingtai shuzi	Static figure (does not include inflation)	静态数字
Jinqian anzhi	Settle nearby	近迁安置
Jiudi houkao	On the spot resettlement, i.e., in the vicinity of their old homes; pushed up the hills	就地后靠
Jiujin houkao	In the neighbourhood	就近后靠
Jiujin houkao anzhi	In the vicinity of their former homes	就近后靠安置
Jizhong anzhi	Settlement in groups	集中安置
Jueding	Decisions	决定
Jueyi	Resolutions	决议
Kaifaxing yimin fangzhen	Developmental resettlement scheme	开发型移民方针
Kouliangtian	Grain ration field	口粮田

Romanization	English equivalent	Chinese characters
Kuaikuai	A 'piece' (refers to the territorial authority among governing bodies)	块块
Liang kong qu	Control of acid rain and SO_2 in two regions	两控区
Lindi	Forest land	林地
Lingdao guanxi	Leadership relations	领导关系
Lingyang	Adopt	领养
Mingling	Orders	命令
Minjian zuzhi	Popular/non-governmental organisation	民间组织
Mu	Traditional unit of area (0,0667 ha)	亩
Nongcun renkou	Rural population	农村人口
Renzhi	Rule by man	人治
San da fa	Three great cuttings (of trees)	三大伐
San he	Three rivers (the Huai, Hai and Liao)	三河
San hu	Three lakes (Tai, Chao and Dianchi)	三湖
Santongshi	Three simultaneouses policy	三同时
Sanxia gongcheng	Three Gorges project	三峡工程
Sanxia gongcheng jianshe weiyuanhui	Three Gorges Project Construction Committee	三峡工程建设委员会
Sanxia gongcheng yimin kaifaju	Three Gorges Project Resettlement and Development Bureau	三峡工程移民开发局

Romanization	English equivalent	Chinese characters
Shangmian pai renwu, xiamian chou ren qu, zhong shu bantian hou, sihuo quan bu zhi	The authorities assign tasks, subordinates whip people to go, after having planted trees for a while, whether they live or die we do not know	上面派人物，下面抽人去，种树半天后，死活全不知
Shanqu nongcun	Mountain areas	山区农村
Shehui tuanti	Social organisations	社会团体
Sheji	Design	设计
Shengchan anzhi fei	Production resettlement fund	生产安置费
Shenghuo anzhi fei	Livelihood resettlement fund	生活安置费
Shengtai yu huanjing baohu xietiao xiaozu	Co-ordination Small Group for the Ecology and the Environmental	生态与环境保护协调小组
Shengtai yu huanjing lunzheng zhuanjia zu	The Ecology and Environmental Verification expert group	生态与环境论证专家组
Shi	Clan	氏
Shidian yimin	Trial resettlement	试点移民
Shigong	Construction	施工
Shiqian, shizhong, shihou	Before, during and after the event is implemented	事前，事中，时后
Shixing	For trial implementation	实行
Shuiping hen di	Low level	水平很低
Shuitian	Paddy fields	水田
Suanyu kongzhiqu he eryang hualiu wuran kongzhiqu huafen fangan	Plan for control of acid rain and SO_2 problems in two regions	酸雨控制区和二氧化硫污染控制区划分方案

Romanization	English equivalent	Chinese characters
Suzhi di	Translates directly as 'low quality'; points to low education	素质低
Tiaotiao	A 'line' (refers to the vertical system among governing bodies)	条条
Tiaozheng	Adjustments	调整
Tongbao	Administrative circulars	通报
Tongzhi	Notifications	通知
Toumingdu	Transparency	透明度
Touqin kaoyou	To be resettled through relatives or friends	投亲靠友
Tuigeng huanlin	Restore land to forest or grassland	推耕还林
Tuitian huanhu	Restore land to lakes	推田还湖
Tuzhi mofen	Make things look better than they are	涂脂抹粉
Waiqian	Resettlement out of the reservoir area	外迁
Wanshan	Improvement	完善
Weifa	Break the law	违法
Weigui	Break regulations/rules	违规
Weiji	Break discipline	违纪
Wushi, shi shenme jiu shi shenme	Deals with concrete matters	务实，是什么就是什么
Xian juece, hou lunzheng	Make the decision first, and then verify	先决策，后论证
Xibu da kaifa	Development of the Western region	西部大开发
Xitong	System	系统

Romanization	English equivalent	Chinese characters
Xuqing paihun	Retain clear water (in the dry season) and flush out the silt (in the flood season)	蓄清排浑
Yewu guanxi	Professional/business relations	业务关系
Yicixing peichang	'Lump sum' type compensation	一次性赔偿
Yimincheng	Relocatee towns	移民城
Yimincun	Resettlement villages	移民村
Yiminlou	Resettlement buildings	移民楼
Yimin houqi fuchi jijin	Later Stage Support Fund	移民后期扶持基金
Yimin zijin shiyong de guanli he jiandu	Supervision and management of the usage of the resettlement funds	移民资金使用的管理和监督
Yi nongye wei jichu	Taking agriculture as a basis	以农业为基础
Yinggai tuichu huichang	Must leave the meeting	应该推出会场
Yong da jiangyou de qian mai le cu	Use the soya sauce money to buy vinegar (i.e. spend resettlement funds on unintended projects)	用打酱油的钱买了醋
Yuandi	Garden plot	园地
Yuanqian anzhi	Resettlement at a distance	远迁安置
Yu fang wei zhu	Focus on prevention	预防为主
Yulun jiandu	Supervision by public opinion	舆论监督
Yutang	Fish ponds	鱼塘
Zanxing banfa	Provisional measures	暂行办法
Zhengzhi genju	Power base	政治根据

Romanization	**English equivalent**	**Chinese characters**
Zhengzhe lingdao kexue jishu	Politics take charge of science and technology	政治领导科学技术
Zhijie yanmo	Directly affected by reservoir inundation	直接淹没
Zhishi	Directives	指示
Zhongda qingkuang rang renmin zhidao	Inform the public about important events	重大情况让人民知道
Zhongda wenti jing renmin taolun	Discuss important events with the people	重大问题经人民讨论
Zhong gongcheng, qing anzhi	Emphasis put on construction of the project, rather than on resettlement	重工程，轻安置
Zhongguo Changjiang Sanxia gongcheng kaifa zonggongsi	The China Yangtze River Three Gorges Project Development Corporation	中国长江三峡工程开发总公司
Zhongyang	The Centre	中央
Zhongyang tongyi lingdao, fensheng fuze, xian wei jichu	The central authorities' unified leadership, responsibility according to province, and taking counties as a basis	中央统一领导，分省负责，县为基础
Zhubu	Gradually	逐步
Zhu Rongji, mei you ta jiu bu zhidao you shenme yangzi	If it were not for Zhu Rongji, it would be impossible to know the state of affairs [for the project]	朱镕基，没有他就不知道有什么样子
Zichan jieji ziyouhua	Bourgeois liberalism	资产阶级自由化
Zuo sixiang gongzuo	Convince (peasants to move)	作思想工作

Chapter 1

Introduction

The topic of the book is the Three Gorges dam project (*Sanxia gongcheng*),[1] which is currently being constructed on the Yangtze River in China. Few dams in the world are as well known. The controversy surrounding the project has contributed to international campaigns against the dam as well as articles and books about the project both inside and outside China. This book attempts to contribute to increased knowledge and understanding of the Chinese domestic resettlement and environment issues related to the dam, and portrays the official policymaking for these issues as well as the academic discourse in China.

Background

The National People's Congress approved the dam project in 1992. Construction of the dam began in 1994, and its completion is scheduled in 2009. The project was approved after decades of debate among bureaucrats, scientists and journalists. The Three Gorges dam is part of China's development and modernisation efforts. The purpose of the Three Gorges project is electricity production, flood control, and improved navigational facilities, which may be beneficial for the population living in the areas surrounding the Yangtze River. Nevertheless, the dam is not perceived by everyone as a symbol of national development, and arguments against constructing the dam have been numerous. Opponents state that the perceived benefits will not materialise, and the negative impact may be greater than anticipated. One of the issues of contention is the main topic of this book: the resettlement issue. The Three Gorges project will displace approximately 1.13-1.2 million people according to Chinese official figures.[2] Reservoir resettlement involving such a large number of people is unprecedented in China, as well as in the world. Furthermore, the official figure is disputed by opponents of the project, who believe that the number of people to be resettled is much higher (1.4 to almost 2 million).[3] In addition, there are potential socio-economic problems that are

---

1 The term *sanxia* means three gorges. The Yangtze River cuts through three gorges on its way to Shanghai, known as the *Sanxia* (Three Gorges): Qutangxia, Wuxia and Xilingxia. The English translation, Three Gorges, will be used in this book.

2 The official figure sometimes varies between 1.13 and 1.2 million people. 1.13 million will be used as reference in the book as it is used most frequently.

3 See 'Li Rui da, Dai Qing wen' (Li Rui answers Dai Qing's questions) in Dai (ed. 1989), p. 58; Dai (1998a); and Li, Heming (2000).

associated with the resettling of such a large population. The opposition is also related to the mixed previous experience with dam resettlement in China, where many of the resettled people still live in poverty. Moreover, while Chinese experts and officials anticipate that the Three Gorges dam and resettlement will have a certain environmental impact on the area, they disagree on the scale of the impact. The dam itself may influence the area in a number of ways, and water pollution is expected to increase when the fast-flowing river becomes a reservoir. Likewise, the resettlement of the large population is expected to have great impact on the environment in the reservoir area, i.e. the region that is affected by the reservoir, and aggravate the existing environmental problems.[4] The Three Gorges reservoir area has a fragile eco-system, as it is densely populated and agricultural practices have created serious erosion problems. The environment issue and environmental policymaking will receive special attention in the book due to its close link to the resettlement issue.

Objectives of the Book

The book focuses on the three following objectives: the dynamics in the decision-making process seen in relation to the changes in Chinese society, the Chinese discussion about the resettlement and environment issues and the validity of the analytical approach used. These three are all interrelated and to some extent overlapping, as they concern the changes in Chinese society and how these have affected the resettlement policymaking for the project.

Dynamics in the Decision-making Process in Light of Changes in the Chinese Society

The book aims to analyse the decision-making process for the resettlement of the Three Gorges dam project, and will link the development of the Three Gorges project with developments in the Chinese society at large. Chinese society is undergoing rapid change, including changes in decision-making. The book will shed light on how these changes may have influenced the resettlement policymaking for the Three Gorges project. The major players in the policymaking process will be introduced, and their role in the process discussed. The discussion within China about the resettlement policy involves mainly the leadership,[5] bureaucracy, some intellectuals, scientists and the media, and is not a grassroots debate.

---

4 The land area of the reservoir area is 54,200 km^2, of which 74 percent is mountainous region, 22 percent are hills and 4 percent are flatland. The *anzhiqu* (the settlement area), the area where the resettled population is to move to, is 12,300 km^2. Zhu and Zhao (eds. 1996), pp. 7-8.

5 The term leadership in this book points to the leadership of CCP, State Council and the Three Gorges Project Construction Committee (TGPCC), depending upon the context.

Thus, understanding the politics and bureaucracy for the project and the interaction between these actors is necessary, as well as gaining knowledge of the relations within the Three Gorges project bureaucracy and their impact on the resettlement policy. Numerous studies have focused on actual environmental problems in China, and research of the implementation of environmental policy in China has been carried out.[6] Fewer studies concern the relationship between resettlement and environmental policymaking. By analysing the resettlement and environmental policymaking for the dam project in relation to the developments that have taken place in China in general, this study will contribute to increased insight and understanding of the resettlement and environmental policymaking in China.

The Chinese Discussion

The Chinese Three Gorges resettlement discussion is not well known outside China, although there is great interest and concern for the dam project in a number of countries. A goal of the book is therefore to shed light upon the Chinese resettlement perspectives for this dam project. It is important to let the Chinese discussion on the resettlement issue come to the fore, as the discussion outside of China is mainly dominated by Western[7]/non-Chinese perspectives. Furthermore, the social and economic costs of the resettlement issue are beginning to receive increased attention in China, which makes a study of these issues relevant. Chinese perspectives on resettlement are diverse, as in other countries, and the intention of this book is not to represent one voice in the discussion. Rather, it intends to shed light on as well as analyse the discourse about the Three Gorges resettlement in an unbiased fashion.

Little research is being carried out about the resettlement policymaking for the dam project, in particular in relation to the environment., In addition to a few long-term analyses for the dam project,[8] the majority of the existing information about the dam includes articles that are published in Western journals and newspapers regarding the different aspects of the Three Gorges project at large, including the resettlement and environment issues. Reporting about the resettlement in relation to the dam project often focuses on specific isolated incidents or topics, and does

---

6 See for instance Jahiel (1994); and Sinkule (1993). One important study about environmental policymaking is Ross (1988).

7 The West is defined as the countries of Europe and North America, in contrast to those of Asia, such as China; the culture and civilisations of these regions as opposed to that of the Orient. *The New Shorter Oxford English Dictionary* (1993). In addition, Western perceptions would also be found in some countries that are not situated in the Western hemisphere, such as Australia and New Zealand.

8 One example of recent research about the Three Gorges resettlement is the Ph.D. dissertation by Li Heming, *Population Displacement and Resettlement in the Three Gorges Reservoir Area of the Yangtze River Central China,* a valuable study about the conditions for the resettled rural population in their new home areas. Li, Heming (2000). Yin (1996) is another example that discusses the historical process leading up to the final approval of the dam.

not discuss policy developments. This may have to do with the nature of the articles, limited space in newspaper articles, or that specific focus has a certain news value. Few articles penetrate deeply into the political processes surrounding the Three Gorges dam, and few look at the diversified discussion that is going on in China. This book aims to provide insight into both official policy and unofficial viewpoints (scientific and others) in order to obtain a deeper understanding of the project's resettlement developments, as well as how these developments relate to the developments in Chinese society at large. It will also seek to identify measures that are being carried out by Chinese authorities, and the basis for these decisions. Moreover, views of the scientific world and the nature of the problems in the reservoir area in relation to the resettlement and the environment will be examined.

The Research Question: Why Resettlement Policy Change?

Before the Three Gorges dam was approved, trial resettlement had been going on for some years in order to find the best resettlement plan for the dam project. The trial resettlement was perceived as successful by Chinese authorities, and was the basis for the 'real' resettlement that was initiated following the approval of the dam in 1992. The final resettlement plan that was intended to guide the large-scale resettlement for the project stressed the reliance on agriculture for the rural population, as well as resettlement of the rural population in the vicinity of their old homes (*jiudi houkao*). In the first period following approval, the resettlement proceeded smoothly according to Chinese authorities. In 1998 the second phase (1998-2003) of the construction was initiated, during which a great challenge would be to resettle a large group of people in a short period. In May 1999, a departure from the original resettlement policy, i.e. policy change, was introduced, indicating that the process may have been less smooth than portrayed in the state media.

This book takes as its starting point the resettlement policy change that was introduced in May 1999, announced by Premier Zhu Rongji. The policy change itself is only the culmination of policymaking carried out behind the scenes; thus, the developments beforehand that made the change possible as well as those afterwards are the topics that make up the study. The resettlement policy change involves moving a large number of people completely out of the reservoir area that originally were intended to be resettled within the reservoir area. The stated reason for the policy change was the weak environmental and ecological condition in the area. Consequently, the environmental policymaking for the dam project receives much attention in this book, as it is directly linked to the resettlement change. The resettlement policy change mainly affects the rural population;[9] the study therefore does not include the urban population.

---

9 The rural population (*nongcun renkou*) is defined as the agricultural population that relies completely on the land for a living, as well as the population engaged in other activities in addition to agriculture.

Since Chinese authorities have given the impression that they prefer to resettle people within the reservoir area, why and how did the resettlement policy change take place? By studying the decision-making process for the resettlement policy, this book attempts to describe and explain the causes for policy change. Since the project was approved in 1992, important developments in Chinese society have appeared. It is necessary to link these general developments with the Three Gorges resettlement. Thus, the research question concerns the concrete policy change seen in relation to developments in the Three Gorges project as well as the dynamics in Chinese society at large. The developments over time may have made the resettlement policy change possible, and three possible explanatory factors for the policy change will be discussed in the book:

1. *The environment*—The stated reason for the resettlement policy change is the environmental degradation in the reservoir area, and the lack of capacity to settle the large population. In order to understand how the environment could become the decisive factor for the resettlement policy change, it is necessary to go through: a) China's environmental developments in general, and b) the specific environmental developments for the Three Gorges project.
2. *The resettlement* problems—A number of problems have emerged in the resettlement process. One assumption in this book is that these may have been important in bringing about the resettlement policy change, and examples from the resettlement implementation process are presented and discussed. In addition, a debate about the possible consequences for the resettled population as well as the environment has been going on in academic journals that have portrayed actual and potential problems. The actual and anticipated problems of the resettlement process will be discussed as possible factors for policy change.
3. *A changing Chinese society*—China is a changing society where information and science are increasingly important for decision-making. One assumption in this book is that this ongoing change has been important for the resettlement policy change, and that information from the academic world may have increasingly been taken into consideration. The book attempts to give an analysis of the factors in Chinese society that may have enabled the policy change, such as the change of leadership (the Premier and the Three Gorges leadership); the new leadership and the interaction with the state media; increased reliance upon think-tanks in decision-making; and the importance of diversified backgrounds of leaders.

Thus, one may say that the explanatory factors are on three different levels: the stated and immediate cause (the environment), the assumed and unsaid cause (the problems of implementing resettlement) and a slightly more indirect cause (developments of the Chinese society) that constitute the framework in which the decisions are made. These factors will be further introduced below, in relation to the chapters of the book.

Organisation of the Book

In order to give the reader some background to the very long debate about this dam project, the book will start, in Chapter 2, with an empirical introduction to the Three Gorges project. The chapter includes facts about the project, the purposes for constructing the dam, the history from the time it was introduced in 1919 until present, as well as the present status of the project. The long history of the project as well as the many attempts to have the project approved signify the complexity of the project, and its political significance is commented on in this chapter. This historical chapter gives the necessary background for the resettlement and environment discussion, as the chapter introduces the issues involved in the debate, the many attempts to launch the project, and a number of the actors and interest groups.

The analytical framework for the book is described in Chapter 3. The chapter begins with an introduction to three theoretical approaches that have been employed in studies of policymaking in general. It continues with an introduction to approaches that have been applied in studying policymaking in China. Finally, an introduction of the approach applied in this book is given, i.e. the fragmented authoritarianism approach. This approach assumes that decision-making in China involves building consensus among equal units, and in the process of reaching consensus, bargaining takes place. This approach has been applied on the Three Gorges project decision-making process more than a decade ago, and was an important tool to explain the ways of the Chinese bureaucracy before the dam was approved. It is therefore useful to apply this approach in the study of the resettlement policymaking, in order to assess its validity more than a decade later.

The following five chapters, 4-8, shed light on the resettlement policymaking and implementation and the environmental policymaking for the dam project. Chapter 4 is an introduction to China's general resettlement history and experience, and is intended as background for the discussion in the resettlement Chapter 5. It begins with an introduction to the international dam building debate in order to link the Three Gorges project to the international discussion. It continues with an introduction to the resettlement experience in China in general, and looks into the background for the development of the Three Gorges resettlement policy. The problematic resettlement history has prompted Chinese leaders to think of alternative methods for the Three Gorges project, and the new resettlement scheme for the Three Gorges project is also presented here.

Chapter 5 is an analysis of the resettlement policy change. It begins with the introduction of the resettlement policy change and why it is interesting to study the change. References will be made to implementation problems (actual or potential) of the Three Gorges resettlement process in order to illustrate the potential impact these have had on the policy change issue. The stated reason for the policy change was the environment; a brief introduction is therefore given to the environmental condition in the Three Gorges reservoir area. The chapter continues with a presentation of the resettlement issue, with references to World Bank reports about dam building in China as well as specific dam projects that demonstrate a positive picture of the resettlement process in China in general. The chapter questions if this is the case in this specific Three Gorges project. In order to structure the discussion of the resettlement problems, the Impoverishment Risks and Reconstruction model (IRR) is applied. The model consists of eight impoverishment risks and reconstruction points, and a selected number of points that were deemed relevant for the topics discussed have been applied in this book. After having analysed the different resettlement problems according to selected points of the IRR model, a discussion of the Chinese government's response to the problems is included. Comments on the new resettlement regulations from 2001 and some major changes are given. Finally, the relevance of the IRR model in connection with the Three Gorges project is commented on.

Chapter 6 is intended as background for the discussion in Chapter 7 regarding the Three Gorges project environmental developments. Chapter 6 gives an overview of China's environmental developments from 1972 until the 2001, with a greater focus on the last two decades. Increased focus on environmental issues in China in general has positively influenced the environmental focus on the Three Gorges dam. Selected elements of environmental developments are discussed in order to give a general overview of the major environmental developments in China in order to illustrate the increased attention paid to this issue. This background chapter looks into the institutional settings, laws and regulations, the increased focus on environmental issues in later years and public participation as well as the development of environmental NGOs. Certain topics will be highlighted, such as water pollution, floods, forests and erosion, since they are particularly relevant in the environmental discussion for the dam project.

Chapter 7 is the specific analysis of how the environmental focus in general in China has manifested itself in the Three Gorges project. The chapter discusses the environmental decision-making process as well as the role of some of the major actors in this process. Furthermore, the chapter looks into the increased understanding of the linkage between the lack of environmental capacity in the reservoir area and the resettlement of the large population. In order to understand the gradual development of the link between resettlement and environment, this chapter discusses the environmental development for the dam from 1972 until 2001. One way to assess the increased importance of the environment, has been the growing emphasis on environmental feasibility studies, which is discussed in this chapter. Furthermore, the media's coverage of these issues is an important indicator of policy developments, which will also be discussed. In addition, a discussion will take place about a few selected institutions involved, as well as the scientific focus before and after approval of the project.

The changing Chinese society and the changing shape of decision-making is a third way to explain how the resettlement policy change could take place. Chapter 8 attempts to illustrate that the change of Premier and leadership in the Three Gorges project, the role of Zhu Rongji, and the interaction between the leadership, media and scholars have all been important factors in the process leading to resettlement policy change. It also discusses the increased importance of information as one factor that has made the Chinese leadership aware of the concrete as well as potential problems in the resettlement process. One assumption is that Zhu Rongji's leadership style has been important for the policy change. Examples of arguments and information from academic journals about the resettlement and environmental problems are given in order to illustrate the type of information which may be available for the leadership. These may be important sources of information in the decision-making process. The state media coverage of the dam project is an important indicator of the political attitude towards the dam project, and the change in reporting after Zhu became Premier is used to illustrate the effect of the leadership change on the Three Gorges resettlement policy. The assumption is that the factors described above are all interlinked and have been important for the policy change.

Chapter 9 is the conclusion chapter, which is divided into two parts. Part one is the summary and conclusion of all the chapters and the findings in the chapters. Part two contains comments and viewpoints regarding the findings in the book. These comments are divided into two sections. Section one concerns conclusions made about the analytical framework for the book and the value of the fragmented authoritarianism approach in relation to the Three Gorges resettlement and environmental policymaking. Section two discusses the following two issues: the level of importance paid to the environmental issue compared with the resettlement issue in the policymaking and the application of the IRR model in China.

Data and Methodology

The methodology followed has been to study Chinese language public materials regarding the resettlement policy and environmental developments for the project. The Three Gorges project is a controversial and political project in China, and obtaining materials other than official material about such a project is difficult. The resettlement issue in particular is a difficult topic to study, since resettlement is perceived as the key to success for the project. Furthermore, much prestige is involved and there are interests to be protected on all levels: leadership, bureaucracy and the scientific sphere. The main sources of information for this book have therefore been public reports, yearbooks, articles in newspapers and academic journals, and books. Since the Three Gorges dam is currently being constructed, an enormous amount of articles about the project are published in the media. These articles are mainly very positive and contain descriptions of the latest developments of the project, including resettlement and environment. The articles are useful for obtaining information about meetings convened for the dam, Chinese leaders' statements about the project, policymaking, policy changes and general project developments as well as official figures for the dam project. Nevertheless, it is necessary to supplement the official information with additional sources in order to obtain a broader picture of the situation. For this book, articles in academic journals have been a very important supplement that have provided additional and more diversified information about the resettlement and environmental processes. The articles, often written from a different angle than the newspaper articles that are more propaganda oriented, aim to discuss actual problems as well as measures to solve these problems. These articles may be critical towards the ongoing resettlement, and they often point to problematic issues. They appear to give a more realistic picture of problems and challenges than do the articles in the state media. The above mentioned materials have been supplemented by interviews of Chinese officials, academics and others. Furthermore, English language material is also applied, mainly for the discussion of the analytical approach and changes in Chinese society, as well as the IRR model.

The figures obtained in the materials used need to be commented on, as they often differ in various texts describing the same topic. This may be due to various definitions of terms, as discussed in Chapter 5 in relation to types of land being inundated. Sometimes official figures provided by authorities are low, while figures found in scholarly journals may be higher. This may be due to the figures having been adjusted over the years in some texts and not in others (but this is not indicated in the text), or that scholars have obtained different figures through research carried out on their own. Also, the authorities may for various reasons wish to use lower figures.[10] The different figures complicate the understanding of the situation, and it is necessary to be aware of the discrepancy between the figures that are applied in the different texts.

Below is an introduction of the principal materials applied in this book. The section is divided into three parts, Chinese-language materials, English-language materials and interviews.

Chinese-language Materials

The Chinese-language materials may be divided into different groups, such as yearbooks, reports, newspaper articles, articles in academic journals, research reports and books.

Yearbooks and reports One source of information is the *Zhongguo Sanxia Jianshe nianjian* (China Three Gorges Construction Yearbook), first published in 1995 by the Three Gorges project authorities. The *Yearbook* gives the official version of the developments in most of the areas linked to the construction of the dam project, including the resettlement and environment. Despite its very positive portrayal of the project developments, it is nevertheless useful as it contains speeches by China's leaders on various occasions, and laws and regulations that have been passed. The speeches, for instance, give an impression of topics of concern for Chinese authorities. Furthermore, the *Changjiang Sanxia gongcheng shengtai yu huanjing jiance gongbao 2000* (Bulletin on Ecological and Environmental Monitoring of the Three Gorges Project on the Yangtze River), which is the annual

10 Different figures exist for instance for agricultural land, such as the difference between the 1997 *Abstract of the First National Agricultural Census on China* and the 1997 *China Agricultural Yearbook*. I am grateful to Eduard Vermeer, Leiden University, for providing information about this topic.

environmental report for the dam project, has been useful for this book. The *Bulletin* is edited by the China Environmental Monitoring Centre under the State Environmental Protection Administration (SEPA), which approves this annual report. The contents of the *Bulletin* have been provided by a number of sources that participate in the environmental monitoring work for the dam, such as the various line ministries. The *Bulletin* provides some information about the environmental developments for the project, such as figures. However, it is quite general in its description of the environmentally related issues. A third source is the *Zhongguo huanjing nianjian* (China Environmental Yearbook), that gives an indication of the environmental developments in general in China, including some information about the developments in the Three Gorges project related to the environment.

Newspapers Important newspapers that were used in this study include *Renmin ribao* (the People's Daily), *Renmin ribao haiwai ban* (the People's Daily Overseas edition, also available in China), *Guangming ribao* (Enlightenment Daily), and *Jingji ribao* (Economic Daily). The *People's Daily* is named the mouthpiece (*houshe*) of the party, which indicates that important decisions made by the Chinese leadership would appear in this paper. Much of the information regarding the resettlement policy change has been found in this newspaper. In addition to the above newspapers, other newspapers such as *Beijing qingnian bao* (Beijing Youth Daily), *Beijing wanbao* (Beijing Evening News), *Zhongguo huanjingbao* (China Environmental News), etc., have been useful in providing information on certain issues of the dam. The Internet has simplified collection of information from China, and has been an important tool for obtaining information for this book. Information from press agencies has also been useful, such as the *Xinhua she* (New China Press), the official state press in China, and *Zhongxin she* (China News Agency), a press agency that focuses on informing the *huaqiao* (overseas Chinese) around the world.

Academic journals and research reports In addition to studying the official information regarding resettlement policymaking for the dam, it was necessary to supplement this information with views from the academic field. Therefore, articles from Chinese academic journals have been an equally important contribution to this book, for obtaining a clearer picture of the actual issues and challenges that are being discussed in the resettlement debate. One of the journals that has been important for the book are *Zhanlüe yu guanli* (Strategy and Management), a journal that has close links to the Chinese government and focuses on policy issues. Other journals are *Sichuan Sanxia xueyuan xuebao* (Journal of Sichuan Three-Gorges University), *Changjiang liuyu ziyuan yu huanjing*, (Resources and Environment in the Yangtze Basin), *Renkou yanjiu* (Population Research), and *Huanjing baohu* (Environmental Protection). The articles in these journals are compiled by academics (and others) who discuss policymaking, actual or potential problems for the resettlement and the environment. A few of the articles are openly critical of the resettlement process, while the majority point to problems and suggest counter-measures. Research reports by CAS, such as *Sanxia gongcheng dui shengtai yu huanjing de yingxiang ji duice yanjiu* (Research on the Three Gorges Project Impacts on the Ecology and the Environment and Counter Measures)[11] have also been useful, as they have given ideas about the research agenda and research results in addition to the research carried out by the project authorities.

Books Books published both on the mainland and in Hong Kong have been important sources of background information on developments and problems in Chinese society. These books have provided useful information regarding the political developments and the changes in society, and have been an important source of background regarding the Three Gorges decision-making. Some of the books are: He Qinglian (1998), *Xiandaihua de xianjing* (The Pitfalls of Modernisation), which includes discussions of the environmental problems in the country, linking poverty and environmental degradation. Furthermore, Zheng Yisheng and Qian Yihong (1998), *Shendu youhuan, Dangdai Zhongguo de kechixu fazhan wenti* (Profound Hardship, Sustainable Development Problems in Modern China), looks into the sustainable development challenges for China. A few books also discuss the Three Gorges project specifically, such as Ling, Zhijun (1998), *Zhongguo jingji gaige beiwanglu (1989-1997)* (A Memorandum Book of China's Economic Reform), who writes about the resettlement issue and the discussion. In addition, the Ma Jun (December 1999), *Zhongguo shui weiji* (China's Water Crisis) discusses the Yangtze River valley at length, including the Three Gorges project in relation to both water, floods, siltation and deforestation. Books that have been useful with regard to understanding developments for decision-making in China in general, and in particular regarding Zhu Rongji and his aides, are Xiao, Zhengqin (1999), *Zhu zongli zhinang qunying* (The Think Tank of Premier Zhu Rongji), and Zheng, Yongnian (2000), *Jiang Zhu zhi xia de* Zhongguo—gaige, *zhuanxing he tiaozhan* (China under Jiang and Zhu—Reform, Change of Pattern and Challenges).

English-language Sources

English-language sources include books on Chinese politics and decision making. One that has provided the analytical framework is Lieberthal, Kenneth G. and Michel Oksenberg (1988), *Policymaking in China: Leaders, Structures, and Processes*. This volume has been useful in the approach to studying the decision-making for the Three Gorges resettlement, as it was applied in analysing the decision-making for the project in general in the late 1980s. A second model that was applied in this book is the Impoverishment Risks and Reconstruction model as found in the Cernea, Michael M. and Christopher McDowell (eds. 2000), *Risks and Reconstruction: Experiences of Resettlers and Refugees* and Cernea, Michael M. (1997), 'The Risks and Reconstruction Model for Resettling Displaced Populations'. Books regarding Chinese society that have been important for the

11 Chen, Xu and Du (1995).

book are Li, Cheng (2001), *China's Leaders, the New Generation* and Fewsmith, Joseph (2001), *Elite Politics in Contemporary China*. In addition, *The China Quarterly* and *The China Journal* have provided important articles regarding resettlement and environmental issues as well as Chinese policymaking, which will be referred to throughout the book.

Interviews

Given the nature of this study, the primary source of information has been the written sources described above. Nevertheless, interviews have been an important source of information, in particular with regard to understanding the political processes surrounding the project and obtaining a more diverse view on the issues that are discussed, as well as confirming some of my own observations of the process. The majority of the interviews were carried out during the 1990s, and the interviewees included representatives from the Three Gorges resettlement bureaux on the national, provincial and local levels, the China Yangtze Three Gorges Project Development Corporation, Chongqing municipality, SEPA and Chongqing EPB, as well as officials in ministries and commissions, academics, intellectuals/journalists and peasants. The interviews are intended to be complementary to the above-mentioned written materials and were significant in obtaining a deeper understanding of the developments over time. The method applied for the interviews was a set of questions that were prepared in advance, in order to ask a number of the same questions to most interviewees. However, the interview method was qualitative; thus, the interviews were loosely structured and the author/interviewee was not bound by the questions. In most cases the conversations continued beyond the set of questions. The number of people participating in the interviews varied from one to several. I have chosen to retain the interviewees anonymity due to the controversy surrounding the Three Gorges project, unless the person interviewed explicitly has stated that references may be made. The interviews for this book were carried out over a period of 3-4 years.

In 1997, several interviews were carried out in Beijing over a period of several months. Interviewees were officials at the Resettlement bureau and academics from various institutions. A field trip in the reservoir area was taken the same year together with a Chinese academic, where I visited Yichang, Wanxian and Wanzhou,[12] Fengjie and Chongqing. Interviews were conducted with resettlement officials in these places. Peasants who had been resettled from Zigui to the Yichang area were also interviewed, as well as peasants who prepared to move in the Wanxian area. During this period (1994-97) I was working for the UN in Beijing with environment related issues,[13] and I often had an opportunity to discuss

12 Wanzhou was previously called Wanxian city. The area is since the late 1990s called Wanzhou kaifaqu (Wanzhou development district), which also includes Wanxian county.

13 During 1994-1995 I worked for UNIFEM (the United Nations Development Fund for Women) in relation to the Fourth World Conference on Women that was held in

the project with officials and academics in informal settings, which has been useful in achieving a deeper understanding of the environmental and resettlement issues of the dam project.

In 1999, two months of fieldwork were carried out in China. This is the period when the core interviews for the book were conducted with officials and academics at major institutions in Beijing as well as in the reservoir area and other cities. Places visited were Wanzhou, Chongqing, Yichang, Nanjing and Wuhan. Interviews during the trip to the Three Gorges were conducted with resettlement officials, persons from municipal governments, environmental bureaux and agricultural organisations, as well as journalists, peasants and academics.

In 2000, during a two-week visit to Beijing, I conducted follow-up interviews with academics, officials and journalists in Beijing.

To conclude, the interviews were important and a necessary supplement to the information obtained from the written sources. As expected, officials were careful about their statements, while academics seemed more open and frank about their own views. During the course of working on this book the author has had numerous conversations with Dai Qing. These conversations have been essential in shedding light on the issues being discussed for the project, as they have provided an unofficial and different version of these issues.

Beijing in 1995. From 1995-1997 I worked in the Energy and Environment division of UNDP (United Nations Development Programme).

Chapter 2

The Three Gorges Project: A Story of Development and Modernisation

The purpose of this chapter is to give a brief introduction to the Three Gorges project in general in order to provide a necessary background for the resettlement discussion.The chapter begins with a brief introduction of the facts about the project such as the main purposes for constructing the dam. Then follows a summary of the most important events of the dam project's history that describe and discuss the developments of the Three Gorges project until it was approved in 1992. The historical account is divided into several periods that describe the initial period when the dam was first proposed, the initiation of a debate under Mao, the frustrations for the dam proponents, the debate in the new economic reform period, the final period before the dam project was approved and the political significance of the project. A brief introduction is then given of the present state of the project.

Finally, the Three Gorges leadership and actors are introduced. One trait of this dam project is the number of actors involved. This section will provide an overview of the actors and interest groups that are relevant to the Three Gorges resettlement and environmental policymaking only, and will not include the entire number of ministries and organisations that are involved in this project.

Facts about the Three Gorges Project

The Three Gorges project is unique both in China and the world due to its great size. The purposes of the project are, power generation, flood control and improvement of navigational facilities. The mere facts about the Yangtze River, China's longest river,[14] underscores the challenges of this project: the river stretches 6300 km long from the Tibetan plateau in the west, and runs into the ocean by Shanghai in the east. The river has a historical record of flood catastrophes, and the flood problem still remains unresolved here as in other rivers in the country. Devastating floods that occur during the rainy season each year, create huge economic losses to the area.

---

14 The Chinese name for the Yangtze River is Changjiang, which means the Long River.

Map 2.1 The Three Gorges Area

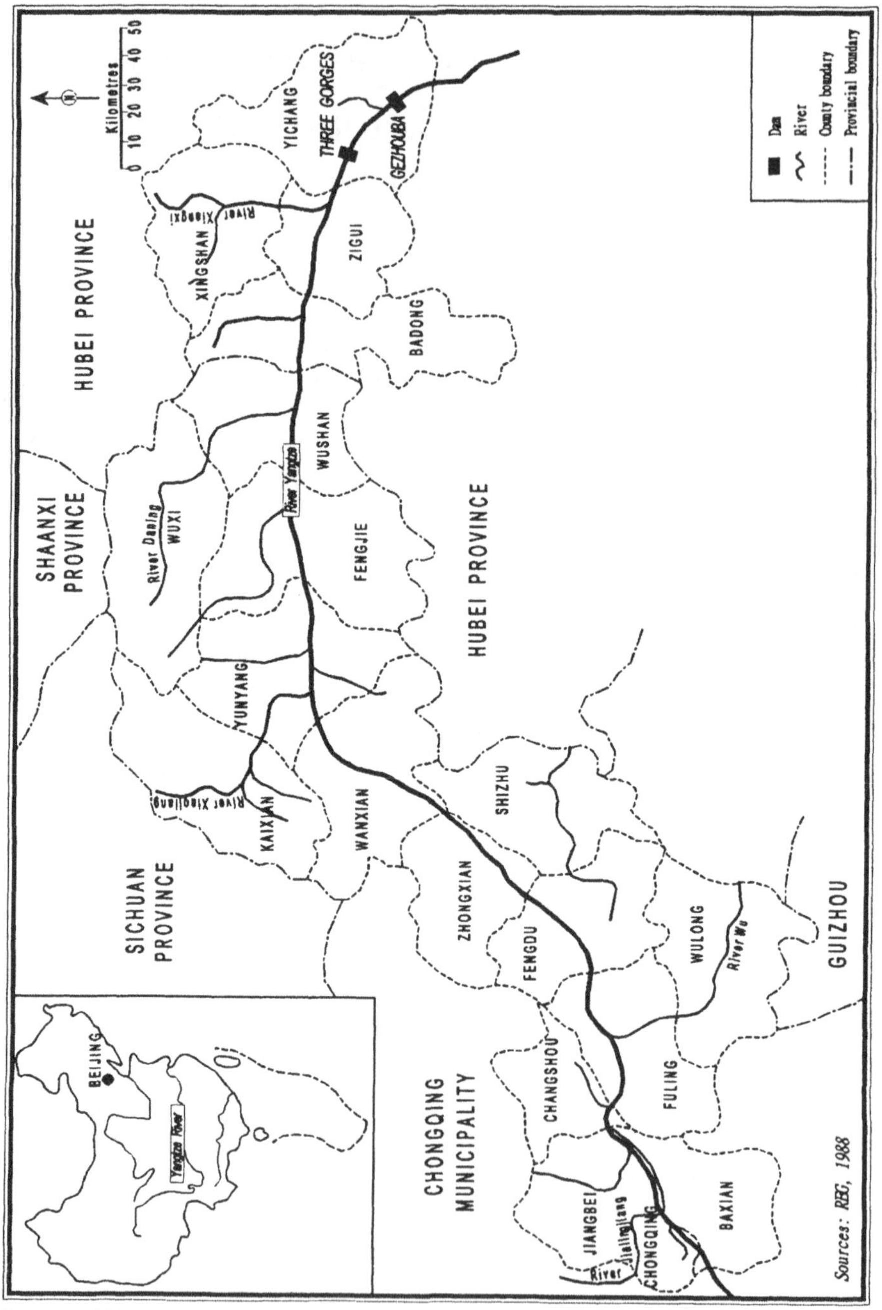

Source: Li Heming (2000), p. 111.

The Three Gorges project began official construction in 1994 and will be completed in 2009. The dam, situated at Sandouping near Yichang in Hubei province, stretches 1,983 metres across the river. The reservoir will be 600 km long and reach Chongqing municipality. Hubei is one of the main beneficiaries of the project with regard to flood protection and electricity. It is responsible for resettling 15 percent of the relocatees. Chongqing, formerly part of Sichuan province, became an administrative municipality directly under Beijing in 1997 in order to administer the Three Gorges resettlement;[15] it is responsible for relocating 85 percent of the population to be resettled. Chongqing then became China's largest municipality, with a population of approximately 30.59 million inhabitants, of which the agricultural population constitutes 79.93 percent and the non-agricultural population 20.07 percent.[16] The reservoir inundation will affect an area of 1084 km^2, of which 632 km^2 is land area.[17] The project is one of the largest in the world when it comes to the actual size of the dam, power production and the number of people that need to move. The large-scale resettlement of the Three Gorges project is undoubtedly the greatest challenge for the Chinese authorities. 20 counties and cities will be partly or entirely inundated by the reservoir.[18] Official figures of the number of people to be resettled is 846,200, of which 361,500 belong to the rural population and 484,700 to the non-agricultural population (cities and towns).[19] The total figure, which includes the population increase expected to take place in the area, is 1.13-1.2 million people. The dam will inundate 390,000 *mu* of cultivated land.[20]

---

15 Chongqing was declared China's fourth municipality with its own government by the State Council, and approved by the National People's Congress at their annual session in March 1997. Chongqing's role in the resettlement and the new administrative position was comfirmed by Wang Yunnong, Vice-Secretary General, Chongqing municipal people's government, interview no.18, Chongqing municipality 14 September, 1999. Before this Chongqing was one of five major river cities (Shanghai, Wuhu, Jiujiang, Wuhan and Yueyang) that were opened to foreign investment by Deng Xiaoping in 1992. Spence (1999), p. 710.

16 The figures for the population are 24.45 million and 6.14 million respectively. See Xibu da kaifa zhinan bianji weiyuan hui (ed. 2000), p. 341. The new municipality became an 82,000-square-kilometre region.

17 Changjiang shuili weiyuanhui (ed. The Yangtze River Water Conservancy Commission) (1997), p. 4. The remaining 452 km^2 is the Yangtze River and its tributaries.

18 These are Yichang, Zigui, Xingshan and Badong in Hubei province; Wushan, Wuxi, Fengjie, Yunyang, Kaixian (Kai county), Wanxian (Wan county), Wanzhou municipality, Zhongxian, Shizu, Fengdu, Fuling municipality, Wulong, Changshou, Jiangbei and Baxian, as well as part of the river reaches in Chongqing municipality. See Wang, Rushu (2000). See map 2.1.

19 Yangtze Valley Water Resources Protection Bureau, MWR and NEPA (eds. 1999), p. 42.

20 Xinhua she (7 March 2001). One *mu* is 0.0667 ha.

Flood Control

Harnessing the great rivers has been an important factor in the lives of Chinese for centuries, starting with the Great Yu, the semi-mythical founder of the pre-historical Xia dynasty, and the father of China's early water control system. Flood control still plays an important role in Chinese lives today. Due to the serious recurrent flood problems, improving the flood control system in the Yangtze River is one of the main purposes for constructing a dam in the Three Gorges. Presently, the 180 kilometre-long Jingjiang Dike in the middle reaches of the river is the critical component of the flood protection system in the Yangtze Valley. If the dike were to break during a flood, it could take the lives of 100,000 people and inundate major urban centres. In addition, there are designated flood overflow diversion areas[21] and low-lying areas along the river side of the main and secondary dikes, susceptible to flooding. The importance of the Yangtze River Valley is emphasised by the fact that it is the home to approximately 35 percent of the Chinese population, who produce about 40 percent of the country's agricultural output and 40 per cent of the industrial output.[22] Each year large numbers of people are engaged in flood control work due to the serious flood problem in the valley. For flood protection purposes, the Three Gorges project involves constructing a dam with a height of 185 metres. At completion, the normal water storage level will be 175 metres. The water level in the reservoir is to be reduced during the flood season to make room for flood water, and the flood restriction level during this season would be 145 metres. The total water storage capacity of the reservoir would be 39.3 billion cubic metres, of which 22.15 billion would be for flood control purposes.

Power Generation

When the Three Gorges project was debated in the 1950s, the power part of the project plans played a minor role.[23] However, China's modernisation process has created a need for increased electricity production, and today the generation of electricity has become the top priority of the dam. The power to be generated at the Three Gorges is assumed to greatly contribute to the electricity needs of Central and Eastern China. The annual power production at the Three Gorges would reach

---

21 The largest diversion areas are the Jingjiang diversion area, with one-sixth of the flood storage capacity of the Three Gorges reservoir, used only during the severe 1954 floods; the Honghu diversion area with over half the of the storage capacity of the Three Gorges reservoir, developed in 1974 and yet to be used; the Dongting and Xilianghu diversion areas. Barber and Ryder (eds. 1990), p. 90.

22 Lieberthal and Oksenberg (1988), p. 271.

23 In 1956 one of the main supporters of the dam wrote a lengthy report that barely mentioned power generation. However, the situation changed and according to Dai (ed. 1989), p. 200, the hydroelectric aspect in the late 1980s accounted for 75 per cent of the total budget, flood control for 21 per cent and navigation only 4 percent.

Table 2.1 Floods in the Yangtze River

Flood year	Number of people killed	Number of people homeless/affected	Inundated land
1870[24]	–	–	–
1931	145,500	14 million	51.51 million *mu* (3.43 million ha)
1935	142,000	–	–
1954	33,000	18.8 million	47.5 million *mu* (3.17 million ha)
1981	1,358	1 million 20 million	17.56 million *mu* (1.17 million ha)
1991	2,628	3.2 million	–
1998	3,656	14 million	206 million *mu* (13.8 million ha)

Sources: Chen (1999g), Dai (ed. 1989), UNEP (2001), Spence (1999), Van Slyke (1988).

Table 2.2 Type of Floods

Year	Type of floods
1870	A 1000-year flood; Flood in the upper reaches which caused flood in the middle and lower reaches
1954	Floods in both upper and lower reaches
1981	Flood in the upper reaches; little or no flood in the middle and lower reaches
1980, 1983, 1988, 1991	Flood due to rain in the lower and middle reaches
1998	Long term floods in the entire Yangtze River (from June to August)

Source: Chen (1999g).

84 terawatt-hours (TWh) annually, with an installed capacity of 18,200 MW.[25] Chinese authorities also emphasise the environmental benefits of the project, as the electricity production would substitute 40 million tonnes of coal annually.[26]

24 Apparently no data exists from this thousand-year flood (occurs once in a thousand years). Chen Guojie states that this flood has occurred only once, however, no details of casualties or inundated land are provided. See Chen (1999g), p. 2.

25 Bureau of Resettlement and Development (1995), p. 3.

26 China relies mainly on coal (75 percent of all energy), and consumes approximately 1.4 billion tonnes annually. See Zhang (2000). The major coal resources of the country are mainly situated in the north-east provinces of Shanxi, Shaanxi, the autonomous regions of Inner Mongolia and Ningxia, while the hydropower potentials are found in the south-western provinces of Sichuan and Yunnan. 40 percent of the railway cargo capacity is earmarked for coal transport.

Altogether 26 turbines will be installed for power generation, beginning in 2003 when 14 of the turbines have been installed.

Navigation

The Yangtze River is called the 'Golden Waterway', a name which characterises the river's importance as a transport system between east and west.[27] It carries 80 percent of China's internal waterborne traffic.[28] Navigation on the Yangtze River, including the Three Gorges, has always been dangerous due to shoals and currents in the river. Navigation by engine power was not attempted until the last years of the 19th century, and the first passage was made in 1899.[29] River crafts became powerful enough to traffic the river a decade later. However, passage on the river continued in the traditional manner. After 1949, the Chinese Government has made great efforts to improve the navigational facilities on the river by blasting and dredging the river, and by establishing more advanced beacon and buoy systems.

The Three Gorges project will improve the navigational facilities by making the river deeper and expanding the width. In some parts of the river, such as a 200 kilometre-long stretch from Fengjie to Yichang the width of the river is reduced to about 100 metres at its narrowest during the dry season, which makes navigation difficult since the largest amounts of dangerous shoals are found in this section. On the other hand, during the flood season the water level can rise by ten metres per day in this section. Existing rapids and shallows will be inundated by the reservoir and raise the safety on the river in general. The transport capacity in the reservoir will increase from the present 10 million tonnes annually (one-way capacity) to 50 million tonnes per year.[30] The construction of the dam with a 175-metre high water level would ensure the passage of 10,000 tonne fleets up to Chongqing. One ship lift and five sluices are planned for the project; the ship lift is intended for smaller vessels, and the sluices for large boats. Improving the navigation facilities of the Yangtze River is expected to bring economic development to the areas along the river.

Cost of the Project

The project authorities differentiate between the *jingtai* figures (based on 1993 figures, which do not take inflation into consideration) and *dongtai* figures (including inflation). The *jingtai* figure is 90 billion yuan (approximately US$ 11

27 The river flows from Qinghai, where it is known as the Tongtian River; on its course eastwards it becomes the Jinsha River when it reaches the Xizang (Tibetan)-Sichuan border. The Jinsha river flows through the north of Yunnan, and then into Sichuan. In south Sichuan it becomes the Changjiang (Yangtze) River. From here it flows through Hubei, Anhui, Jiangsu and enters the east China Sea at Shanghai.

28 Van Slyke (1988), p. 16.

29 Ibid. p. 20.

30 Sanxia gongcheng lunzheng lingdao xiaozu bangongshi (1988), p. 96.

billion). Of this, a total of 40 billion yuan (US$ 4.8 billion) will be used on the resettlement work, which means approximately 40,000 yuan per capita. The *dongtai* figure for the project is estimated to reach 200 billion yuan (US$ 24 billion) on completion in 2009. Project authorities outline the following sources for obtaining the 200 billion yuan needed for the construction of the dam and the resettlement:[31] i) The Three Gorges Construction Fund, which will collect funds, approximately 100 billion yuan, from other power stations in developed areas of the country; ii) 10 billion yuan from the power production at the Gezhouba dam;[32] iii) When the 14 turbines go into operation in 2003 at the Three Gorges dam, power production will provide 40 billion yuan; iv) Loans over a ten-year period from the China Development Bank are estimated to reach 30 billion yuan (interest rates were not yet fixed);[33] v) 1 billion in Government bonds.[34] These funding sources will provide 181 billion yuan, and the remaining 19 billion yuan will come from sources such as issuance of Government bonds, international bonds, loans, export credit (in connection with purchase of foreign equipment), etc.

In sum, the Chinese authoritiees see three major benefits from the Three Gorges dam: electricity production, flood control and improvement of the navigational faciltites on the Yangtze River. The cost of the project is estimated to reach 200 billion yuan (US$ 24 billion).

The History of the Three Gorges Project

The Three Gorges project has a history of more than 80 years, and the account is divided into the following different periods: the initial period; a debate begins; frustrations for dam proponents; new economic reform period; leading to project approval; the political significance of the project; the present state of the project; and an introduction of project leadership and actors.

---

31 Based on information by Xu Keda, Director, International Cooperation Department, China Yangtze Three Gorges Project Development Corporation, Yichang April 1997.

32 This dam project was begun in 1970, and completed in 1989. It is regarded as the forerunner to the Three Gorges project, and was meant to be the testing ground for some of the technical issues of the much larger Three Gorges project. See 'The history of the Three Gorges project', for further details.

33 US investment banks, such as the Morgan Stanley, have been criticised by Probe International and International Rivers Network for helping to underwrite bonds for China Development Bank. Its largest loan commitment is the China Yangtze Three Gorges Project Development Corporation (*Zhongguo Changjiang Sanxia gongcheng kaifa zonggongsi*). See Knight (31 May, 2001); and Behn (6 April, 2000).

34 Government bonds for 1 billion were sold out in five days time in 1997. In 1999, the China Yangtze Three Gorges Project Development Corporation planned to raise five billion yuan through a stock market floatation in order to finance the second phase of the project. Wang (23 February, 1999).

Initial Period

The project was first suggested in 1919 when Sun Yat-sen, the first president of the Republic of China proposed to build a dam in the Three Gorges. The earliest reference to building such a dam is found in Sun's *Plan to Develop the Industry* in the section on 'Improvement of Navigable Rivers and Canals'. Sun proposed 'to build a dam to store water, so that ships can sail upstream against the flow of the river, and the river can be harnessed for electric power'.[35] Under the Nationalistic government, deliberations on the construction of a dam in the Three Gorges area continued. A number of foreign experts were also invited to visit the area and to write reports on the feasibility of a dam in the gorges. One of them, John L. Savage, chief design engineer at the US Bureau of Reclamation (USBR) visited the Three Gorges area in 1944 and wrote the 'Preliminary Report on the Yangtze River Three Gorges Plan',[36] formally presenting a concrete plan to construct a dam. This was the beginning of a number of plans to construct a dam in the Gorges, and a formal co-operation between the USBR and the Nationalist Government's Resources Committee was initiated in 1946. Nevertheless, the Nationalist Government decided to terminate the design work on the project in May 1947.[37]

After the People's Republic of China was established (1949), the discussions about this project resumed in the mid-1950s. The nation had started paying attention to the flood problems of the country in the beginning of the 1950s, including the Yangtze River Valley where serious floods had occurred in 1931 and 1949. In 1953, Mao Zedong emphasised the importance of studying the Three Gorges project for flood control purposes. The 1954 flood, one of the worst floods in the last century, took the lives of more than thirty thousand people and rendered millions of people homeless (see Table 2.1). China had recently completed the construction of the Jingjiang flood diversion project, which made an important contribution to limiting damage from the flood in 1954. However, it was considered unrealistic to rely on the Jingjiang section alone for flood control in areas along the middle and lower reaches of the river, and they would also have to depend upon the construction of dams. The Central Committee (CC) of China's Communist Party (CCP) therefore decided to commence planning for dam projects on the river, and Soviet experts arrived in 1955 to assist China in this work.[38]

A Debate Begins

Two individuals emerged in the technical debate on the Three Gorges that followed the renewed focus on the construction of a dam on the Yangtze River. One was the

---

35 Luk and Whitney (eds. 1993), p. 43.

36 Ibid., p 44.

37 Ibid. This might have been due to the national situation at the time, when the National Party and the Communist Party were struggling against each other for power in China.

38 Luk and Whitney (eds. 1993), p. 48.

head of the Yangtze Valley Planning Office (YVPO),[39] Lin Yishan, an ardent supporter of the project. Lin Yishan had no formal training in water management. However, he became a driving force behind the water conservancy work on the Yangtze and helped shape the development of the YVPO (established in 1956). By the 1960s, the YVPO had a university of its own and employed thousands of people who hoped to make a career by participating in the preparations for and the construction of the Three Gorges dam. The second person, Li Rui, became a vice minister of Electric Power in the mid-50s and was head of the Ministry's General Bureau of Hydroelectric Construction.[40] He represented the hydropower interests in opposition to the project. When the Ministry of Water Resources merged with the Electric Power Ministry, he continued to serve as vice-minister until he was purged at the Lushan meeting in 1959.[41] He has continued to be an opponent to the Three Gorges dam until the present day.

The first disagreement between Lin Yishan and Li Rui occurred in 1954, when Lin Yishan personally briefed Mao on the catastrophic flood and the potential of a dam in the Three Gorges. Li Rui at the time was in a group of officials on a study trip to the Soviet Union[42] when they received a telegram stating that the Ministry of Water Resources was considering the construction of a dam in the Three Gorges. Li Rui quickly responded to the telegram, and stated that taking China's economic situation into consideration it would be too heavy a burden on the nation to engage in such a project. Li Rui stated that the proposal by Lin Yishan and the YVPO was too focused on flood control only, and he believed that the Three Gorges project should be a multi-purpose project with hydropower and navigation taken into consideration. As for the timing of the project, the Ministry of Electric Power suggested a staged approach to the flood control of the Yangtze, by strengthening dikes along the river as well as constructing smaller dams on the main tributaries. The economic and technical conditions were the main reasons for this; the additional time would provide time for the nation to acquire technical know-how and funding for the multi-purpose Three Gorges project.

During the Hundred Flowers Campaign in 1956-1957, a campaign where Mao had invited intellectuals to engage in public discussion over various issues, a debate seemed to evolve in the Chinese media about the project.[43] The debate

---

39 Since 1990 is known as the Yangtze River Water Resources Commission (YRWRC). YVPO will be used throughout the text for the period up to the name change in the early 1990s.

40 Lieberthal and Oksenberg (1988), p. 293.

41 Li Rui was purged as a rightist in the Anti-Rightist Struggle, an attack against the 'bourgeois rightists' that began in 1957. The Central Committee of the CCP held its 8th plenum at Lushan, Jiangxi province, in 1959.

42 Li Rui was in the Soviet Union together with Liu Lanbo, the person in charge of electric power in the Ministry of Fuel Industry; Liu was Minister of Electric Power from 1979-1981.

43 An article about the Three Gorges project had been published in the People's Daily on 1 September 1956, sounding as if the project were about to be launched any moment. Li Rui wrote a response that was accepted by the paper, but stopped by Premier Zhou,

revealed differences regarding the technical issues such as the large locks (that had never been built before), problems of navigation during the construction period, and developing the world's largest generators. Engineers critical of the project criticised the proponents for not taking the technical issues seriously.[44] In 1957 the Hundred Flowers turned into an Anti-Rightist Campaign, which prompted an 'atmosphere of scorn for technical issues',[45] and provided an opportunity for the proponents to strike back.

In 1958 several meetings took place that were important for the dam project. At a 1958 Central Committee meeting in Nanning, Mao was briefed by both Lin Yishan and Li Rui regarding the pros and cons of the project. Mao had initiated this meeting since he had continued to receive conflicting views on the project.[46] Upon completing the reporting on the Three Gorges, Mao ordered the two to compile their opinions in writing to be discussed at a meeting later that month. At that meeting, Mao and others criticised Lin Yishan for being unrealistic, and agreed with Li's opinions that the time was not ripe for the Three Gorges project.[47] Following the Nanning meeting, Mao appointed Premier Zhou Enlai to take personal charge of the planning of the Yangtze River Basin, including the Three Gorges project.[48] Premier Zhou then organised a boat trip for key advisors and decision makers involved, including representatives from the relevant State Council ministries,[49] provinces and municipalities. Zhou's report from the boat trip

who did not wish to have a public debate on the topic in the paper at that time. However, the project was eventually discussed in magazines and newspapers, and Li Rui states that a public debate about the Three Gorges began in 1956. Li (1985), pp. 1-2.

44 Articles by proponents of the project were often less technical and advocated the Three Gorges role in the historic effort of controlling floods in the Yangtze River, and the benefits to be derived from the project without addressing the issues raised by the critics. Lieberthal and Oksenberg (1988), p. 296.

45 Ibid.

46 Li Rui stressed that Lin Yishan spoke for two hours, while Li spoke only for half an hour. Lin stressed the importance of the Three Gorges project in the flood protection work of the Yangtze River, and claimed that, looking at the history of floods in the past two thousand years, there was an increasing tendency for serious flooding in the area. The Three Gorges would be able to solve these problems, and Lin suggested construction to begin in 1963. The suggested dam height was 200-210 metres. Li Rui's brief response stressed that the Three Gorges project alone cannot solve the flood problems of the Yangtze River. His main point concerned the recent strengthening of the dikes and development of flood diversion areas, which were meant to withstand a flood such as the flood disaster in 1954. Thus, he didn't see the need to embark on such an ambitious project, both financially and technically. Li (1985), p. 8.

47 These are Li Rui's own words. Other critics of Lin Yishan were Bo Yibo and Hu Qiaomu. Li, (1985), p. 9.

48 Luk and Whitney (eds. 1993), p. 50.

49 It is interesting to note that the Ministry of Water Resources and Ministry of Electric Power were merged into one ministry in early February 1958. The reason for this is not clear. However, Lieberthal and Oksenberg (1988), p. 299, states that the merger occurred at a time when there was a favourable attitude towards developing

and a meeting held in Chongqing,[50] provided the base for the discussion that followed at the Chengdu Conference of high officials on 28 February. At this Conference, Zhou presented a plan for the Yangtze River Valley including the Three Gorges project. The Conference adopted Zhou's resolution, 'Opinion on the Three Gorges Project and Planning of the Changjiang River Basin', and concluded that the construction of the Three Gorges project was necessary and feasible; however, not until the following seven contradictory relationships had been worked out: long term and short term; main channel and tributaries; upper, middle and lower reaches; large, medium and small scale; flood control, power generation, irrigation and drainage, and transport; hydropower and thermal power; and power generation versus power consumption.[51]

Following the decisions made at the Chengdu Conference, the first Three Gorges Scientific Conference was held in June 1958 to review the technical elements of the project. About two hundred units and 10,000 scientific personnel participated in research on the Three Gorges project over the next two years, indicating the importance placed on hydropower projects in that period. The first feasibility study of the project was begun in 1958 and completed in 1960. Further work on the Three Gorges project came to a halt due to the national economic crisis caused by the Great Leap Forward.[52] The Danjiangkou dam on the Han River, which was decided upon at the Chengdu Conference in 1958, was hastily built with inadequate planning and substandard concrete, which led to delays and reconstruction.

Frustrations for Dam Proponents

According to Lieberthal and Oksenberg (1988), the period from 1960 to 1978 proved to be a frustrating time for the dam proponents. 'They adjusted their arguments and approaches to the prevailing national priorities', in order for the

---

hydropower projects and several major projects were being planned for China's major rivers.

50 Luk and Whitney, (1993), p. 50, mentions that Li Rui at the Chongqing meeting had praised the Three Gorges project, and found it feasible. Li Rui himself states in Dai (ed. 1989), p. 60, that if he had once agreed to the project (150-metre dam), it was because he had heard that the project had been approved by the Central Committee (CC) of the Party, which made it difficult to hold a diverging opinion. In the historical account in Li (1985), p. 10, the Chongqing meeting is only briefly mentioned.

51 The request at the Chengdu meeting to look into these seven relationships is described in a number of books; Tian, Lin and Ling (1987), p. 18; Lieberthal and Oksenberg (1988), p. 300; Luk and Whitney (1993), p. 51; Chen (ed. 1992), p. 12.

52 The Great Leap was an attempt to achieve an economic revolution by increasing the production of steel and grain in a short period. The grain production figures were disastrously over-inflated, and the local backyard furnaces produced inferior quality steel. The result was a human and an environmental disaster. In the period from 1959 to 1962, 20 million or more people died from the effects of the Great Leap. See Spence (1999), pp. 550-53. For the environmental effects, see chapter 6.

Three Gorges project to be included in each subsequent Five-Year plan.[53] Li (1985) goes quickly through the period of the 1960s and 1970s, giving examples of the attempts by the YVPO to launch the project.[54] Although it does not give an analysis of the reasons why the project was rejected, it nevertheless underscores the difficulties the dam proponents had in launching the project in that period. It is also clear that leaders such as Mao Zedong and Premier Zhou Enlai were personally engaged in the Three Gorges project, and were very much involved in the decision-making for the project.[55] National conditions in the 1960s and 1970s did not permit the planning of such a huge and complicated project. The failed policy of the Great Leap that brought several years of starvation, hindered the project from moving ahead in the early 1960s. Moreover, the Cultural Revolution (1966-1976) produced such chaos, especially from 1966-1968, that the necessary organisational, financial and technical conditions for commencing the construction of a dam in the Three Gorges were all in abeyance.

In 1964, Mao decided that the third Five-Year plan (1966-1970) must provide resources for a 'third line' of industries in the south-western parts of the country in case of war with the United States. The YVPO and Lin Yishan therefore submitted a 'Report on Design Issues on the Yangtze Three Gorges' directly to Mao and the Party centre, suggesting a staged construction of the Three Gorges with different dam heights to be initiated in 1968. This would permit installation and operation of turbines, and ensure income from electrical generation before the completion of the final dam. However, due to the deteriorating relationship with the Soviet Union in that period, the dam was perceived as a potential target if a war were to break out. As two bloody border clashes between China and the Soviet Union had taken place in March 1969, Mao concluded that the war preparations made planning for the Three Gorges dam impossible.

In 1970 Lin Yishan and the YVPO submitted a report to upgrade the Three Gorges project from the *guihua* stage to the *jihua* stage.[56] Later that year the

---

53 Lieberthal and Oksenberg (1988), p. 305.

54 Premier Zhou seemingly was not in a hurry to launch the dam project, and had stated in 1961 during a visit by a Vietnamese hydropower delegation that the Three Gorges project would be realised, but that it was not a matter of urgency and the construction should not be rushed. Li (1985), p. 10.

55 In 1963 when Mao was briefed about the project, he commented that aspects regarding the sediment problem must be clear before launching the project. In 1969 he stated that due to the war preparations it would be impossible to plan for the Three Gorges project. In 1970 Premier Zhou proposed in a letter to Mao to first construct the Gezhouba dam, and wait until the PRC had developed anti-aircraft capability to think about the Three Gorges. Construction of the Gezhouba began in 1970. Li (1985), pp. 10-12. According to Li Rui in Dai (ed. 1989), p. 55, the Gezhouba project was dramatically rushed into operation on Mao's birthday, with the consequence that the construction was halted and a new design drawn up.

56 Lieberthal and Oksenberg (1988), describe five major stages of decisional process for construction projects: *guihua* (set long term goals), *jihua* (concrete project planning), sheji (design), *shigong* (construction) and utilisation. Three Gorges was accepted as a

Ministry of Water Resources and Electric Power (MWREP) suggested to first start the construction of another dam project on the Yangtze River, the Gezhouba dam. Premier Zhou agreed with the MWREP to postpone the construction of the Three Gorges until the international situation and China's anti-aircraft capability permitted it. Furthermore, it was desirable to start the Gezhouba construction first since the project was regarded as the testing ground for resolving the technical issues in connection with the much larger Three Gorges project. Gezhouba was hastily initiated on Mao Zedong's birthday in 1970 and was scheduled to be completed in 1975, but experienced delays that moved the completion date to 1989. Due to the enormous cost overruns and a prolonged construction period of the Gezhouba project, the Three Gorges project was not included in the fifth Five-Year Plan (1976-1980).[57]

New Economic Reform Period

In the 1970s and 1980s several major changes occurred that would have great impact on the decision-making process of the Three Gorges project. The YVPO lost its direct reporting relationship to the Premier Zhou Enlai, and became subordinate to the MWREP. Zhou Enlai and Mao Zedong, who had both played important decisive roles in the decision-making of the project, passed away in 1976. Furthermore, the reform movement in the 1980s diminished the role for personal leadership and created a greater regard for consensus in decision-making. Lieberthal and Oksenberg state that the system allows top leaders to exert enormous pressure to have a project endorsed. Nevertheless, with the reforms, top leaders became more reluctant to employ Mao's methods to overcome bureaucratic resistance, such as ideological campaigns and broadsides and deification of top leaders. Top leaders now prefer active co-operation of all major parties concerned to prevent long delays of a project.[58]

With the need for electricity in relation to the new policy adopted in 1978, the so-called Four Modernisations of agriculture, industry, national defence, and science and technology, the Three Gorges project received serious attention in the preparation of the sixth Five-Year plan (1981-1985). Lieberthal and Oksenberg speculate that there was a connection between the attention paid to the Three Gorges project and the division of the MWREP into two separate ministries again in 1979.[59] Traditionally, the electric power part of the ministry had been against the Three Gorges dam and favoured smaller dams on the tributaries. The electric

---

guihua plan in the mid-1950s, and the *jihua* stage was formally completed in April 1984. However, the Three Gorges project is so large and controversial that several of these stages have overlapped simultaneously. p. 287-90.

57 The construction took 18 years instead of five and costs reached RMB 4.8 billion instead of 1.35 billion. Heggelund (1993), p. 100, and Dai (ed. 1989), p. 55. 26,000 were resettled. McCully (1996), p. 322.

58 Lieberthal and Oksenberg (1988), p. 23.

59 The water resources and power ministries were merged twice, in 1958 and in 1982.

power part had also been the more dominant part of the two, which could be due to the fact that the electric power part had exercised operational control over subordinate units and produced revenues for the state budget, while the resources part played more of a policy role.[60] Splitting the ministry into two would enable the water resources part to push the project.

In April 1979,[61] a meeting was held in the State Council regarding the Three Gorges project, where it was decided that Lin Yishan and the YVPO should organise a site selection meeting, which eventually was held in May 1979. At the meeting different opinions were expressed; several CC leaders had expressed support,[62] while a number of participants expressed in private their fear of rushing the project, as well as the need for additional feasibility studies.[63] The two potential sites were visited by professors, specialists and technicians in a number of fields (water resources, hydropower, environmental protection, shipping and communications, etc.).[64] During the early 1980s, several groups of American experts visited China, including a group representing the US Department of Interior's Bureau of Reclamation (USBR). They made appropriate site visits, and held detailed talks with the YVPO and MWREP regarding the Three Gorges project. This group was in general positive to the plans of constructing a dam in the Gorges, but advised the Chinese to carry out feasibility studies that would indicate costs and benefits of the various aspects of the dam. According to Lieberthal and Oksenberg (1988), foreign ideas were for the first time since the 1950s starting to play a role in the Chinese debate on the project. For instance, the Ministry of Finance was asked for the first time in 1982 to prepare a financial feasibility study of the project, which most likely was a result of the USBR's recommendations.

In 1982, the MWREP was reconstituted out of the Ministries of Water Resources and Electric Power, as part of State Council's reorganisation of several ministries. As two separate ministries, the conflicts would have reached the State Council for settlement. Lieberthal and Oksenberg (1988) state that Chinese familiar with the process consider the merger as an attempt by the State Council to force the water resources and electric power parts to resolve their disputes internally and present a consensus view to the State Council. The new minister of the merged MWREP, Qian Zhengying, was an ardent supporter of the Three

---

60 Lieberthal and Oksenberg (1988), p. 96, footnote 63. Lieberthal and Oksenberg also assert that had electric power been the main purpose of the dam, it is likely that the project would have ended up in the Electric Power Ministry, which probably would have rebuked the dam proposal. Ibid., p. 311, footnote 117.

61 Lin Yishan and the YVPO had submitted a report on the project to the State Council in April 1979, where they once again proposed an early construction. Li (1985), p. 12.

62 Ibid. Li Rui does not mention the name of the other leaders expressing support for the project.

63 They did not dare to express this openly at the meeting in fear of being referred to as opponents of the project. Ibid.

64 The two sites were Taipingxi and Sandouping; Sandouping was chosen as the site.

Gorges dam.[65] Li Peng, then promoted to Vice Premier,[66] received responsibility for the energy portfolio at the State Council level, which was formalised by the State Council in 1984, and he was instructed to set up a 'leadership small group' to oversee the preparatory work for the Three Gorges project. The following year, in 1983, a conference was convened by the State Planning Commission to evaluate a feasibility study carried out by the YVPO the year before. A group of 350 experts participated and discussed the various aspects of the dam such as dam height, flood control capacity, electrical generation, relocation of people, environment, silting, etc. A formal decision was made to support a 150-metre high dam, although several experts preferred a higher dam.[67] Li Rui praises this meeting as the first in the history of the Three Gorges debate to thoroughly discuss all aspects of the dam project. Furthermore, it was the first meeting where such a great number of scientists participated.[68] The conference concluded that some parts of the YVPO report were satisfactory, such as matters related to the direct costs of constructing the dam, while other elements required additional research, such as relocation of people and enterprises, environmental impact, technical issues, etc. The Conference recommended that the State Council adopt the YVPO report in principle; which it did in 1984 (a dam of 175 metres with maximum pool level of 150). Furthermore, the State Planning Commission agreed that the time had come to build a dam in the Three Gorges. The YVPO then carried out several of the studies that were recommended at the 1983 conference, such as the potential environmental effects of the dam.

The positive results of the conference in 1983 and the developments in 1984 (adoption by the SC) seemed to indicate that the Three Gorges project would be included in the Seventh Five Year plan (1986-1990) and that construction would begin in that period. The State Council and Politburo approved the 150-metre plan in 1984; this provoked a heated debate and concern in Chongqing.[69] If the 150-metre plan were to be implemented, the accumulation of sedimentation at the backwater region (where Chongqing is situated) would gradually create obstacles

65 Qian Zhengying has been a leader of the pro dam faction for the Three Gorges project, and for the past few decades has been a leading person in developing water-engineering programmes in China. Qian Zhengying was formerly the minister of Water Resources and a minister of MWREP before it was split up in 1979. Following this second merger in 1982, the MWREP was later split into two ministries again in 1988. Qian is now a vice-chairperson of the Chinese People's Political Consultative Conference (CPPCC). Qian has reportedly said that 'If I am kept from damming the Three Gorges, I will not rest even in death'. Dai (1994), p. 248.

66 Li Peng, the adopted son of the late Premier Zhou Enlai, had been the former Vice Minister of Electric Power, and also Vice Minister of MWREP. Li has background in hydropower; he was educated at the Moscow power institute in the 1950s. Lieberthal and Oksenberg (1988), p. 319.

67 This was not the final decision on the dam height, as it has ended up at 185 metres, with a normal pool level of 175 metres.

68 Li (1985), p. 14.

69 Yin (1996), p. 499.

to navigation. Chongqing then suggested a 175-metre plan that was delivered to the leadership. When Deng Xiaoping learned about the background for the suggestion he preferred the 175-metre plan.[70] In May 1986 the CCP Central Committee and the State Council issued a document to annul the Preparation Group of Sanxia Province and instead established the Three Gorges Economic Development Office. In addition, the Chinese economy had encountered major problems, as the too rapid growth had created shortages of capital and reduced foreign reserves substantially. Consequently, the Three Gorges project was not included in the Seventh Five Year plan.

Leading to Project Approval

Preparations for the project went nevertheless ahead. In June 1986 the Central Committee of the Chinese Communist Party and the State Council instructed the MWREP to do a third feasibility study of the project. A leading group was established under the leadership of the MWREP, and 400 experts were engaged in a two-year comprehensive study of the Three Gorges project. The group was divided into smaller groups according to topic.[71] The leading group concluded in 1988 that the Three Gorges project was a feasible project, and that construction should start as early as possible. Only ten experts did not sign their names to the report, indicating that they disagreed with the conclusion. At the same time, the Canadian government had signed an agreement with the Chinese government to do a feasibility study of the project with a grant from the Canadian International Development Agency (CIDA). Their report was completed in 1989, recommending the construction of a dam in the Three Gorges.

Within China, opposition to the project became more vocal as the preparations for the dam project intensified. Representatives from the Chinese People's Political Consultative Committee (CPPCC) conducted a field trip in 1986 to the Three Gorges area and held meetings with ministries and bureaux in a number of cities to be affected by the project to gather information about the project. They submitted a report to the State Council recommending that the project should not proceed in the near future. Furthermore, prominent citizens, journalists, scientists and bureaucrats gathered to criticise the plans for the project, and tried to influence the decision-makers by publishing books and articles about the project. Dai Qing, a former journalist in the *Guangming ribao* (Enlightenment Daily), edited *Changjiang, Changjiang*, which was one of the important books holding divergent views on the

---

70 The CCP secretary of Chongqing went to Beijing to present the appeal to Hu Yaobang, Party Secretary General and Zhao Ziyang, Premier, who took it to Deng Xiaoping. Yin (1996), p. 503.

71 Minister Qian Zhengying set up the leading group with herself as the leader. Under it, altogether 14 groups were organised to study the following topics: flood control, power generation, navigation, ecology and environment, hydrology, geology, earthquake, sedimentation, resettlement, electrical and mechanical equipment, comprehensive planning, construction, investment and economy.

project that was published in early 1989.[72] The massive opposition towards the dam project within China and the lobbying among the decision makers had a definite impact on the Three Gorges process. In April 1989, the representatives to the National People's Congress (NPC) were scheduled to vote over the fate of the Three Gorges project, and whether China should go ahead with such an ambitious project. The project was not voted on at that session of the NPC, as 270 representatives to the NPC who opposed the project requested that the project be postponed until the next century. These protests resulted in the then vice premier Yao Yilin announcing that the decision on the project would be postponed until 1995. Nevertheless, after the student demonstrations were brutally ended on 4 June 1989, opposition against the project died down and preparations for the project continued in 1990. In July 1990, the State Council held a meeting to discuss the Three Gorges project verification report in order to listen to the different opinions of the expert groups. It was decided at the meeting to establish the Investigation Committee under the State Council, whose task was to study the feasibility report that had been compiled by the experts. Ten expert groups were established to go over the work, which consisted of 163 experts as well as relevant representatives from the various ministries. They completed their work in May 1991, and reports from the different expert groups were submitted to the State Council. On 17 September 1991, the State Council studied the report of the Investigation Committee, agreed to construct the Three Gorges project and decided to put it on the agenda of the National People's Congress.[73]

At the same time as preparations were going on behind the scenes, the media also started paving the way for project endorsement by publishing articles that strongly favoured a dam at the Three Gorges. In addition, the serious floods during the summer of 1991 once again put the need for flood control in the Three Gorges area to the foreground. A national water resources campaign was initiated, and the project proponents once again were given a pretext to call out for the immediate launching of the Three Gorges project. In August 1991, Wan Li, the chairman of the Standing Committee of the NPC, proposed a go-ahead for the project at the NPC's upcoming session and proposed that the project should be listed among the key projects for the ten year development programme (1991-2000).[74]

Li Peng, then Premier, on 16 March 1992 submitted the approval by the State Council, *Guowuyuan guanyu tiqing shenyi xingjian Changjiang Sanxia gongcheng de yi'an* (Proposal submitted from the State Council regarding the review of the

72 Dai (ed. 1989). Dai Qing was at the time a prominent journalist in the newspaper. Dai Qing is the adopted daughter of one of China's eight famous marshals, Ye Jianying. She has written a series of controversial articles on the history of the CCP. After 4 June, 1989 she publicly resigned from the CCP, and was kept in detention for approximately 10 months. Dai Qing stated to this author that this was due to her involvement in the publication of the book *Changjiang Changjiang*, which was a collection of essays and interviews by officials and scientists who were critical towards the dam project published in 1989.

73 Shuilibu Changjiang shuili weiyuanhui (1992), p. 3.

74 See Heggelund (1994), pp. 38-9.

construction of the Yangtze River Three Gorges project). Consequently, a few days later on 3 April, the plans for the Three Gorges project were endorsed by the NPC, and were listed in the country's ten-year programme for economic and social development. One thousand seven hundred and sixty-seven representatives voted for the project, while 177 voted against and 664 abstained. As the NPC traditionally is known to be a rubber stamp organ, the number of representatives voting against the project signifies the great opposition to the dam project within China.

The Political Significance

The Three Gorges debate started in the 1950s as a technical debate where main participants were engineers and bureaucrats in the water resources and electric power administrations. Eventually, the debate evolved and became wider, and in the 1980s a more diversified group was involved in the debate. The political significance of the Three Gorges project debate is that it was not only a debate about the technical issues for a dam project, but also a cloak of legitimacy to criticise the way decision-making in general is carried out in the PRC. This criticism comes mainly from parts of the bureaucracy itself, as well as from scientists and journalists.[75] The criticism concerns issues such as lack of democratic and scientific decision-making methods, one-sidedness on the part of the ministry responsible for the project (MWREP), manipulation of information, lack of press freedom and freedom of speech. The attitude among officials seemed to be that scientific issues should be debated among scientific experts, the results of scientific research should be provided to leaders in order for them to make the decisions. The discussion about the project should not be used to further political purposes. Science would therefore be a tool for authorities in decision-making. Finally, it was significant that the debate about the Three Gorges dam indicates that it was possible to engage in some sort of discussion about social and political issues in the late 1980s.

The project was approved in 1992 despite widespread opposition which culminated at the NPC. The number of people who voted against the dam or abstained illustrates the great opposition against the dam project in China. The crackdown on the student demonstrations a couple of years earlier might have created a favourable condition for the approval of the project, as all debate had been silenced at that time, and a harsher political climate had emerged with the changes at the central power structure.[76] Furthermore, the floods during spring and

75 The Three Gorges debate was not a so-called 'grassroots debate'. People in general do not show much interest in this project. One reason could be that they lack divergent information about it, since the only information accessible is mainly the official pro-project propaganda. Another reason concerns the the problems people would encounter if they were to question the decisions made by the central authorities.

76 The then Secretary General of the CCP, Zhao Ziyang, who was an advocate of economic and political reforms, was removed from his position and dismissed from all his party posts. He was replaced by the present Secretary General Jiang Zemin. See Nathan and Link (2001), pp. 256-64, pp. 268-72.

summer of 1991 on the Yangtze River had created increased attention to the need for flood control in the area, with much positive coverage of the project's flood control ability in the media. These factors, in addition to the fact that top leaders were perceived to have given consent to the project, made the approval of the project possible.

Present State of the Project

Preparatory work on the project was initiated in 1993, a year after the adoption of the project at the NPC in April 1992. In 1994, a ceremony was held to mark the official beginning of the actual construction of the dam with the participation of Li Peng.[77] The construction of the dam is divided into three major phases, the first one covering the period from 1994-1997. The closure of the river that took place 8 November 1997 signifies the end of the first phase. During this phase a diversion canal for navigation was constructed. This canal opened for navigational services on 6 October 1997, meaning that all ships have stopped using the main river for navigation. The work in the first phase has mainly involved preparations for the construction of the main dam. The second phase, 1998-2003, is the crucial period for the project, when two coffer dams (each 76 metres high) must be constructed in order to facilitate the work on the high dam. During this stage the first turbines (14) on the north bank of the river will be installed. In December 2002 a second river closure took place in the diversion channel. During spring 2003 the river will be closed for navigation for about two months. Power generation is scheduled to begin in 2003 when the water levels will reach 135 metres. In the third phase, 2004-2009, the final 12 generators will be put in place in order to achieve the goal of 18,200 MW installed capacity. The permanent ship lifts and sluices will also be completed in this phase. Since the construction of the project began in 1993, until the end of 2002, 645,200 people had been resettled.[78] As the Three Gorges resettlement is discussed in Chapters 4 and 5, further mention will not be necessary here.

The Three Gorges Project Leadership and Actors

In order to understand the decision-making structure for the project, let us take a brief look at the institutional picture for the dam project. Central Commissions (*weiyuanhui*) and leading groups (*lingdao xiaozu*) that are directly under the Party apparatus or the State Council are important in the shaping of policy in China. A member of the Politburo leads these organs or its standing committee. The *Sanxia*

77 Li Peng's speech at the *Changjiang Sanxia gongcheng kaigong dianli* (Ceremony for the construction start of the Yangtze River Three Gorges project) on 14 December, 1994, is found in Zhongguo Sanxia jianshe nianjian bianjibu (1996), pp. 3-4.

78 Fu, Jing (8 February, 2003).

gongcheng jianshe weiyuanhui (The State Council Three Gorges Project Construction Committee) is such a committee. The State Council Three Gorges Project Construction Committee (TGPCC) is the highest policy-making body under the State Council established for the project. The TGPCC is headed by Zhu Rongji whose title is chairman of the committee. Zhu became the chairman in May 1998 when he took over the premiership from Li Peng. There are seven vice-chairmen of the committee. In addition, there are 19 committee members representing the relevant commissions, ministries and other institutions. The TGPCC co-ordinates all work for the dam project, including environmental and resettlement work.

The chairman and vice-chairmen of the committee are among the most powerful people (men) in China (see Table 2.3 below). At the time of this study, there were two politburo members, former Premier Zhu Rongji, and Wu Bangguo, the former vice-Premier.[79] The three Central Committee members are Jiang Zhuping, who is also the deputy secretary of the Hubei CCP,[80] Pu Haiqing[81] and Zeng Peiyan, the Chairman of the SDPC.[82] One alternate Central Committee member is Guo Shuyan. Gan Yuping is the vice-mayor of Chongqing as well as deputy secretary of the Chongqing CCP. Lu Youmei is the head of the Three Gorges Development Corporation, the proprietor of the project.[83] All members of the TGPCC took over their posts in the committee in May 1998, except for Guo Shuyan who has been the General Manager of the Three Gorges project General Office since November 1993. Provincial interests which are affected by the project in varying degrees are represented by both Hubei and Chongqing.

79 Zhu Rongji stepped down as Premier during the 10th National People's Congress in 2003. Wen Jiabao has taken over the position of Premier and will most likely take over the chairmanship of the TGPCC if previous procedure is followed. However, at the time of writing this has not yet occurred. Wu Bangguo took over the chairmanship of the National People's Congress after Li Peng stepped down.

80 Jiang Zhuping is the former governor of Hubei province.

81 Pu Haiqing is the former mayor of Chongqing. See Zhongguo Sanxia jianshe nianjian bianjibu (1998), p. 23.

82 The State Development Planning Commission (SDPC), formerly called the State Planning Commission (SPC until the government restructuring in 1998) has been an important actor in Chinese politics and the development of the economy, for instance in relation to the Five-Year plan. SDPC also co-ordinates work between the various ministries. Following the structural changes during the National People's Congress in in March 2003, the SDPC will be renamed the the State Development and Reform Commission (SDRC). SPC will be used in the period before 1998, SDPC in the period up to 2003. The new head of the SDRC is Ma Kai. Zeng Peiyan is now a vice-Premier.

83 Lu Youmei is a senior engineer in hydrology. He was Vice-Minister of the Ministry of Water Resources and Electric Power between 1984 and 1988, and Vice-Minister of the Ministry of Energy between 1988 and 1993.
See www.china3gorges.com/HTML/Information/hottalks/inht0708.htm.

Table 2.3 The State Council Three Gorges Project Construction Committee[84]

Name	Three Gorges Project (TGP)	State Council, Commissions, Ministries, Bureaux	CCP	Provincial positions
Zhu Rongji	TGPCC Chairman	Premier	Politburo member	–
Wu Bangguo	TGPCC Vice-chairman	Vice-premier	Politburo member	–
Jiang Zhuping	TGPCC Vice-chairman	–	Secretary, Hubei CCP; Central committee member	–
Pu Haiqing	TGPCC Vice-chairman	Director, State Metallurgical Industry Bureau	Central committee member	–
Zeng Peiyan	TGPCC Vice-chairman	Chairman, SDPC	Central committee member	–
Gan Yuping	TGPCC Vice-chairman	–	Deputy secretary, Chongqing CCP	Vice-mayor, Chongqing municipality
Guo Shuyan	Director, TGP General Office	–	Alternate central committee member	–
Lu Youmei	General manager, China Yangtze Three Gorges Project Develop-ment Corporation	–	–	–
Qi Lin	Head, Resettle-ment and Development Bureau	–	–	–

84 Table 2.3 illustrates positions held by these leaders during the course of the study; several of them have left their posts indicated in the table, such as Zhu Rongji, Wu Bangguo, Zeng Peiyan, GanYuping, etc. The author is aware that these 8 people have several other official positions such as heads of Leading Groups etc, but has chosen to limit the table to the CCP, State Council and its ministries, commissions, bureaux and provinces.

The 19 members of the TGPCC are high-level officials such as ministers, vice-ministers, bank directors, vice-governors and vice-mayors. They represent the following commissions, ministries and institutions: the State Economic and Trade Commission (SETC),[85] Ministry of Science and Technology (MOST);[86] Ministry of Finance; Ministry of Land and Natural Resources; Ministry of Communications; Ministry of Water Resources; Ministry of Agriculture; People's Bank of China; China Development Bank; Hubei province; State Environmental Protection Organisation (SEPA); Chinese Academy of Sciences (CAS); State Forestry Bureau; National Relics Bureau; State Power Corporation; Yangtze River Water Resources Commission (YRWRC); TGPCC Supervision Bureau; and the Vice-director of the Three Gorges project general office.[87]

There are several institutions subordinate to the Three Gorges Project Construction Committee (see Figure 2.1): One of them is *Yimin kaifaju* (the Resettlement and Development Bureau) headed by Qi Lin,[88] which is in charge of implementing the resettlement for the project. The governments of Chongqing municipality and Hubei province are in charge of concrete resettlement in their own areas, and have set up municipal, provincial, county and township resettlement bureaux.[89]

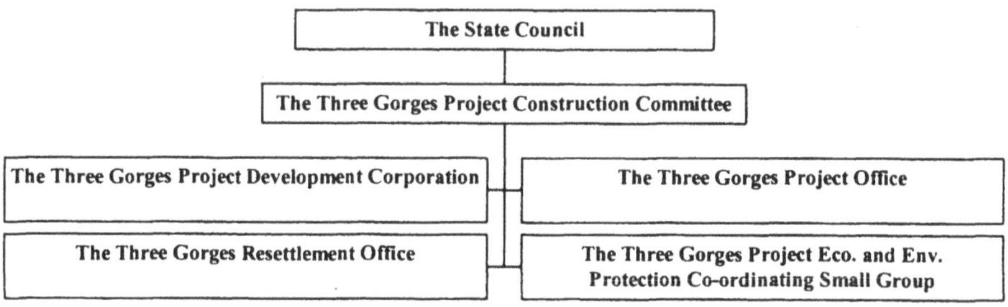

Figure 2.1 Three Gorges Project Organisational Chart

Sanxia bangong shi (The Three Gorges Office), headed by Guo Shuyan, is in charge of the day-to-day activities for the project. *Zhongguo Changjiang Sanxia*

85 Following the NPC in March 2003, the SETC is to be merged with the Minstry of Foreign Trade and Economic Co-operation into a new agency, the Ministry of Commerce.

86 MOST was previously known as the State Science and Technology Commission (SSTC), but was demoted to a ministry during the restructuring that took place in 1998.

87 Zhongguo Sanxia jianshe nianjian bianjibu (1998), p. 23.

88 Qi Lin became a vice-chairman of the TGPCC in September 2002. His entry into the TGPCC illustrates the increased importance placed on the resettlement issues for the dam project.

89 Each county has an overall plan for the resettlement with concrete tasks to carry out in the different areas. These plans are based upon the overall national resettlement policy of the central authorities.

gongcheng kaifa zong gongsi (The China Yangtze Three Gorges Project Development Corporation, CYTGPDC, shortened to TGPDC), the proprietor of the project, headed by Lu Youmei, allocates all funding related to the project, such as construction and resettlement that are first approved by the TGPCC. The TGPDC is responsible for the construction of the dam; it is situated in Yichang, Hubei province. One important leading group under the TGPCC, is *Guowuyuan Sanxia gongcheng jianshe weiyuanhui shengtai yu huanjing baohu xietiao xiaozu* (The State Council Three Gorges project Construction Committee Ecological and Environmental Protection Co-ordinating Small Group), which has been established to co-ordinate the ecological and environmental work for the project. The Ecological and Environmental Protection Co-ordinating Small Group distributes work to the various ministries and holds annual meetings to discuss important issues that have been raised during the year. Key members of the committee are the State Environmental Protection Administration (SEPA), the Ministry of Water Resources (MWR), Ministry of Agriculture, State Forestry Administration and the Chinese Academy of Sciences (CAS).[90]

The book will not look into the role of all actors listed above. However, it will focus on a few institutional and individual actors that are deemed important for the resettlement and environmental policymaking for the Three Gorges project. Their roles and backgrounds will be discussed in several chapters throughout the book.

Summary

This historical account illustrates the long discussion about the Three Gorges dam project, first suggested in 1919 by Sun Yat-sen. The purpose of the project is electricity production, flood control and improved navigational facilities. The project is estimated to cost RMB 200 billion (US$ 24 billion). The discussion in the PRC began in the mid-1950s and continued until 1992, when the project was approved by the NPC. The long history illustrates the complexity of the Three Gorges project and the interests of a number of people who were involved in the discussions since the beginning. Discussion between individuals in the Ministries of Water Resource and Electric Power took place in the 1950s. Mao Zedong chose to postpone the construction of the dam, as the time was not yet ripe for such a large undertaking. The 1960s and most of the 1970s were a frustrating period for the dam proponents due to other political priorities for the Chinese CCP. The Great

90 The committee is headed by the vice-director of the TGPCC office, Li Shichong. In addition, members are the State Meteorological Administration, the Health Ministry, Ministry of Construction, Ministry of Communications, the Yangtze River Water Resources Commission, administrations for geology and mineral resources (since 1998 under the Ministry of Land and Natural Resources) and China Bureau of Seismology, and members from the TGPCC, the Three Gorges Project Resettlement Bureau and the TGPDC. For a comprehensive list of all members (including individual names) of the Co-ordinating Small Group, see Zhongguo Sanxia jianshe nianjian (1998), p. 26.

Leap Forward (1959) and its consequences, the Cultural Revolution (1966-76) as well as border clashes with the Soviet Union influenced the plans for the Three Gorges. The project was not included in the fifth five-year plan (1976-1980). Mao's role had been important in the process, with his death in 1976 this changed. The new economic reform policy was initiated in the late 1970s and diminished the role for personal leadership in the 1980s. It created a greater regard for consensus in decision-making, and active co-operation instead of Mao's methods of ideological campaign and broadsides and deification of top leaders to overcome bureaucratic resistance.

In the 1980s numerous discussions took place about the dam project, and topics ranged from dam site selection and resettlement to technical issues. The project was in principle adopted by the State Council in 1984, and was expected to be included in Seventh Five-year plan (1986-1990). However, obstruction from Chongqing as well as economic recession hindered the approval of the dam project. Preparations for the project went nevertheless ahead, and feasibility studies were carried out by both Chinese and international experts. Apart from a few Chinese experts, the conclusion was positive and the reports recommended the construction of a dam in the Three Gorges. Nevertheless, opposition to the project became more vocal within China as the preparations for the dam project intensified. Two hundred and seventy representatives to the NPC opposed the project, and requested that the project be postponed until the next century. These protests resulted in the then vice premier Yao Yilin announcing that the decision on the project would be postponed until 1995. Nevertheless, after the student demonstrations were brutally ended on 4 June 1989, opposition against the project died down and preparations for the project continued in 1990. The floods in the summer of 1991 once again put the need for flood control in the Three Gorges area to the foreground. The project was finally approved by the NPC in 1992, with one-third of the representatives voting against or abstaining, signifying the great opposition to this project in China. The political significance of the Three Gorges project debate is that it was not only a debate about the technical issues for a dam project. It also gave a cloak of legitimacy for the bureaucracy itself as well as the scientific sphere to criticise the way decision-making in general is carried out in the PRC. The construction of the dam began in 1994, and is now in the second phase of the three-phase construction period. By the end of 2002, 645,200 people had been resettled.

The institutional organisation is important for understanding the resettlement and environmental decision-making for the dam project. The *Sanxia gongcheng jianshe weiyuanhui* (The State Council Three Gorges Project Construction Committee, TGPCC) is the highest policymaking organ for the dam project, and is headed by Premier Zhu Rongji. The Committee members are all high-level officials representing commissions, ministries, banks, Chonqing municipality and Hubei province. There are several institutions subordinate to the TGPCC that are in charge of construction of the dam, resettlement and environment.

Chapter 3

Analytical Framework

The purpose of this chapter is to establish a framework for the analysis of the Three Gorges resettlement and environment policy that will structure the empirical material obtained during the course of the research. I will present different models, which will be used to systematise the materials. There are many ways to analyse political processes in China, and scholars contend on the methods most appropriate for the country. One of the issues that arises when studying policymaking in China is whether or not it is possible to use theories established by Western researchers when studying a country that appears so different from the Western world. During the past decades several approaches have been employed to study the political system and how policy decisions are made in China. Some of these theories pay more attention to the cultural traditions of the country and how those would influence policymaking. Other studies of decision-making in China draw on Western perceptions of bureaucratic processes that aim to shed light on the bureaucratic processes taking place in the country. This chapter does not aim to identify the appropriate approach for research of policymaking in China, which would be impossible, as one model cannot capture all aspects of decision-making in China. The approaches must rather be regarded as complementary. Thus, this chapter attempts to indicate to what extent one specific approach is suitable to explain the dynamics in the process surrounding the resettlement policy change for the Three Gorges project. A key point in the conclusion chapter will be to draw a conclusion regarding the relevance of the analytical approach described below.

The chapter begins with a very brief introduction to three general approaches that have been employed in the studies of policymaking. Then, a brief introduction to some selected central analytic approaches and theories that have been employed to discuss politics and decision-making in China by Western scholars. Next comes an introduction of models that may be applicable in the analysis in the book, the fragmented authoritarianism and the reasons for choice of analytic approach.

Three General Approaches to the Study of Policymaking

There are a number of possible approaches in the study of domestic policymaking. Three basic models that have been common have been introduced by Graham T. Allison in his classic analysis, using three conceptual models in relation to the Cuban missile crisis in 1962.[91] The text below is a summary based on the

[91] Allison (1969).

definitions of Allison's three models, (I) the rationality policy model, (II) the organisational policy model and (III) the bureaucratic politics model as presented in Allison's article on the crisis.

Rationality Policy Model

Analysts using this model attempt to understand events as the more or less purposive acts of unified national governments. Each analyst 'assumes that what must be explained is an action, i.e. the realisation of some purpose or intention'.[92] Nations and governments are believed to have rationally chosen a specific action. According to Allison's organisational concepts the actor is the nation or government. The actor is conceived as a rational, unitary decision-maker, and has specified goals. More importantly, the actor has 'one set of perceived options and a single estimate of the consequences that follow from each alternative', i.e. providing exact and neutral information to decision-makers is a prerequisite for making the correct judgement. When a problem appears, action is chosen in response to this. In selecting the response to solve the problem, action to be carried out by the nation or the government is conceived as a steady-state choice among alternative outcomes.

Allison states that action as rational choice consists of the following factors:

1. Goals and objectives—national security and national interests are the principal categories in which strategic goals are conceived.
2. Options—various courses of action provide the spectrum of options.
3. Consequences—enactment of each alternative course of action will produce a series of consequences, each having benefits and costs.
4. Choice—rational choice is value-maximising, i.e. selecting the choice that ranks highest in terms of goals and objectives.

From this, one can conclude that the rational choice model would mean to consciously look at a number of alternative actions, then make a selection of the most favourable alternative, i.e. an increase in the value of the set of consequences which will follow from that action.

Organisational Process Model

Allison states that a government consists of 'a conglomerate of semi-feudal, loosely allied organisations, each with a substantial life of its own'.[93] The organisational process model implies that choices made are not deliberate choices of leaders, but are more outputs of large organisations functioning according to standard patterns of behaviour. Governments consist of large organisations that have responsibility for specific areas. Each organisation takes care of a special set

92 Allison (1969), p. 693.
93 Ibid., p. 698. See also Allison (1971).

of problems. Few problems fall exclusively into the area of one organisation. Therefore, government behaviour reflects the independent output of several organisations, partially co-ordinated by government leaders. Government leaders cannot substantially control the behaviour of an organisation, although they are able to exert some influence. The behaviour of these organisations regarding particular events depends upon the set routines established in the organisations prior to that event. Organisation may also change, especially in response to a major crisis.

The actor in model II is a constellation of loosely allied organisations on top of which government leaders sit (and not a monolithic nation or government). Detail and nuance of actions are determined mainly by organisational routines, not government leaders' directions. Actions according to standard operating procedures (SOP) do not constitute far-sighted, flexible adaptation to an issue. Standard operating procedures constitute routines for dealing with standard situations. Problems are factored and power fractionated, i.e. problems are cut out and parcelled out to various organisations. Each organisation perceives problems, processes information and performs a range of actions in quasi-independence within broad ranges of national policy. 'Primary responsibility for a narrow set of problems encourages organisational parochialism', states Allison.[94] Organisations develop partiality regarding operational priorities, perceptions and issues, due to procedures within the organisation.

Pre-established routines are important in regard to action as organisational output. Allison lists several general points which sum up aspects regarding the activities of each organisation in producing outputs:[95]

1. The operational goals of an organisation are seldom revealed by formal mandate, but emerge as a set of constraints defining acceptable performance. The constraints emerge from expectations and demands of other organisations, interest groups, and from bargaining within the organisation.
2. The organisation deals with the problems in a sequential way, i.e. when they arise.
3. Organisations perform their duties according to standard operating procedures (SOP); SOP do not change easily. Without the procedures it would be difficult to carry out certain tasks.
4. Programmes and repertoires are important as determinants of organisational behaviour.
5. Negotiated environment is arranged in order to regularise actions of other actors with whom they have to deal.
6. When situations are not standard, organisations engage in search; the style of search and the solutions are largely determined by existing routines.
7. Organisational learning and change basically follow existing routines.

---

94 Ibid., p. 700.
95 Ibid., pp. 700-703.

Bureaucratic Politics Model

Government behaviour can be understood as outcomes of bargaining games. Model III claims that political leaders 'pull and haul' in order to reach solutions to an issue. 'Each organisation tries to extend its own influence, guard and/or expand its own resources, and preserve its own interests to the best possible extent'.[96] The representatives from various bureaucracies form alliances, and 'participate in a bargaining process that has little to do with the grand plans of national development or ideological advancement'. In contrast with Model I, the bureaucratic politics model observes no unitary actor, but rather many actors as players. These actors do not focus on one issue, but on many intra-national problems as well. They do not have one single goal but rather national, organisational and personal goals. Political leaders at the top of the national apparatus government and persons in charge of organisations form the circle of central players. In policymaking, the result is often different from what anyone intended due to different groups pulling in different directions. The power and skills of the proponents and opponents determine the course of action.

Policy as political outcome occurs where a solution is chosen not as a solution to a problem, but which results from compromise, coalition, competition and confusion. Policy outcome is the result of 'bargaining among players with separate and unequal power over particular pieces and with separable objectives in distinguishable subgames'.[97] The players in action are individuals, and not a unitary nation nor a conglomerate of organisations. Individuals become important players by occupying critical positions in the administration. Positions define what players must do. One person has many hats, and performance in one area affects his credits and power in the other areas. The reaction to a problem is coloured by the person's position. The success of each player depends upon his/her power; power is effective influence on policy outcome. Each player must pick the issues that he/she is most likely to succeed with. The player focuses on the decision that has to be made now, and not on a total strategic problem. The issue to be is coloured by the position of the player ('Where you stand, depends on were you sit').[98]

To sum up, Allison states that Model I has been useful in providing a standard frame for analysis, a model which implies that important events have important causes. Nevertheless, he continues, it is necessary to supplement the model by frames of reference that focus on the large organisations and political actors involved in political processes. These aspects are represented in Allison's Model II and III. Model II implies output of large organisations functioning according to certain regular patterns of behaviour, while in Model III an action performed by a nation is the outcome of bargaining among individuals and groups within the government. The three models are described by Allison:

---

96 See Christiansen and Rai (1996), p. 19.

97 Allison (1969), p. 708.

98 Ibid., p. 711.

> Model I concentrates on 'market factors', pressures and incentives created by the 'international strategic market place'. Models II and III focus on the internal mechanism of the government that chooses in this environment.[99]

The three models above are general approaches to domestic policymaking, and constitute the background for the specific models that have been employed in the study of Chinese policymaking from the 1960s until today to be described in the next section. However, the models have been further developed by scholars in the course of their study of Chinese politics.

Western Approaches to the Study of Decision-making in China

China began its reform process and 'open-door policy' in the end of the 1970s. Due to the reform process and the consequent opening up of the country to the outside world, the amount of materials accessible to Western scholars increased, which made it possible to obtain a better understanding of the political processes going on in China. Thus, a development has taken place with regard to the models used to analyse Chinese politics and decision-making. Below I will describe and discuss briefly some of the central approaches that have been employed by Western scholars in studying China, set in a historical framework of the developments within the country.

The Mao Era and Top-down Approach

One of the earliest models used to study Chinese politics was *the totalitarianism model*, represented by the view of Doak A. Barnett,[100] which was dominant in the 1950s and 1960s. According to this model, Chinese politics was based upon centralised power of the CCP, where a small handful of leaders maintained control. At the time when this model was dominant, Mao Zedong was China's undisputed leader. 'The structure of the Party, with its network of cells, its strict code of obedience, and its organisational hierarchy, was seen as symptomatic of a 'totalitarian' authority tightly controlling every aspect of Chinese society'.[101] In the late 1950s and the 1960s, however, events such as the Great Leap Forward, the ideological clash between the then Soviet Union and China as well as the Cultural Revolution (1966-76), made it clear that the totalitarian model was inadequate in understanding Chinese politics. Chinese politics were more complex than previously perceived.

Therefore, *the two-line struggle* and *the class struggle approach* became the dominant approach in the 1970s. This model was based on the political facts in China, where it was officially stated that Cultural Revolution was a struggle

---

99 Ibid., pp. 716-17.

100 Barnett (1964).

101 Christiansen and Rai (1996), p. 3.

between two lines. On the one side was Mao Zedong and his mass line policy (proletarian-revolutionary line), and on the other side Liu Shaoqi's bureaucratic politics (bourgeoisie-reactionary politics).[102] From the late 1960s the two-line approach was adopted by Western scholars as the tool to analyse Chinese politics and policymaking. In the class struggle approach, where Mao's line would take China on a new road to socialism, one factor was the workers' gaining control over their own work units. One important exponent was Charles Bettelheim,[103] and his analysis of the struggle between workers' control over management of their own work units and the manager-cadres controls. It is thought that the reason for the decline of the class struggle approach model in the mid-1970s was due to the emergence of new empirical data regarding the Cultural Revolution. This information disclosed the darker sides of the Cultural Revolution, and 'revealed not the triumph of the workers' management but the manipulation and coercion of workers and Party members by contending groups in the CCP leadership that led to great injustices, corruption and even terror'.[104] The reason that the two-line approach lasted until the end of the 1970s was the limited materials available to Western researchers that gave an impression of the political conflict that was taking place in the Chinese society.[105]

The Post-Mao Era—Pluralistic, Present Approaches

A new era began following Mao Zedong's death in 1976, and the arrest of the Gang of Four.[106] Hua Guofeng had been chosen by Mao as his follower. Deng Xiaoping had been ousted in 1976 by the Gang of Four, but was rehabilitated in 1977. It became apparent that political conflicts existed between Hua and Deng, as Deng wished to reform, while Hua was determined to continue Mao's line. Deng succeeded in the power struggle and eventually introduced the policy of the four modernisations and economic reform.[107]

In response to the new developments in China, Western scholars have come up with a number of models, of which the most important are: the clientelist model (factionalism); interest group model; cultural approach model; and bureaucratic

---

102 Brødsgaard (1989), p. 302.

103 Bettelheim (1974).

104 Christiansen and Rai (1996), p. 11.

105 The sources for research were mainly: i) red guard material; ii) refugees interviewed in Hong Kong, and iii) articles from the official press. Brødsgaard (1989), p. 305.

106 Jiang Qing (Mao's wife), Zhang Chunqiao, Yao Wenyuan and Wang Hongwen were the radicals and the instigators of the Cultural Revolution.

107 Hua Guofeng was labelled a believer in 'the two whatevers': 'obey whatever Mao had said and to ensure the continuation of whatever he had decided'. Hua wished to continue the politics of Mao Zedong. Deng wished for reforms; he had more powerful connections in the party and in the army, as well as among the leading intellectuals. See Spence (1999), p. 640.

politics model.[108] It is important to emphasise that these models stress different aspects of the Chinese policymaking process and should be regarded as complementary to each other rather than mutually excluding. Below is a brief introduction to some of the models that were developed following the new political reality in China and that are being used on present Chinese politics.

Factionalism and elite conflict was one model used to analyse the politics in China in the early 1970s. This model was based upon the fact that disagreements existed within the CCP, as had been reflected in the 1950s and 1960s through various political campaigns and purges. The most well known analyst representing this model is Andrew Nathan.[109] Nathan describes a faction as based on clientelist ties, i.e. a maze of individual personalised ties between a leader and his political supporters.[110] The system was not based on friendship or representation but on patron-client relationships. The political process was based on establishing consensus between the leaders of the various fractions. This model analyses mainly elite-level politics, and has been regarded as artificially separated from the social reality of the country, where the social and economic dynamics are not the focus.

The clientelist model thus developed from this model of factionalism and elite conflict. Both models, the *factionalism and elite conflict* and *clientelism*, focus on patron-client relationships, albeit with different emphasis. The clientilist model seeks to analyse 'how patron-client relationships affect the interaction between the state elite and the citizens, and how that effects policy implementation',[111] and not only the patron-client relationships that exist among the elite. The emphasis of this model is less on the formal ties and institutions, and more on the informal, personal relationships. Major exponents for this model are Jean Oi and Andrew Walder.[112] The clientilistic approach is an important shift in focus, as it tends to focus more on the local level dynamics of politics and gives a different view of the individual citizens' potential to influence the implementation of centrally decided politics. The shift in analytic focus on policymaking in China must be seen in relation to the changes in the country in general, where the reforms have contributed to a more pluralistic society. Nevertheless, this approach has also been criticised for forgetting the group, and focusing entirely on the individual. It does not take into account the changing political reality of the country.[113]

The interest groups model is based on the study of Soviet politics first put forward by Gordon Skilling,[114] which springs out of the notion that in socialist societies as anywhere else, different groups have different outlooks and preferences. The model 'focuses on the differences between people and the degree to which they assert specific interests and seek to influence the political

---

108 Brødsgaard (1989), pp. 306-7.
109 Nathan (1973).
110 Christiansen and Rai (1996), p. 4.
111 Ibid., p. 6.
112 Oi (1989); Walder (1986).
113 Christiansen and Rai (1996), p. 9.
114 Skilling (1966), pp. 435-51.

establishment'.[115] Skilling defines an interest group as 'an aggregate of persons who possess certain common characteristics and share certain attitudes on public issues, and who adopt distinct positions on these issues and make definite claims on those in authority'.[116] Much of the research on Chinese policymaking in the post-Mao period has taken its point of departure in the notion that there are different groups with their own interests that determine the outcome of policymaking in China. Michel Oksenberg was the first to try out the interest groups model in relation to China,[117] and made a study of the interest and influence by seven selected groups in Chinese society. David Goodman (1984) and Victor Falkenheim (1987)[118] applied further the interest groups model, where broad occupational groups were seen to have different interests, such as economists, provincial party secretaries, workers, cadres, peasants, military and intellectuals.

The culturalist approach represents the view that China is different from other countries due to the historical past; 'it is assumed that in various ways the historical events and structures of thought in the past determine the present'.[119] The scholars representing this view think that Chinese traditional values are dominant in policymaking, i.e. seeking consensus based on political views (not based on which bureaucracy one represents) as well as building on personal relations. Traditional cultural values and norms are important in understanding the political developments of the country and Lucian Pye[120] is the foremost exponent for this model, stressing the importance of *guanxi* (relations) in Chinese politics. Tang Tsou[121] is another scholar that has stressed the importance of the informal aspects of Chinese politics. He is known for his idea of one side winning all and/or the other side losing all. Tsou states that '[t]his is a feature not only of elite CCP politics but of Chinese politics throughout the twentieth century'.[122] He believed that it was necessary to develop concepts and approaches to study Chinese politics, including the CCP politics, and was not convinced that the concepts developed in American political science could be applied directly to the study of China.[123] This approach also stresses personal relations, *guanxi,* as being the key factor in Chinese politics.[124] This leads on to the importance of *informal politics* in Chinese politics,

---

115 See Christiansen and Rai (1996), p. 13.
116 Ibid., p. 13.
117 Oksenberg (1968). See Brødsgaard (1989).
118 Goodman (ed. 1984). See also Falkenheim (1987).
119 Christiansen and Rai (1996), p. 21.
120 Pye (1968); and Pye (1985).
121 Tsou (1995); Tsou (1976), reprinted in Tsou (1986).
122 See Tsou (1995), p. 97.
123 See Fewsmith (2001a), p. xv.
124 Tsou gives an example of Hua Guofeng and Deng Xiaoping; Hua held formal positions as chairman of the Central Committee as well as the Party's Military Commission, and was Premier of the State Council. Hua's authority was not backed by informal power, and he therefore lost to Deng Xiaping whose informal power was

of which Joseph Fewsmith and Lowell Dittmer are advocates.[125] Dittmer concludes that 'informal politics can be defined on the basis of a combination of behavioural, structural and cyclical criteria'.[126] Fewsmith states that informal politics have been important at all political levels, 'both at the local level where personal relations could protect reform inititatives, and at the central level, where new ideas could be conveyed with surprising rapidity to the highest levels'.[127] Fewsmith stresses the need to study relationships and interaction between formal and informal politics, and states that while formal structures have been important, the informal politics were predominant. Fewsmith argues that studies of political systems in China have neglected the interaction between formal and informal poitics, and have mainly drawn upon Western understandings of bureaucratic processes, of which the *bureaucratic politics model* is one.

The bureaucratic politics model focuses on processes through which decisions are arrived at and the implementation of policy. The model claims that political leaders 'pull and haul' in order to reach solutions to an issue, and bargain with each other to reach solutions. Bureaucratic leaders seek to maximise the interests of their own organisations. Political decisions are a result of bargaining, and decision-making is fragmented. The bureaucratic politics approach in relation to China has developed in the light of scholars doubting whether or not it is possible for interest groups outside the state and party bureaucracy to independently influence policymaking.[128] There is almost a consensus that this bureaucratic politics approach fits the study of bureaucratic politics in China well. This is due to the fact that the country has one of the oldest bureaucratic systems in the world, and the bureaucratic political structure has been strengthened by the communist system. Scholars who have developed the bureaucratic politics model further in relation to the study of Chinese policymaking are David Lampton, Doak A. Barnett, Kenneth Lieberthal and Michel Oksenberg.[129] These scholars believe that in formulating policy, the parties concerned (bureaucratic actors) bargain in order to obtain consensus. Even in the implementation phase, the bargaining continues so that even if a decision has been made at a higher level, it is re-bargained due to influence from lower levels.[130] This implies that bargaining and consensus building may reach down to the grassroots level, implying that bureaucracies or institutions involved in the process may actually influence the outcome of policymaking at the

---

eventually manifested in formal channels of expression when he acquired formal positions. Tsou (1995), p. 154.

125 Fewsmith (2001b); Fewsmith (2001a); Dittmer (1995). Dittmer (2001); and Dittmer, Haruhiro and Lee (eds. 2000).

126 Dittmer (1995), p. 33.

127 Fewsmith (2001a), p. xx.

128 Brødsgaard (1989), p. 314.

129 Lampton (1987); Lieberthal and Oksenberg (1988); and Lieberthal and Lampton (eds. 1992). Barnett (1985).

130 The Three Gorges project is an example of this, as it has been approved in principle twice, but both times the discussion continued.

highest levels. The scholars representing this view call the approach *fragmented authoritarianism.*

To sum up, the totalitarianism and the two-line struggle models have more or less been abandoned by scholars as a method to explain Chinese politics. Two important reasons for this are the changes on the political arena in China and the opening up of the country to the outside world, which have increased the availability of new research materials to Western scholars and and have increased access to Chinese people in general. These two factors have contributed to a higher level of sophistication in the analysis of Chinese politics. The approaches described above have developed in the period following Mao's death and until today. They must be regarded as mutually complementary in portraying the political reality in today's China, as they describe different components of the Chinese policymaking process. These approaches are more suitable today than earlier models, since we have more knowledge about political developments within China that are based on Chinese written sources and interviews.

The Fragmented Authoritarianism Approach

As mentioned above, important studies using the bureaucratic politics model in relation to Chinese politics have been carried out by Lampton, Lieberthal and Oksenberg, which they call the *fragmented authoritarianism model.*[131] Lieberthal states that 'the fragmented authoritarianism model seeks to put into better perspective two well-developed groups of literature concerning policy making in post-1949 China'. One is the rationality model, the other a 'cellular' conception of the political system. According to Lieberthal, the fragmented authoritarianism model adds a third dimension of the political system: 'the structure of bureaucratic authority and the realities of bureaucratic practice that affect both the elite and the basic building blocks of the system'.[132] The models introduced in the previous section may all be useful in the study of modern Chinese society. However, some models may be more appropriate for the study of Chinese bureaucracy and policymaking and the organisations involved. On the general level, Allison's organisational process model and the bureaucratic politics model are both relevant, as they describe the internal mechanism of the governments. With regard to the approaches that have been employed in the study of China, the fragmented authoritarianism approach stands out as particularly relevant in the analysis of Chinese policymaking in relation to the Three Gorges project and will be the starting point for the study of the resettlement policy change of the Three Gorges project. Below is a brief summary of the fragmented authoritarianism approach which is based on Lieberthal and Oksenberg (1988), *Policymaking in China* and

131 Lampton (1987); Lieberthal and Oksenberg (1988); and Lieberthal and Lampton (eds. 1992).

132 Lieberthal and Lampton (eds. 1992), pp. 10-11.

Lieberthal and Lampton, (eds. 1992) *Bureaucracy, Politics, and Decision-Making in Post-Mao China.*

Policymaking in China portrays the following bureaucratic structure where authority consists of four tiers. i) The core group of twenty-five to thirty-five top leaders who articulate national policy; ii) the layer of staff, leadership groups research centres, and institutions which link the elite to and buffer them from the bureaucracy; iii) State Council commissions and ministries that have supra-ministerial status and co-ordinate activities of the line ministries and provinces; and iv) line ministries which implement policy, the provinces and corporations. Consensus building among these is important to the policy process, although power traditionally has been vested more in individuals. The shaping and implementation of policy take place within the bureaucracies at the national level, which consists of commissions and ministries. At this level feasibility studies are conducted, appropriate funding is allocated and energy policy decisions are implemented.

The fragmented structure of authority has seemingly produced increased bargaining in the Chinese bureaucracy. Lieberthal[133] states that 'bargaining involves negotiations over resources among units that effectively have mutual veto power', and encourages consensus building among various organs in order to initiate and develop projects. According to Liebertehal and Oksenberg, the structure for the energy sector highlights the fragmented structure of authority,[134] which requires that any policy suggestion of project proposal obtain the co-operation of many bureaucratic units with different authority. Consequently, issues tend to go to the highest decision-making bodies such as the SPC and the State Council. Their role would then be to co-ordinate and mediate between the institutions. This means that one single ministry or province would seldom be able to push a project alone; it would need to seek co-operation from other units as well. Furthermore, the fragmentation of authority would demand great efforts at all stages of decision-making to maintain or create consensus in order to push the proposal forward. The fragmented structure of authority thus makes consensus building crucial to the policy process. Chinese leaders are able to apply pressure in order to advance a project over the objections of participants. Nevertheless, their pressure alone is not sufficient to move the proposal, as it easily could get stuck in the bureaucracy. Lieberthal and Oksenberg also state that top leaders are more reluctant to use the methods from the Mao era, such as political campaigns and ideological broadsides. They think that the reforms have increased the opportunities for units to use resources to their own benefit instead of concentrating activities on state-mandated activities. The reforms have therefore increased the need for consensus building in decision-making for large energy projects.

Due to the bargaining and consensus building, the policy process is diffuse according to Lieberthal and Oksenberg:[135] It is protracted, meaning that policy is

---

133 See 'Introduction' by Lieberthal in Lieberthal and Lampton, (eds. 1992), p. 9.

134 Lieberthal and Oksenberg (1988), p. 22.

135 Ibid., p. 24.

being shaped over a longer period, acquiring history that is well known to the participants. For instance, searching for the timing of a particular decision is not meaningful, as the process is protracted as consensus for it is built. The policy process is also disjointed, where key decisions are made in a number of different and loosely co-ordinated agencies. This influences the policy results, which are distinct and often contradictory (in particular in relation to energy policy). Finally, the policy process is also incremental and policy changes gradually.

What then determines whether a particular policy will be decided within the State Council bureaucratic apparatus or by the top 25 to 35 leaders? Lieberthal and Oksenberg state the following five factors that can propel an issue onto the agenda of the top leaders:[136]

1. The particular interests of individual top leaders explain many instances of problems being brought to this level.
2. Chinese bureaucrats can try to force an issue onto the agenda of the highest leaders.
3. The emergence of a critical problem may capture the attention of the top leaders and force decisions to be made.
4. Foreigners may force an issue onto the agenda of the highest level leaders.
5. Procedural Requirements (such as five-year plans, etc).

According to Lieberthal and Oksenberg, the formal table of organisation in China gives an impression of a system that is hierarchically clearly defined, with a core leadership at the top and individuals linked to vertical ministries that control units from the centre to the local level. On the top of the ministries are the commissions. However, there are factors that need to be taken into consideration, since this hierarchical chain of organisation actually is fragmented. Some of the organisational concepts that need to be paid attention to in the study of Chinese policymaking are:[137]

- *Centre versus locale.* One hierarchical divide is between the national level (the Centre, *zhongyang*) and the provinces, municipalities, and counties (locale, *difang*). The Centre includes the State Council and its commissions, ministries and leadership groups in Beijing, as well as the Party Politburo, Secretariat and the organs of the Central Committee. The relationship between the Centre and the local levels changes constantly. One important aspect of central-provincial relationship is finances: budgetary process and the national banking system. The budgetary reform has permitted provinces to retain increasing portions of revenues they generate. This budgetary change has increased the ability to make independent decisions at the local levels.

---

136 Ibid., pp. 30-31.
137 Ibid., pp. 138-51.

- *Xitong*—The meaning of xitong is system; this refers basically to vertical functional hierarchies and stretch from Beijing to local units. Each central ministry or State Council commission heads its own xitong. One major issue is the relationship between the vertical functional systems and the horizontal territorial governing bodies. The Chinese express this as *tiaotiao* (a 'line', refers to the vertical system) and *kuaikuai* (a 'piece', refers to the territorial authority such as provincial, municipal or county government). One issue is whether *tiaotiao* serves *kuaikuai* or vice versa. These various hierarchies find it difficult to co-operate, as formal lines of authority and communication are handed down within the vertical chain of authority (for instance from a ministry to the provincial subordinate) or the territorial chain of command (such as between a provincial government and a coal department). Thus, the system does not invite to co-operation across these vertical and horizontal lines, as they are part of different *xitong*.
- *Bureaucratic ranks*—bureaucratic rank is important. A ministry, a provincial government, and certain State corporations have the same rank. This means that a ministry cannot tell a provincial government what to do. Likewise, a provincial government department and ministerial bureau also have the same rank.
- *Individual rank and stature* of the leader of a unit is also a factor that influences a unit's place. Most high level officials have three formal ranks: their formal government and/or party title; their civil service grade; and their position in the Communist party. 'These together determine their real stature when they deal with others'.[138]
- *Interagency relations*—the system of overlapping authority (for instance a department in a provincial government is usually subordinate both to the provincial government itself and the central ministry) has developed a vocabulary that describes the various relationships between the agencies. *Lingdao guanxi* (leadership relations) and *yewu guanxi* (professional/business relations) are applied. Leadership relations apply when the higher unit can issue a binding order to its subordinate. Professional/business relations apply mainly for interrelated areas of activity.

These five points illustrate the complexity of the the Chinese bureaucratic apparatus, and the potential conflicts that may arise between them. These five points should be kept in mind throughout the discussion of the resettlement policy change, and will be commented on in the conclusion.

The Fragmented Authoritarianism Model and the Three Gorges Project

There are several reasons for choosing the fragmented authoritarianism approach as a starting point for the analysis in the book:

---

138 The civil service system includes more than twenty grades. Each grade specifies a salary range. Ibid., pp. 145-6.

- *Previous study.* An important analysis of the decision-making process for the Three Gorges project in general is found in *Policymaking in China*, by Lieberthal and Oksenberg, which employs the fragmented authoritarianism model as a tool to explain the decisons for the project. The case study illustrates that the fragmented authoritarianism approach has been a relevant approach for the decision-making process for the Three Gorges project. The study concerns discussions about whether or not to launch the dam project in general, and an analysis is made of the bureaucratic apparatus involved, the various interest groups and the attempts and failures to launch the dam project are discussed in great detail. Their conclusion to the case study is that bargaining took place, and the actors involved were relatively stable in their position over the years. Other factors that were important for the decision-making process were a diminishing role of a single leader, and the widening range of actors who played a role in the dam project. Furthermore, their study also illustrates the great opposition that certain actors demonstrated, and how they bargained for changes even after decisions were made.[139] These are topics that also will be investigated in the analysis of the resettlement policy change for the dam project. *Policymaking in China* was published a few years before the approval of the dam project in 1992, and does not have a conclusion as to how and why the project finally was approved. Nevertheless, their study is valuable in understanding the bureaucratic processes leading up to the project approval, and to obtain an impression of the interest groups involved in the process. This book will use the study and the fragmented authoritarianism model as a point of departure for the analysis of the Three Gorges resettlement policy change, and the relevance of the model will be checked and discussed in the conclusion chapter of the book.
- *Bargaining.* This book is a study of the resettlement policymaking change and the causes for the change. According to the fragmented authoritarianism approach, policy decisions are announced at the centre, but there are usually long processes beforehand due to bargaining. With regard to the Three Gorges resettlement policy change, input may have come from actors outside the core leadership before the decision was made to move the large number of people. In this case, provincial and local governments as well as other institutions may be likely candidates as they do not always support the decisions from the centre, and may attempt to influence the decision-making or the implementation process. With regard to resettlement, some of the questions that are being asked in the book concern the political processes and reasons for policy change. The relevance of the fragmented authoritarianism approach in relation to the resettlement policy change will be discussed in the final chapter. Furthermore, the environment is an issue that is closely linked to the resettlement policy change, and the extent to which bargaining takes place

139 Such as Sichuan province and the Ministry of Communication that continued to demand a 180-metre maximum normal pool level after the State Council in principle had accepted a 150-metre dam. See Ibid., p. 336.

regarding environmental issues will be discussed. According to Lieberthal and Oksenberg, the bargaining takes place over resources, and the fragmented authoritarianism model was constructed around large investement projects where bargaining is prone to take place. The environmental area may not appear a likely bargaining arena. Yet funding for environmental purposes may create opportunities for bargaining. One may assume that the people (be they officials, scholars,) representing environmental interests would be interested in obtaining as much funding as possible for environmental protection and research. These actors may be engaged in some sort of bargaining for funding, and would have to compete with other groups for their portion of the project funds. Reasons and motivations may differ among these environmental actors, and one motivation may be increased funding in order to satisfy higher environmental quality and standards in China in general. Moreover, accruing resources for themselves or their unit may be an important reason.[140] The fragmented authoritarianism approach is the basis for assessing the bargaining factor in the Three Gorges project environmental policymaking.

- *Agenda setting and leadership*. Lieberthal and Oksenberg list five factors that determine whether issues appear on the agenda of the top leaders or within the State Council bureaucratic apparatus. One interesting issue to be checked out is why and how the issue of moving out people rose to the leadership level, and to what extent the leadership change in the State Council as well as in TGPCC affected the decision to move a large number of people from the reservoir area.

The fragmented authoritarianism approach will be the framework in which the analysis and discussion of the resettlement policy change and the factor for change take place. However, as this model was developed in the 1980s, it may also be necessary to draw on theories that focus on additional aspects of policy-making. The Chinese society is constantly changing and developing, and the fragmented authoritarianism approach may not be as relevant as it was earlier. For instance, informal aspects may have become increasingly important in policymaking. One indication that this may be the case is the statement from Michel Oksenberg that 'such previous depictions as "totalitarianism", a "Leninist party state", "fragmented authoritarianism" or "bureaucratic pluralism" miss the complexity of China's state structure today'.[141] This conclusion is interesting to note, as Oksenberg was one of the scholars who developed the fragmented authoritarianism approach (in addition to Doak A. Barnett, David Lampton and Kenneth Lieberthal). Oksenberg states that the fragmented authoritarianism captured a static description of how core state apparatus worked during the 1980s, which to a considerable extent still works. Nevertheless, in his opinion, this approach alone is no longer adequate to describe the present, and he believes that it is necessary to

---

140 See Ibid., pp. 160-67 for further details regarding leaders and behavioural consequences in connection with motivations in bargaining for additional resources.

141 Oksenberg (2001), p. 21.

include the state-society interactions in any comprehensive model of the Chinese system.[142]

As stated by Lieberthal and Oksenberg in *Policymaking* in the late 1980s, the reforms had increased the need for consensus building and bargaining for large energy projects, as these would involve large resources. One aspect that may be more important than is portrayed in the fragmented authoritarianism approach is *information*. Leaders depend upon information provided to them by bureaucratic agencies, research institutes and so on, in order to either make decisions or to justify decisions already made.[143] According to some scholars (Naughton, Halpern), control of information is one important factor in the decision-making process in China in general. Naughton claims that the two key resources that structure bargaining positions in relation to large investment projects are control of information and skills and control over resources.[144] The fragmented authoritarianism model has mainly focused on the control over resources, almost to the exclusion of the control of information. According to Nina P. Halpern, little attention has been paid to this source of informal authority within the bureaucracy.[145] As Halpern states, control of information could be an important influence of the information flows within the government that could possibly enhance the leaders' ability to formulate co-ordinated policies and influence their decision-making. Halpern introduces the model 'competitive persuasion', which she suggests captures the bureaucratic process in which the research centres in her study were engaged.[146] These research centres were formally subordinate to the Premier's office or a leadership group. The creation of the research institutions and information provided by 'in-house' research facilities had two effects: the institutes increased the leadership information on policy choices, as well as increased competition in relation to the line ministries/bureaucracies to their policy proposals. It encouraged a broader vision in policymaking that had not been common under Mao.[147] Information and knowledge may have increased importance in the resettlement decision-making process for the Three Gorges project, and will be commented on in the conclusion chapter.

---

142 Ibid., p. 28.

143 The latter was typically true for the Three Gorges project where the decision by leaders to construct the project was first made. Feasibility studies were then carried out to prove that the project was feasible (xian juece, hou lunzheng). For more details on the decision-making process of the Three Gorges project see chapter 6 in Heggelund (1993).

144 Lieberthal and Lampton (eds. 1992), p. 12.

145 Halpern's essay deals with research institutes that were established in the 1980s, mainly under the former Party Secretary Zhao Ziyang to develop policy options for the top leaders. She emphasises in her article that competitive persuasion is intended to apply to normal bureaucratic decision-making, and not as a response to crisis situations. Halpern (1992).

146 Ibid., p. 126.

147 Lieberthal and Lampton (eds. 1992), pp. 14-15.

A second area that is necessary to mention in relation to policymaking, is the study of *informal politics*.[148] The scholars who advocate informal politics as a way to illustrate decision-making, focus on the interaction between formal and informal politics, an area which they believe has been neglected in the study of Chinese politics. The approach builds to some extent on to the cultural approach as it regards the personal relations, *guanxi*, as important in the decison-making process. Joseph Fewsmith[149] states that informal politics have been vital to the reforms, as reform is characterised by rapid change, and cannot be carried out through bureaucratic channels. Furthermore, he states that informal politics 'has been critical for identifying and raising issues, for bringing new information to bear on the analysis of problems, and for proposing new policy recommendations'.[150] This approach may also be useful in shedding light on the decision-making process for the Three Gorge resettlement change, and its relevance will be commented on in the conclusion chapter.

Summary

Three general approaches have often been employed in the studies of policymaking: the rational choice model, the organisational process model and the bureaucratic politics model. The chapter gives an overview of some selected central analytic approaches and theories that have been employed to discuss politics and decision-making in China by Western scholars. The totalitarianism and the two-line struggle models have more or less been abandoned by scholars as a method to explain Chinese politics due to the changes on the political arena in China and opening up of the country to the outside world. This development has increased the availability of new research material to Western scholars and access to Chinese people in general, which has contributed to a higher level of sophistication in the analysis of Chinese politics. The approaches discussed in the chapter were factionalism and elite conflict, the clientelist model, the interest groups model, the culturalist approach, the bureaucratic politics model and the fragmented authoritarianism model. The conclusion from the brief overview is that the approaches described in this chapter must be regarded as mutually complementary in portraying the political reality in today's China, as they describe different components of the Chinese policymaking process.

With regard to the approaches that have been employed in the study of China, the fragmented authoritarianism approach stands out as particularly relevant in the study of Chinese policymaking in relation to the Three Gorges project. Lieberthal and Oksenberg state that authority is fragmented and that in the bureaucratic structure authority consists of four tiers: i) The core group of twenty-five to thirty-

---

148 For a discussion of the importance of the informal nature of politics in China see Dittmer, Haruhiro and Lee (eds., 2000).

149 Fewsmith (2000).

150 Ibid., p. 163.

five top leaders who articulate national policy; ii) the layer of staff, leadership groups, research centres, and institutions which link the elite to and buffer them from the bureaucracy; iii) State Council commissions and ministries that have supra-ministerial status and co-ordinate activities of the line ministries and provinces; and iv) line ministries which implement policy, the provinces and corporations. The fragmented structure of authority has seemingly produced increased bargaining in the Chinese bureaucracy, and the economic reforms have increased the need for consensus building in decision-making for large energy projects. Consensus building among the actors above is important to the policy process. Lieberthal and Oksenberg state five factors that can propel an issue onto the agenda of the top leaders: particular interests of individual top leaders; Chinese bureaucrats can try to force an issue onto the agenda of the highest leaders; the emergence of a critical problem; foreigners may force an issue onto the agenda; and procedural requirements (Five-year-plans). The hierarchical chain of organisation is fragmented and increases the need for consensus building: central versus locale; the *xitong* (the system) and the relationship between the vertical and the horizontal systems; bureaucratic ranks; individual rank and stature; interagency relations and the system of overlapping authority.

The fragmented authoritarianism approach is the starting point for the analysis in this book for the following reasons: the case study carried out by Lieberthal and Oksenberg illustrates that the fragmented authoritarianism approach has been a relevant approach for the decision-making process of Three Gorges project; bargaining over resources is relevant in the fragmented authoritarianism approach, and this will be discussed in relation to the resettlement policy change and the environmental policymaking for the project in the conclusion chapter of the book. In addition, as the fragmented authoritarianism model was developed in the 1980s and changes have taken place in Chinese society, it may be necessary to draw on theories that focus on additional aspects of policymaking such as the importance of information in the desicion-making process, and informal politics and personal relationships (*guanxi*).

Chapter 4

Resettlement Experience in China

This chapter gives an introduction to the resettlement history and experience in China in general, in order to set the stage for Chapter 5, which concerns the resettlement change and the concrete experience from the Three Gorges resettlement. Reservoir-related resettlement in China has often been unsuccessful. Nevertheless, some developments have taken place in this area in China, which will be demonstrated in this chapter. The extent to which this development has been successful in the Three Gorges project will be discussed in Chapter 5.

Before going into the concrete dam developments in China, it is necessary to introduce some facts about what is going on in international dam building in general in order to link the debate in China to what is happening internationally. In the Western/developed world, large-scale dam projects have become increasingly controversial, and there is an ongoing debate in these countries about the need for such large projects, as the negative impacts for people and the environment have received increased focus. Large-scale dam building has seemingly become a developing world phenomenon, with criticism coming from the Western world and from some NGOs in developing countries. This chapter looks into the recent developments in the international dam building debate and relates these developments to the Three Gorges project.

In China, the resettlement experience from early dam projects has laid the foundation for the new resettlement policy of the Three Gorges project. In order to set the background for the development of the Three Gorges resettlement policy it is therefore necessary to give a brief introduction to the general resettlement history in China.

The reservoir resettlement history urged the Chinese authorities to think about new alternatives in resettlement work. The *kaifaxing yimin fangzhen* (developmental resettlement scheme) was developed gradually as a result of this new thinking about reservoir resettlement in general in China. It is also closely linked to the development of the Three Gorges project, which will be illustrated in this chapter.

International Dam Building

The purposes for dam building are usually irrigation, electricity production, flood control and improvement of navigational facilities on rivers involved. Dam construction is often part of a country's development scheme, and has an overall goal intended to benefit the entire nation. Prestige is another important factor, in addition to the above mentioned benefits, that has led many governments in the West and the East to construct large dams in the last decades. 'Big dams have been potent symbols of both patriotic pride and the conquest of nature by human ingenuity'.[151] Dams are often controversial due to large numbers of people who must be involuntarily resettled, as well as negative environmental impacts. Large dams generally displace more people than other types of physical infrastructure projects, as they inundate fertile farmland along rivers. Families from reservoir resettlement are more vulnerable than people who are moved due to other infrastructure projects, as entire villages or townships may be relocated to distant places, where the farmland may not be as fertile. In comparison, urban development projects such as road building is perceived as easier, as people may keep their jobs but are provided with new (and in many cases improved) housing. According to McCully,[152] the dam building industry has continued to build dams over the past decades without questioning the damage done or evaluating whether the promises of water, power and food actually materialised. McCully states that nearly 1200 dams with a height of at least 15 metres were under construction at the beginning of 1994.[153] Furthermore, the dam-building industry now concentrates on the developing world since most rivers have been exploited in the industrialised world, and because there is too much resistance against dam building. In many developing countries, the political systems do not permit the same type of opposition.

Nevertheless, during the last decade, the dam building trend has been turning. The reason for this is that, in recent years a debate has been going on between proponents and opponents of dams, which has resulted in increased negative focus on large dam projects. It has also been suggested by some that the dam building era has come to en an end.[154] The obvious reasons for this are mainly environmental impact, high costs, and social problems due to resettlement. A new argument has recently been added to the dam building debate which challenges the notion that hydropower is clean power; research has found that all reservoirs (not only hydropower reservoirs) emit greenhouse gases due to rotten vegetation.[155] Historically, the social issues surrounding reservoir resettlement have been neglected in dam-building in China as well as in the rest of the world. Restoring people's livelihoods has been, and still is a difficult issue in resettlement due to dam construction. Also, an additional explanation for the increased attention paid

151 McCully (1996), p. 1.

152 Ibid., pp. 236-7.

153 Ibid., p. 22. According to the International Commission on Large Dams (ICOLD) dams above 15 metres in height are defined as large dams.

154 See Probe International (14 November, 2000).

155 World Commission on Dams (2000), p. 75. The report states that an initial estimate suggests that the gross emissions from reservoirs may account for between 1 percent and 28 percent of the global warming potential of GHG emissions. The large scope suggests that further studies are necessary to understand the exact contribution of reservoirs to the GHG emissions.

to dam building and reservoir resettlement is that more information has become available regarding the negative impacts of dams on the environment and the social difficulties that may arise. This is due to efforts by academics and activists who have been collecting data that show the extent to which dam building is damaging the environment and the national economy, and that dams have not fulfilled their promises. Internationally well-known groups such as International Rivers Network and Probe International have contributed to the dissemination of information about the negative impacts of dams.[156] In addition, citizen protests against dams have grown and become better organised and have fought projects on the local, national and international level. One such example is the national movement, the Narmada Bachao Andolan (Struggle to Save the Narmada River) in India.

The establishment of the Inspection Committee by the World Bank in 1994, a three member, semi-independent body whose purpose was to monitor and ensure compliance with dam related policies, was an attempt by the World Bank to clarify rights and responsibilites with regard to borrowers.[157] It eventually became clear that the Inspection Committee would be unable to curb the controversies. The dam building debate resulted in distrust between proponents and critics of large dams, which was unproductive for further communication. Therefore, in 1997 the World Bank and the World Conservation Union (IUCN) sponsored a meeting in Gland in Switzerland between the proponents and the critics of large dams. The purpose of the meeting was to discuss and review dam building, as well as discuss the World Bank 1996 review of the impact and performance of large dams.[158] The establishment of the World Commission on Dams (WCD) in 1998 was one outcome of the Gland meeting. The WCD consists of 12 members representing global expertise on dam building, diversified opinions on the subject, and stakeholders perspectives. The purpose of the Commission was to review the effectiveness of large dams and assess alternatives, as well as to develop internationally acceptable criteria; guidelines and standards for dam building.[159]

In 2000, the World Commission on Dams published the results of their review on international dam building, the *Dams and Development, a New Framework for Decision-Making*.[160] The report concludes that even though dam building may have been justifiable, and has contributed to human development, there is no doubt that both people and the environment have suffered, and in many cases an unnecessary

---

156 See www.irn.org/index.html and www.probeinternational.org/pi/index.cfm?DSP=home.

157 The background for the establishment of the Inspection Committee was the Sardar Sarovar controversy in the Narmada Valley in India, where the World Bank pulled out of the project in 1993, which resulted in a split between the government of India and the World Bank regarding the conditions attached to loans. Bissel (2001), p. 38.

158 See World Bank (1996). A country study of China's experience with dam construction and resettlement was undertaken. See Fuggle and Smith (2000).

159 World Commission on Dams (2000), pp. 27-8.

160 The existence of the World Commission on Dams ends with publication of the review. Information about the WCD's work and the review can be found at www.dams.org/.

price has been paid. The report asserts quite bluntly that dams often have not met their targets.[161] This questions the usefulness of dam building; if performance does not provide the promised benefits, then the price paid by communities is even higher. It is too early to indicate what the impact of the report may be. However, the major developing agencies will have to take the report's conclusions into consideration. The Asian Development Bank (ADB) hosted a regional workshop in February 2001 to discuss implications of the report for the Bank, with representatives from its member states in Asia, and the recommendations were to be reviewed in terms of the Bank's operational guidelines. The World Bank began a formal dialogue on the WCD report during the meeting of its Board in mid January 2001, and was consulting member states in preparation for a full discussion of the report.[162]

Despite the negative attention being paid to dam projects in recent years, a number of countries, in the third world in particular, continue to believe in large dam projects as good for the country. In Asia, Vietnam, Thailand, Indonesia and Laos[163] in addition to China and India, are all examples of countries that still engage in extensive dam building. Two shared perceptions that exist about dam building in these countries are:

1. *National development*: A common feature among the countries in Southeast Asia and China with regard to dam building is that dam projects are perceived as giving a potential for modernisation and national development. It takes place in a country's modernisation process, where the main rationale of building a dam is to contribute to economic growth and to satisfy the rapidly growing demand for electricity. In many countries dam building has come to be a symbol of national development, as well as a barometer of development progress.
2. *National interests versus local interests*:[164] National interests as opposed to regional and local interests is an argument used in many countries, including China, in order to justify a dam project vis-à-vis the public, and especially the local people who must move. The size and influence of individual dam projects puts them at the level of national importance, yet their localisation affects a number of local communities. Large dams are typical examples of appropriation of locally-based resources in the name of national interest and national development, as they inundate large areas of land and forests to

161 World Commission on Dams (2000), p. xxviii and p. 21.

162 See World Commission on Dams (26 January 2001). For information about the ADB workshop see Asian Development Bank (ADB, 2001a). For information about ADB's ongoing and planned responses to the WCD's review see Asian Development Bank (2001b).

163 For more information about dam building in Southeast Asia see Hirsch and Warren (eds. 1999).

164 Although in some cases when the main function is irrigation or power supply, one may also say the dam serves a regional interest. I am grateful to Eduard Vermeer for this comment.

provide power, irrigation benefits and flood control. National interests are prioritised and the local population is expected to give up their own interests in the name of the public interest.

This last point is particularly relevant for the population to be resettled, in particular the rural population which is dependant upon the natural resources in their neighbourhood.

With regard to the dam building in China and large dams such as the Three Gorges project, it is clear that the international debate about the negative effects from dam building and reservoir resettlement has had consequences. As mentioned above, international funding institutions have become more reluctant to fund large, controversial projects such as the Three Gorges project due to social and environmental costs involved. This is also due to these institutions' experience with controversial dam projects in a number of countries. The World Bank, for instance, the largest foreign provider of financing for large dams world-wide,[165] withdrew from the Sardar Sarovar project in India in 1993,[166] and then from the Arun III in Nepal in 1995. The World Bank has also shied away from the Three Gorges Project. Other agencies have also withdrawn their support for the Three Gorges dam project. The US Bureau of Reclamation pulled out of its contract to provide technical support for the Three Gorges Dam in 1993, with its Commissioner saying that the dam was 'outdated and overly expensive'.[167] Furthermore, the President of the Export-Import Bank of the United States stated in 1996 that the Ex-Im Bank could not issue a letter of interest for the project because the information the Bank had received 'fails to establish the project's consistency with the bank's environmental guidelines'.[168] These explanations may be true; nevertheless, reasons may also be related to the global opposition and controversy surrounding certain dams, and the fact that international NGOs put heavy pressure on the funding agencies. Thus, one could say that international NGOs that engage in campaigns against dam building have a certain amount of power in the dam building debate. Therefore, controversial projects are not as easily funded by international agencies as they were earlier. A debate has also been ongoing within China in relation to the Three Gorges project in particular (as is discussed in Chapter 2), where the participants of the discussions mainly were bureaucrats and scientists. Arguments against the plan for such a large-scale dam on the main river were common, and critics instead promoted the construction of smaller dams on the tributaries in order to avoid large-scale resettlement. The final go-ahead for the

165 The World Bank has lent some US$ 58 billion (in 1993 dollars) for more than 600 dam projects in 93 countries. McCully (1996), p. 19.

166 For a critical review of this project see Jarman and Scrivener (eds. 1992).

167 McCully (1996), p. 21.

168 Kamarck, Martin A. Statement of the Board of Directors of the Export-Import Bank of the United States, by Martin A. Kamarck, President and Chairman at a Three Gorges Press Briefing, Thursday May 30, 1996.

Three Gorges dam in 1992 illustrates that the construction of a large-scale dam was accepted at the time as an important step in the country's modernisation process.

Past Resettlement Experiences in China

After the People's Republic of China (PRC) was established in 1949, tens of millions of people have been resettled due to construction of roads, railroads, dams and factories.[169] Hydrological experts from the former Soviet Union assisted China in developing extensive dam building during the first years of the PRC, especially during the Great Leap Forward. An average of more than 600 dams were built every year in China in the three decades following the establishing of the PRC.[170] Many of the projects were rushed into operation, which involved serious delays and cost-overruns.[171] Chinese authorities claim that altogether 10 million people have been resettled due to the construction of water conservancy and hydroelectric projects, while others believe that the number is much higher.[172] It is generally acknowledged even by Chinese authorities that resettlement has been unsuccessful, due to the lack of comprehensive resettlement plans. Officially, about one-third of the resettlement has been declared unsuccessful, which suggests that more than 3 million people have not been satisfactorily resettled and compensated. It is also generally acknowledged that among the three million that have been fairly successfully resettled there are still problems which was stated by former Premier Li Peng in 2000.[173] Finally, the remaining one-third is regarded as successful. The lack of success has been blamed on the fact that emphasis has traditionally been put on the construction of the project, rather than resettlement (*zhong gongcheng, qing anzhi*).[174] China introduced the first resettlement regulations in 1952, which provided compensation for land expropriated for state construction projects. Despite this regulation, resettlement before the 1980s is regarded as quite unsuccessful, even by Chinese authorities. The main focus was on moving the

---

169 For a general introduction to the resettlement strategies and experiences in China, see World Bank (1993).

170 McCully (1996), p. 19.

171 The Gezhouba project on the Yangtze River by Yichang in Hubei is one example. Work began on Mao's birthday in 1970. Due to poor planning the whole project had to be redesigned; it was completed in 1989.

172 See Li (1992a), p. 11. However, World Commission on Dams (2000), p. 104, states that independent sources in China believe the number is much higher, with 10 million people resettled in the Yangtze Valley alone. Critics of the Three Gorges project, such as Dai Qing, believe the number of resettled people in China is closer to 40-60 million people. McCully (1996), p. 67. Three dams are known for large resettlement in China, the Danjiangkou Reservoir 382,000, Xin'anjiang Reservoir 274,000 and the Sanmenxia on the Yellow River 319,000. Zhu and Zhao (eds. 1996), p. 239.

173 Li (11 January, 2000). Li mentions Danjiangkou and Xin'anjiang as examples of resettled population whose livelihoods still have not been restored.

174 Zhu and Zhao (eds. 1996), p. 240.

people away from the areas to be inundated, and, as a result, the relocatees were not successful in restoring their livelihoods. The rural population in particular was neglected in the resettlement process, and their livelihoods and the development of the reservoir areas were seldom or never taken into consideration. In many cases the resettled rural population had to move a second time. If any compensation was given, it was based on administrative decisions, and not on the actual needs of the relocatees.[175] Often the resettlement funding was spent on other things such as the construction of buildings for local governments, etc.

One of the typical dam projects that is often referred to as a failure project both with regard to resettlement and technical issues is the Sanmenxia (Three Gate Gorge) dam on the Yellow River, which was completed in 1960.[176] Approximately 400,000 people[177] were unsuccessfully resettled due to the dam to the western and more arid areas of the home province Shaanxi, as well as to equally arid areas such as Ningxia and Xinjiang, where they did not have ways to make a living. The resettlement took place during the Great Leap Forward (1958-59), when the rural population was resettled without compensation. The resettled population wished to return to their homes in Shaanxi, as they had nothing to eat. Some land that was previously occupied by the relocatees was never inundated. Moreover, due to the siltation problem of the dam, the operating rules for the Sanmenxia had been changed, and a large backwater area became available for farming. At the time, this land was not returned to the relocatees, but was given to state farms that were run by soldiers. The relocatees petitioned the government for years to return to their land, and finally, in 1985, the State Council issued decree no. 29,[178] which allowed 150,000 of the resettled population to retrieve the land that was occupied by state owned farms. The first relocatees began returning to their land in 1986. In the early 1990s the people resettled by the Sanmenxia dam were still living in poverty.[179]

In the late 1970s, political changes took place in China when Deng Xiaoping introduced a new economic policy. This also influenced resettlement policy, as greater focus was put on laws and regulations. A series of laws and regulations were introduced to regulate resettlement, including reservoir resettlement. One important regulation in 1981 required all hydroelectric power stations to allocate RMB 0.001 per kWh for a 'reservoir maintenance fund'. The purpose of this fund was to restore living conditions or improve infrastructure in reservoir resettlement.

175 Ibid., p. 240.

176 See Leng (1996), p. 69. The storage capacity of the Sanmenxia reservoir was quickly reduced in the first years of operation due to taking in too much silt; 93 percent of all silt remained in the reservoir.

177 McCully (1996), pp. 322-3. World Bank (1993), pp. 3 and 5 says 319,000 were displaced first, 430,000 people including the population growth. Leng (1996), p. 60 says 450,000 people, which includes the population increase.

178 Guanyu Shaanxi sheng Sanmenxia kuqu yimin anzhi wenti de huiyi jiyao (Meeting summary regarding the problem of the Sanmenxia reservoir resettlement in Shaanxi province) in Leng (1996), p. 84.

179 See Leng (1996), p. 89. One example is given from Fuping county where more than 80 percent of 1240 families lived in poverty in 1992.

This policy was introduced because of the social instability that existed in the resettled population due to the failure to improve their economic conditions. A survey was made in 1984 which showed that reservoir resettlers were still quite poor, and in 1985 the Ministry of Water Resources and Electric Power (MWREP) instituted new reservoir resettlement design requirements.[180] The significance of these regulations was that they required project design and resettlement planning to take place at the same time. Also, the regulations stipulated that resettlers' real income should be maintained, as well as compensation of lost assets. Furthermore, local governments and design institutes were to share the design responsibilities. Local governments were also to be responsible for planning and budget estimates for local resettlement. In 1986, MWREP required projects to include resettlement funding in the overall project budget. The following year, 1987, a MWREP regulation insisted that the design requirements introduced in 1985 should be followed, as well as 'forbade approval of any design not containing adequate resettlement plans'.[181] The most recent reservoir resettlement regulations for reservoir construction in general were introduced in 1991 by the State Council: *Dazhongxing shuili dian gongcheng jianshe zhengdi buchang he yimin anzhi tiaoli* (Guidelines for compensation and resettlement in large- and medium sized water conservancy projects and hydraulic power projects construction).[182] They contain most of the principles of the 1985 regulations, and state that if they are followed, the resettled population will be able to gradually recover or surpass their former standards. In 1993, the resettlement regulations for the Three Gorges project were promulgated, *Changjiang Sanxia gongcheng jianshe yimin tiaoli* (The Resettlement Regulations of the Three Gorges Project) (see footnote below). On 25 February 2001, the edited Three Gorges resettlement regulations were issued. The new version was issued due to problems that had appeared in the process, which will be discussed further in Chapter 5. The resettlement regulations require both Chongqing municipality and Hubei province to issue resettlement outlines, *Changjiang Sanxia gongcheng shuiku yanmo chuli ji yimin anzhi guihua dagang* (The Outline for managing the Yangtze Three Gorges reservoir inundation and resettlement programme) that are approved by the TGPCC. The resettlement Regulations provide the range of compensation rate, while the Outline sets specific rates for the farmland, entitlement for each resettled person, how many *mu* per person, compensation for cultivated land, etc.[183] In addition to the resettlement

---

180 World Bank (1993), p. 10. Planners were required to: a) minimise resettlement, b) maintain real income of resettlers, c) compensate lost assets at replacement cost, d) attend to special needs of resettling minority peoples, and e) design replacement housing to conform to local styles.

181 Ibid., p. 11.

182 Changjiang shuili weiyuanhui (ed. 1997), p. 63.

183 I am grateful to resettlement expert Zhu Youxuan for the details. See Zhonghua renmin gongheguo guowuyuan ling (1993), article 7; and Zhonghua renmin gongheguo guowuyuan ling (2001), article 9.

regulations, the local governments have issued management methods for the resettlement funding for the reservoir area.[184]

In addition to the country's own experience in reservoir resettlement, foreign influence has been important in improving resettlement regulations in China. The World Bank in particular has played an important role here, as the bank has financed a large number of projects since the 1980s, including dam projects that include reservoir resettlement.[185] One of the most recent projects is the newly completed Xiaolangdi project on the Yellow River that involved the resettlement of 181,000 people. Although China had begun to develop its own regulatory framework for reservoir resettlement, the dialogue between the World Bank and China about this issue has been productive. Since the World Bank's early involvement in a Chinese dam project in 1984, the Lugube project, Chinese regulations have developed rapidly.[186] The World Bank Operational Policies (OP) 4.12 on Involuntary Resettlement[187] for instance, set strict demands for both planning and follow-up for resettlement. The principles in the Chinese Three Gorges resettlement regulations are in many ways similar to the main principles of the World Bank OP 4.12, as they put emphasis on the importance of restoring or improving the livelihoods of the resettled population.[188]

New Resettlement Method: *The Developmental Resettlement Policy*

The principal method of compensation that has been employed in reservoir resettlement in China was the 'lump sum' (*yicixing*) type compensation.[189] This was a simple method of returning to the peasants the amount of money that the land and house were worth, according to established standards set by the central government, with no thought of how the resettled people were to live in the new areas. There was no focus on the production or the development. Often, when short of construction funds, the resettlement money was reduced, and sometimes resettlement funds were used for various other purposes. In many cases the resettled peasants lived in poverty after being resettled, and the peasants often returned to their home areas when they could not sustain themselves in their new

184 *Hubei sheng Sanxia gongcheng baqu yimin zijin caiwu guanli banfa* (Financial management methods for the Three Gorges reservoir area resettlement funding of Hubei province), issued by the Hubei provincial government in 1994. Zhongguo Sanxia jianshe nianjian bianjibu (1996), pp. 30-32.

185 The World Bank has also influenced the development of similar resettlement regulations of other developing agencies, such as the Asian Development Bank, Overseas Development Administration of UK, and the OECD. World Bank (1996), p. 102.

186 Ibid., p. 98-9.

187 These were the former OD 4.30 on involuntary resettlement. See World Bank (2001e).

188 See article 3 of the Regulations. Also, for a discussion about the World Bank OD 4.30 and Three Gorges project resettlement regulations see Heggelund (1994), pp. 31-7.

189 See Li (1992a), p. 9.

areas. Sometimes the central authorities at a later stage have decided to invest in areas where relocated people have settled, in order to improve their lives.[190] Li Boning states that the frequent use of 'lump sum' compensation was due to the lack of experience in resettling people in dam projects, as well as lack of knowledge about the complexities regarding the resettlement issue.[191] In addition, the 'leftist' political climate at the time was the main factor behind the 'lump sum' resettlement scheme and the political mobilisation in order to have the people resettled.

Due to the problematic resettlement history in China, a new awareness of the need to ensure the livelihoods of the relocatees gradually emerged, in particular the rural population. One source of this new awareness is the resettlement officials' personal contact with the resettled population, and their knowledge of continued poverty in the rural population. A second source for the raised awareness must be related to the scientific research and articles published about resettlers' living conditions.[192] Consequently, the requirements issued in the mid-1980s (mentioned above) required incorporation of economic development plans in resettlement strategies, as opposed to the 'lump sum' (*yicixing*) compensation. The Three Gorges project was being debated during this period, and the reconstruction of livelihoods for the relocatees was one important issue. The discussion about the project is clearly linked to the appearance of new resettlement regulations in general in China. A new resettlement plan called 'development type resettlement' (*kaifaxing yimin fangzhen*) was gradually developed for the Three Gorges project. The rationale behind the developmental resettlement scheme is to solve the daily subsistence problems of the resettled population. Thus, as opposed to old resettlement schemes where this was more or less ignored, the resettlement policy initiated for the Three Gorges project intends to develop the economy and infrastructure in the reservoir in order to ensure the livelihoods of the resettled people. This policy was not intended for the Three Gorges project alone, but it is closely related to this dam project as the project was being discussed during the 1980s. As the Three Gorges resettlement discussions influenced the resettlement developments in general in China, the history will be described briefly below.

The history of improving the developmental resettlement scheme goes back to 1965 when one of the ardent proponents of the Three Gorges project, Lin Yishan,[193] put forward the idea of *yimin gongcheng* (resettlement project), which

190 One example is the Danjiangkou reservoir that commenced construction in 1958 and was completed in 1974; in 1984, the authorities invested RMB 300 million in the area in order to improve people's lives.

191 Li (1992a), p. 12.

192 See for instance Chen (1999a), which is a collection of articles published in the 1980s and 1990s on the Three Gorges dam and other environmental issues. In September 2001 the document was still unpublished due to sensitivity over the Three Gorges dam.

193 The former head of the Yangtze Valley Planning Office (now the Yangtze River Water Resources Commission YRWRC). For further details on Lin Yishan see the Three Gorges project history chapter.

also included the concepts of development and production, using the resettlement funding for this purpose.[194] Lin presented his thoughts to the State Council in October 1978 in the report *Guanyu Sanxia shuiku yimin wenti de baogao* (A report about the Three Gorges Reservoir resettlement issue), and to Deng Xiaoping in April 1979, *Sanxia shuiku yimin de baogao* (Report on the Three Gorges Reservoir Resettlement). During the 1970s and 1980s the construction of the Gezhouba and Danjiangkou dam projects were under way, and experience from these two projects was to be included in the preparations for the Three Gorges resettlement. In February 1984, the Central Finance Leading Group (*Zhongyang caizheng lingdao xiaozu*) put forward the idea of moving from a settlement type resettlement to a developmental resettlement scheme. In June the same year, the MWREP convened a meeting in Yantai, Shandong, where the serious resettlement problems of the Danjiangkou project were also discussed.[195] In reality, it was also a meeting to prepare for the Three Gorges resettlement. In September a few months later, The State Council Three Gorges project Preparatory Leading Group convened a meeting where the then Vice-Premier Li Peng stated that based upon 30 years of resettlement experience, it was now necessary to change from a settlement-type resettlement towards a developmental resettlement scheme.[196] He expressed a year later that the main reasons for the resettlement problems were that resettlement policy and plans relied completely on a subsidised-type resettlement. Instead he emphasised that it would be necessary to develop a number of areas in addition to the agricultural sector, such as forestry, fruits, etc. The formal acceptance of the new resettlement concept, developmental resettlement, was included in the *Guidelines for compensation and resettlement in large- and medium sized water conservancy projects and hydraulic power projects construction* issued by the State Council in 1991.[197] The *Guidelines* state that 'The State recommends and supports developmental resettlement', and state further that 'this is the guiding ideology of the legislation and a policy to abide by' The *Guidelines* formally put the developmental resettlement policy into the nation's laws and regulations, and all resettlement projects must abide by this regulation.

Leading up to the acceptance of developmental resettlement is the trial resettlement experience, initiated in 1985. The purpose of the trial resettlement was to explore the best options for the resettlement scheme and to obtain more

194 Changjiang shuili weiyuanhui (ed. 1997), p. 58.

195 Danjiangkou on the Han River began construction in 1958, and was completed in 1973, and displaced 383,000 people. When 100,000 in the first phase were displaced, only a little bit of funds were given them, and little attention was paid to their production potential. Two thirds of the displaced returned to the reservoir area shortly after having been displaced due to insufficient planning and conditions. Li, Heming (2000), p. 75.

196 At the same meeting, the establishment of the Three Gorges special area was discussed.

197 Li (1992a), p. 12. *Dazhongxing shuili dian gongcheng jianshe zheng de buchang he yimin anzhi tiaoli* (Guidelines for compensation and resettlement in large and medium sized water conservancy projects and hydraulic power projects construction).

experience prior to the construction of the Three Gorges project. The Chinese authorities have allocated RMB 460 million[198] for the trial resettlement since it was initiated in 1985. 25.6 percent of the funding was spent on agriculture and soil improvement and establishing orchards (fruit, tea etc); 11.5 percent of the funding was spent on relocating factories in counties and towns (2,275 people); 51.4 percent was invested in infrastructure in towns, including the construction of roads, water and electricity, and postal and telecommunications. Totally these efforts would enable 85,000 people to resettle. In addition, some funding was spent on training of students, peasants and cadres, altogether 20,000 people.[199]

One important part of the authorities' resettlement policy is *duikou zhiyuan* (counterpart support).[200] This policy was initiated in 1992, when the dam project was approved by the NPC.[201] At the time, the State Council issued an appeal for the 26 provinces (regions and cities) to participate in the development of the Three Gorges area. The purpose of the counterpart support is to provide extra funding in addition to the project funding allocated by the government to implement the population resettlement and the relocation of enterprises. The policy is part of the strategy to develop the reservoir area in order to provide livelihoods for the resettled population (both urban and rural). Provinces, cities, ministries, enterprises are obliged to provide either funding or equipment, investment, and training of resettled people in the affected counties, towns and villages of the Three Gorges reservoir area. In 1994, the TGPCC together with Ministry of Internal Trade issued a notice that decided which cities and provinces were to provide support for the counties and cities of the Three Gorges reservoir area.[202] For instance, Beijing municipality would support Badong county, and Shanghai municipality would support both Yichang and Wanxian.[203] The units responsible are sometimes enterprises or bureaus that give support to institutions in the Three Gorges area, such as the Beijing Municipality Food Industrial and Trade Corporation that is the counterpart for Badong County Trade Bureau, and Beijing Municipality Grain Bureau that supports the grain bureaux of Badong and Zhongxian. From 1992-1998, *duikou zhiyuan* in the Three Gorges area amounted to RMB 5.89 billion,[204] and the projects included public welfare projects, economic co-operation projects,

---

198 Zhu and Zhao (eds. 1996), p. 256. In the beginning RMB 20 million was disbursed annually, by 1990 allocations reached RMB 110.2 million. Li (1992a), p. 16.

199 Zhu and Zhao (eds. 1996), p. 256. See this page for more details, such as how many *mu* of land were opened up, how many *mu* of tea orchards, and how many roads that were built.

200 The full name of the 'project' is *Quanguo duikou zhiyuan Sanxia gongcheng yimin gongzuo* (National counterpart support for the Three Gorges resettlement work). Zhongguo Sanxia jianshe nianjian bianjibu (1996), p. 205.

201 Zhongguo Sanxia jianshe nianjian bianjibu (1997b), p. 189.

202 See Zhongguo Sanxia jianshe nianjian bianjibu (1996), pp. 210-11 for the Notice from the TGPCC and Ministry of Internal Trade regarding *duikou zhiyuan*.

203 Ibid., pp. 209-12 for an overview of all the provinces and cities that are involved in the counterpart support.

204 Zhongguo Sanxia jianshe nianjian bianjibu (1999), p. 199.

construction of schools, providing work to migrants, training and capacity building. In 1999, the external funding through the counterpart support system amounted to RMB 1.38 billion[205] on similar projects. Counterpart support meetings are convened annually in order to report on the work carried out in the Three Gorges area. It is anticipated by many that the sums that are provided through the *duikou zhiyuan* system in the end may even reach or surpass the sum allocated for the entire Three Gorges resettlement, RMB 40 billion.[206]

The Developmental Resettlement Scheme for the Three Gorges Project

The developmental resettlement scheme is based upon many years of experience from a number of projects as described above. The Three Gorges resettlement regulations were issued on 19 August 1993, State Council decree no. 126: *Changjang Sanxia gongcheng jianshe yimin tiaoli* (The Resettlement Regulations of the Three Gorges Project), which is the legal basis for the Three Gorges resettlement. Resettlement was to be carried out according to these regulations. It sets the standard for the resettlement, such as Chapter I, article 3, which states:

> The state shall adopt a developmental relocation policy during the construction of the Three Gorges project. The relevant governments shall organise and guide the relocation and resettlement work and shall assume overall control over the use of the relocation funds and the rational development of resources; *taking agriculture as the foundation and combining agriculture with industry and commerce* (author's emphasis), they shall properly resettle those relocated through various channels, various industries, various forms and various methods in order to both enable the living standards of those relocated to match or surpass the original standard of living and to create conditions for long-range economic development around the Three Gorges reservoir and for upgrading the living standards of those relocated.

Four important principles are emphasised in this article, such as the government being in charge of the resettlement organisation and implementation, the production principle, ensurance of living standards remaining the same or higher as well as economic development. There are several points that make the developmental resettlement scheme different from the previous resettlement plans. In addition to the compensation of land and housing, funding is allocated to develop the economy in the area, in order to create production opportunities for the resettled people. It unites the resettlement and economic development of the reservoir area; and the change to non-agricultural activities will speed up 'urbanisation' of the countryside. The host areas in the provinces, cities/towns or villages will receive resettlement funding.

---

205 Zhongguo Sanxia jianshe nianjian bianjibu (2000), p. 182.

206 Interview no 6, Beijing, August 1999.

The regulations state in chapter II, article 10 that:

> Those relocated should be resettled in their original villages, townships, cities and counties. If they cannot be resettled in their original villages, townships, cities and counties, they should be resettled in their original province. If they cannot be resettled in their original province, they should be resettled elsewhere in accordance with economic principles.

In accordance with the regulations (chapter II, article 10), the following concrete resettlement methods that are being used in China, including for the Three Gorges project, are:[207]

The authorities' original plan has been to resettle the rural population in the vicinity of their former homes. *Jiujin houkao* (in the neighbourhood) or *jiudi houkao* (on the spot resettlement) which means that the resettled people move above the inundation line of the reservoir. They are literally pushed further up the hills from their previous homes along the riverside. The reasoning behind this idea of resettling in the neighbourhood is of course that the relocatees would be living close to their original homes, would still be in the reservoir vicinity, customs would be the same. The resettlement funding would be kept in the area, as well as the human resources. Resettlement in the vicinity would also be an opportunity to open up new land in the area; by using resettlement funding and expertise one may change the low-yield fields into stable high-yield fields. A fourth advantage is that the administrative organisational system remains the same. Resettlement funds do not have to be handed over to other administrative units, which would be beneficial to the local governments. Furthermore, from the point of view of the resettled people, they would be happier if they don't have to move far away from their original homes. However, the largest limitation of this relocation in the neighbourhood is the environmental capacity in the higher lying areas.[208]

Jinqian anzhi (settle nearby) is the second alternative for resettling people. This type of settlement is characterised by the lack of conditions in the vicinity to enable them to live there, and refers to the villager groups that are moved a bit further out, in their own village, township or county to other villagers' groups.[209] The *jinqian anzhi* method is similar to the *jiujin houkao* in many ways, but the difference is that it uses the county administrative divisions as the base for resettling. For the resettlers in this group, their county, township, as well as their part of the village is only partially inundated. The advantages of this resettlement method: if the resettled people can be settled in their original county, it will be the same county authorities that will take care of the process, including the funding. Even if the

---

207 See Zhu and Zhao (eds. 1996), pp. 256-61.

208 The issue of resettlement and the environmental capacity is discussed in more detail in chapter 5 regarding the resettlement policy change and the environment.

209 Generally, township/town (*xiang/zhen*) replaced the communes that were dismantled in the early 1980s. The administrative village (*xingzhencun*) replaced the brigade, and the natural village (*zirancun*) or the villagers' groups (*cunmin xiaozu*) replaced the team. See Ho (2001b), p. 405.

peasants leave their village or village group, the administrative units will be the same. Furthermore, when moving to a fairly close place, the resettled people and the host population will be from the same native land. Thus, their customs would be the same, and their social networks will most likely remain identical. The difficulty with this method as well as the above method is the environmental capacity. Conflicts between the resettled population and the host population could arise, as the land given to resettled population is often taken from the host population due to the lack of available land.

The third alternative is *yuanqian anzhi* (resettlement at a distance). This group includes the people who cannot be resettled in their own county, but who may be resettled in their own province or in other provinces. The method of resettling people far away will take place when for instance projects take up farmland, when the natural resources of this area are limited, and where neither agricultural nor non-agricultural opportunities exist. There are different options when resettling people to a distant ocation: i) The organisational system remains the same, and they are moved in groups in order to preserve their original living environment; ii) The second type is scattered resettlement, meaning that resettling occurs in small groups, even according to production teams; the people are 'inserted' into other villages. Resettling groups together enables the community to keep the interpersonal relationships, and maintain the basic production groups. The biggest disadvantage of this method is if the resettled population does not have good living and production conditions, conflicts between the host and guest populations could arise, and they could easily become isolated islands in the host areas; iii) The third option for distant resettlement is to take households as a unit and insert into production teams in the host areas. However, the relocatees are apprehensive about this method, and have often returned to their home areas.

The fourth resettlement method is *touqin kaoyou* (find a way to be resettled through relatives or friends). This method is voluntary and is much encouraged by the authorities. However, only a minority of the resettled population might have the opportunity to do this. The relocatees receive a sum of money when they move to their new home areas, and the authorities do not have the responsibility for finding jobs or homes. The Geheyan reservoir in Hubei province is used as an example where this method was employed successfully.[210]

As described in the history chapter, the authorities plan to spend 40 billion yuan (US$ 4.8 billion) on the resettlement compensation and reconstruction. This means an average of 10,000 yuan per capita. However the funding is not given directly to the population to be resettled. The types of compensation may be divided into three main parts: the production resettlement fund (*shengchan anzhi fei*); livelihood resettlement fund (*shenghuo anzhi fei*); and house-rebuilding fund (*fangwu*

210 433 households, 1,631 people, which constituted 9.3 percent and 8.5 percent respectively of the total population to be resettled; they moved to 7 different provinces, 26 counties, 76 townships and villages, 318 villages (hamlets), and 128 county units, Zhu and Zhao (eds. 1996), p. 260.

huanjian fei).[211] The production resettlement fund is compensation for land to be inundated by the reservoir, such as farmland, fruit orchards, mountain hills, and fish ponds. The money is given to the regions or individual units, and not to the individual.[212] The amount of compensation depends on the quantity and quality of the land. The livelihood resettlement fund is intended for infrastructure, removal and life subsidy in the transition period.[213] These sums vary according to region, and include various subsidies for the relocated population. The house-rebuilding fund is intended to compensate for house building. The compensation depends on the size of the house and the materials used, such as bricks, concrete, wood, and so on. The house-rebuilding fund is the largest portion of the resettlement compensation for the relocatees.

Summing up

Dam building and reservoir resettlement in general are controversial issues due to the costs inflicted, and in particular due to the social and environmental impact. Dam building now seems to be more of an issue in the developing world than in the developed world, as most areas in the developed world have been developed, and there is much resistance. A debate has been going on in the international community about the need for dam building. The establishment of the Inspection Committee by the World Bank in 1994 was one attempt to clarify rights and responsibilites with regard to borrowers. The founding of the World Commission on Dams in 1998 following the 1997 Gland meeting sponsored by the World Bank/IUCN was an initiative to create understanding for the arguments of the stakeholders on both sides of the fence. It was an attempt to create a dialogue between the opponents and proponents of dam building in order to reach an agreement on the way ahead with regard to water access and energy services. In 2000, the WCD report concluded that even if dams had been useful for human development, dams had also created misery and suffering. Thus, a new approach is needed regarding the building of dams.

In China, reservoir resettlement is regarded by Chinese authorities (and others) as having been unsuccessful until the beginning of the 1980s. There are several reasons for this:

- Emphasis has been put on the construction of the dam project and technical aspects while little importance has been given to the resettlement.
- The lack of regulations at the time, and the lack of comprehensive planning.

211 This information is taken from Li, Heming (2000), pp. 192-5.

212 The exception is the self-employed, who are able to obtain 70 percent of the money. Ibid., p. 194.

213 The self-employed may obtain the full livelihood fund, while those employed by enterprises can obtain the removal fund and subsidy only. *Ibid.*, p. 194.

- The method employed: basically a lump sum compensation was given to the resettled people, with no thought of how they were to sustain themselves in their new environments.
- A lack of sufficient funding.

Gradually, the Chinese authorities have increased their understanding of the social needs and potential social instability if these are not met. In 1985, 1986 and 1991 important resettlement regulations in general were issued that intended to improve the situation for the relocated population. Specific regulations for the Three Gorges project were issued in 1993 and 2001. The 1993 regulations state the need to take agriculture as the foundation for reconstructing the rural population's lives.

The developmental resettlement scheme (*kaifaxing yimin fangzhen*), which seeks to address the subsistence problems experienced in earlier dam projects, was compiled over a long period of time, and is closely connected with the planning of the Three Gorges project. One important part of the authorities' resettlement policy is *duikou zhiyuan* (counterpart support), which is financial support provided by provinces, cities, enterprises and units as an additional source of resettlement funding. The trial resettlement which took place in the 1980s, was an important experience for the authorities in relation to the resettlement challenge of the project. The intention is to move people within their original villages, townships cities and counties. However, moving within the province is an option, as is moving them farther away to other locations. The resettlement funding includes compensation for land to be inundated by the reservoir, for the construction of infrastructure, for removal and life subsidy as well as for house building. The majority of the (per capita) resettlement funding is not given directly to the resettlers, but is handled by the local governments and resettlement authorities.

Chapter 5

The Three Gorges Project Resettlement Policy Change: Implementation Challenges

The chapter will discuss the Three Gorges resettlement policy change and the possible causes for policy change. This chapter therefore begins with an introduction of the facts surrounding the policy change in May 1999, such as the stated reason for the policy change, and why it is interesting to study this change. It continues with an introduction to the environmental capacity in the area, as this was the stated reason for moving out such a large number of people.

Furthermore, a brief introduction of the Impoverishment Risks and Reconstruction (IRR) model developed by Michael M. Cernea and used by the World Bank is given followed by a discussion of the problems which emerged in the resettlement process. This section looks into the concrete problems that have been described in the media or in articles. These problems have likely been instrumental in bringing about the resettlement policy change in May 1999. The arguments are structured according to selected points of Cernea's IRR model.

Finally, the chapter includes a discussion of several issues: The Chinese authorities' response to the problems that have emerged; whether or not a risks and reconstruction perception exists in the planning of Chinese authorities; and the relevance of the IRR model for the Three Gorges project and in China.

Changes in the Rural Resettlement Policy

On 19-20 May, 1999 Premier Zhu Rongji was present at a Three Gorges project resettlement working meeting that was organised by the State Council,[214] at which time Zhu emphasised the importance of the second phase of the construction during which 550,000 people would be resettled (before 2003). At the meeting two 'adjustments' (*tiaozheng*) were announced. One was an adjustment and improvement (*wanshan*) of the resettlement plan, which involves resettling a large number of the rural population out of the reservoir area. The second adjustment was in relation to moving enterprises, which is not the focus of this book.[215] The main

214 Lu and Jiang (24 May, 1999).

215 However, it will be necessary to touch upon the enterprises in relation to the rural relocatees finding new occupations.

change involved shifting from a policy where the rural population were to be resettled in the vicinity of their former homes (*jiujin houkao anzhi*), i.e. literally pushed up the hills along the river, to resettlement away (*waiqian*)[216] from the reservoir area, mainly to other provinces for a large portion of the relocatees. The main reason for this as stated by Zhu Rongji, was the weak and limited ecology and environment in the Three Gorges area. Thus, the environmental capacity in the area was given as the direct cause for the policy change.

Even though *waiqian* (moving out) is included as an alternative in the resettlement regulations,[217] resettlement in the vicinity has always been stressed as the favoured alternative for the majority of the rural population.[218] Therefore, the policy adjustment introduced in May 1999 must be regarded as an important change and deviation from the original and stated plan. Article ten of the Three Gorges resettlement regulations states that the relocatees should be resettled in their original townships, villages, cities and counties. If this is not possible they should at least be resettled within their own province. The last resort is to move people out of their original provinces. Nevertheless, moving the rural population out has been an unpopular choice, and not an official alternative. Discussing this alternative in relation to the Three Gorges project has always been a very sensitive issue. The reason for this is the unsuccessful resettlement in earlier dam projects in China, when moving people out of their counties and provinces was the common way of resettling the rural population. As was described in Chapter 4, the rural population has in most cases been unable to improve their living conditions after moving to new areas. The problems and issues regarding distant removal are well documented in literature over the years.[219] During the planning period for the Three Gorges project, the rural population has been moved to distant places in trial projects. For instance in the 1990s, rural populations were moved to Xinjiang, Heilongjiang, Shandong, and Hainan, but the peasants did not like it in these areas due to the vast difference in life styles and climate, so they returned.[220] These moves have not been discussed in the media. Nevertheless, the disastrous floods in 1998 along the Yangtze River were an eye opener with regard to the weak ecological condition in the areas along the Yangtze. It is therefore likely that following the floods, Chinese authorities began discussing the possibility of

216 The term *waiqian* should be interpreted as moving out of the reservoir area; *waiqian* implies both moving within the same province or to other provinces. See Lu and Jiang (24 May 1999).

217 Zhonghua renmin gongheguo guowuyuan ling (1993).

218 For the discussion about environmental capacity between the '*leguanpai*' (optimists), i.e. government officials such as Li Boning who stated that there is enough capacity to resettle the population, and '*beiguanpai*' (pessimists), scientists, Chinese People's Political Consultative Conference (CPPCC) representatives, who believe that the environmental capacity is exhausted in the reservoir area, see for instance Ling (1998), pp. 344-52.

219 See Leng (1996); Jing (1997); World Bank (1993).

220 Xia (1999).

Map 5.1 Areas to be Inundated by the Three Gorges Reservoir

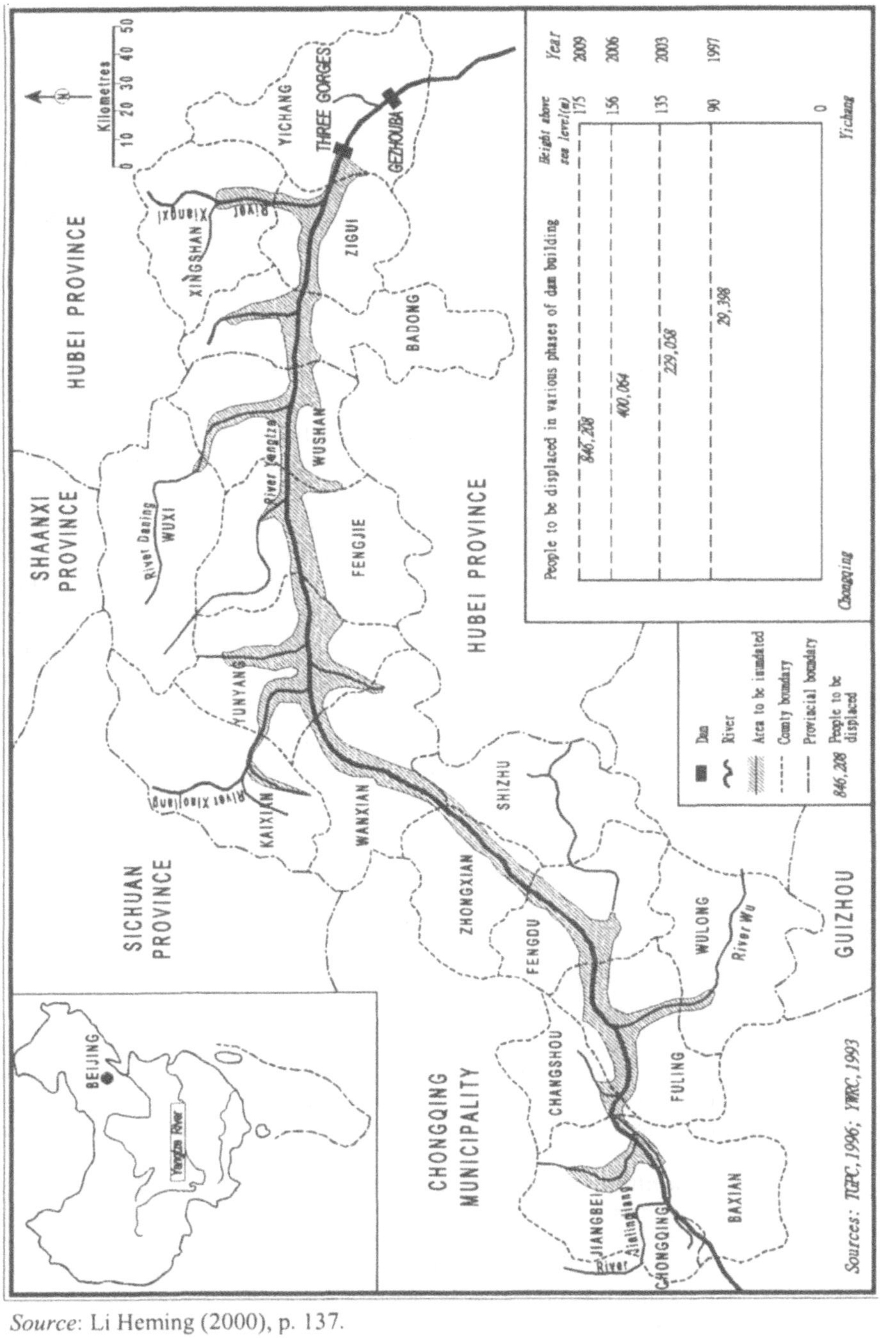

Source: Li Heming (2000), p. 137.

moving some of the people out of the reservoir area, which was made public in May 1999.

At the May resettlement meeting, Zhu stressed that due to the erosion problems as well as lack of farmland, it would be necessary to carry out the resettlement of the rural population in a number of ways. His main message was to encourage more people to move out of the reservoir area, and he encouraged creativity in the ways this was to be carried out. The main points were the following: when moving people out of the reservoir area, one should attempt to move them into the same province or municipality. Should this not be possible, move them into the neighbouring province. Furthermore, Zhu emphasised that it is important to try to move the relocatees in clusters into the host areas outside of the reservoir area, although it may be necessary to separate them. The rural population should be encouraged to find ways to resettle through relatives or friends, instead of the usual way where the authorities are in charge.[221] Zhu also stressed the importance of providing the population moving out with subsidies according to the resettlement plan. The local governments in the host areas shall be responsible for taking care of land contracts for the resettled population, purchase of or construction of housing, and the formalities in relation to the change of permanent residence (*hukou*).

As mentioned in the history chapter, the population that is directly inundated by the Three Gorges reservoir amounts to 846,200 people. Among them the non-agricultural population amounts to 484,700 people and the agricultural population to 361,500 people (see Table 5.1).[222] The 125,000 to be moved out of the reservoir area thus makes up more than one-third of the total affected agricultural population.[223] The remaining agricultural population was to be resettled within the reservoir area. The final number to be resettled, which includes the population increase in the area, will according to Chinese authorities reach 1.13-1.2 million.[224] Chongqing municipality, with a large rural population (nearly 80 percent) has the largest resettlement burden.[225] The municipality will have to move out 100,000 by 2003. 25,000 from Hubei province will be moved out of the reservoir area, but within the province borders.[226]

---

221 Lu and Jiang (24 May, 1999).

222 Zhu and Zhao (eds. 1996), p. 239 and p. 3.

223 It was first was announced at the working meeting of the Resettlement Bureau (under the Three Gorges Project Construction Committee) on 12 November 1999 that between 83,000 and 120,000 people of the rural population were to be resettled by the Three Gorges project out of the reservoir area. The number was later adjusted to 125,000.

224 References to the official resettlement vary between 1.13 and 1.2 million people. 1.13 million will be used as reference in the book.

225 However, according to a scholar involved in research for the dam project resettlement, the funding is divided in half between Chonqing municiplaity and Hubei province, as more funding was needed in Hubei. Interview no. 24, Nanjing 1999.

226 Qu (4 January, 2001).

Table 5.1 Population to be Resettled from Chongqing and Hubei

	Total *dongtai* figure*	Total *jingtai* figure**	Non-agricultural population	Agricultural population
Reservoir area	1.13 million	846,200	484,700	361,500
Chongqing municipality	–	719,398	418,133	301,265
Hubei province	–	126,802	66,567	60,235
Wanxian municipality***	–	570,874	315,118	255,756

* including the estimated population increase to take place before 2009 when the dam is completed

** not including the population increase

*** Wanxian municipality, part of Chongqing, is one of the municipalities with the heaviest resettlement loads; the rural population constitutes over 90 percent.

Sources: Zhu and Zhao (1996) and Wanxianshi Sanxia gongcheng yimin bangongshi (1996).

Destinations have been identified for the majority of the people to be moved out. For instance, approximately 70,000 people from Chongqing municipality will be moved to altogether 11 provinces and cities (see Table 5.2).[227] The trial work of moving the 70,000 people will begin in 2001, and be completed in the second half of 2003.

Table 5.2 Total Number of People to Move Out (*waiqian*)

Moving out from Chongqing municipality to other provinces	70,000
Moving out of the Chongqing reservoir area but within Chongqing municipality	20,000
Rely on relatives and friends to move out from the Chongqing reservoir area	10,000
Moving out of Hubei reservoir area within the Hubei borders	25,000
Total number of people to move out of the reservoir area	125,000

Source: Dong (18 November, 1999).

227 The 10 provinces and one city are Anhui, Fujian, Guangdong, Hubei, Hunan, Jiangsu, Jiangxi, Shandong, Shanghai, Sichuan, and Zhejiang. Dong (18 November, 1999).

Table 5.3 Resettlement Distribution of 70,000 from Chongqing Municipality

Sichuan province	7,000
Shandong province	7,000
Jiangsu province	7,000
Zhejiang province	7,000
Guangdong province	7,000
Fujian province	5,500
Shanghai municipality	5,500
Hunan province	5,000
Anhui province	5,000
Jiangxi province	5,000
Hubei province	7,000
Total	68,000[228]

Source: Dong, Jianqin (18 November, 1999).

The Environmental Capacity of the Three Gorges Area

The environmental carrying capacity in the Three Gorges area was given as the direct reason for the resettlement policy change as stated by Zhu Rongji in May 1999. It is therefore necessary to look into the environmental condition of the Three Gorges area, before moving on to some of the resettlement problems that may have influenced the policy change in the next section.

Environmental capacity is the key to a successful resettlement of the rural population, as resettlement and environmental capacity are closely linked. In an area that already is densely populated and where the land is considerably developed, protecting the environment becomes essential. This is related to sustainable livelihood for the resettled population, and the ability to settle and to produce are mutually dependent. Thus, the environmental capacity is a central issue in the resettlement controversy for the dam project. The definition of environmental capacity is given in *Questions and Answers on Environmental Issues for the Three Gorges Project,* an official guide to the dam project published by the Ministry of Water Resources and NEPA:[229] 'The environmental assimilative capacity for resettlement means the maximum relocatees number which can be borne or accepted by a given region, in the precondition of assuring the normal circulation of nature ecosystem and maintaining certain production ability, living standard and environmental quality'. A report by the Chinese Academy of Sciences

228 This adds up to 68,000 people. The article does not mention where the remaining 2,000 people will be resettled.

229 Yangtze Valley Water Resources Protection Bureau, MWR and NEPA (eds. 1999), pp. 48-9.

(CAS), *Sanxia gongcheng dui shengtai yu huanjing de yingxiang ji duice yanjiu* (Research on the Three Gorges project impacts on the ecology and the environment and counter measures) points further to the fact that the study of the environmental capacity in relation to resettlement links together several issues, such as resettlement, economic development, and ecological and environmental protection of a fixed area.[230] A third study, the *Sanxia gongcheng yimin yanjiu* (Resettlement Research for the Three Gorges Project) by the Yangtze River Valley Commission, defines environmental carrying capacity as the level of productive forces, living standard and environmental quality of an area.[231] This study also divides the environmental capacity into the actual existing environmental capacity and the potential environmental capacity that can be developed in the reservoir area.

From the above definitions, the contradiction between resettlement and environmental capacity becomes evident in the Three Gorges project. As one scholar said: 'the resettlement issue is not only an economic or social issue, but it is also an ecological issue'.[232] A number of studies have been carried out regarding the environmental capacity in the reservoir area in order to find out how it is possible to resettle the large number of people. Some scholars are quite outspoken in their conclusion '...when it comes to the understanding of the environmental capacity, we believe that the Three Gorges reservoir area already surpasses [the capacity], and that pushing the resettlers up the hills is to make matters even more difficult than they are'.[233] From this viewpoint it is clear that settling such a large number of the rural population would be a difficult and nearly impossible task.

In addition to the above-mentioned studies, a number of environmental studies have been carried out for the project.[234] Many of the studies have a common conclusion, namely that the environmental capacity in the area is questionable. The final environmental impact statement for the project, *Changjiang Sanxia shuili shuniu huanjing yingxiang baogaoshu* (*Yangtze River Three Gorges Project Water Conservancy Project Environmental Impact Statement*) [235] was approved by the National Environmental Protection Agency (NEPA)[236] in February 1992. The *Statement* was compiled by the Environmental Impact Assessment Department,

---

230 Chen, Xu and Du (1995), p. 5.

231 Changjiang shuili weiyuanhui (ed. 1997), p. 79.

232 Chen (1999b), p. 55.

233 Ibid., p. 55.

234 EIAs were completed in 1980 and 1983 by the Yangtze Valley Water Resources Protection Bureau (YVWRPB). CAS conducted a study that was submitted in 1987. In 1988 a report was compiled by the Yangtze River Water Resources Commission. In China, the construction units frequently write the EIA themselves, such as the YVWRPB, which is under the MWR, the main proponent of the Three Gorges project. The units involved assess themselves, which is considered a loophole in the environmental policy of the country and regarded as a big problem by scientists interviewed by the author.

235 The term *Statement*/EIA will be used throughout this book.

236 NEPA became the State Environmental Protection Administration (SEPA) in March 1998. Prior to 1998, NEPA will be used in the text.

Chinese Academy of Sciences (CAS) and the Research Institute for the Protection of the Yangtze Water Resources, Ministry of Water Resources (MWR). Before the approval by NEPA, an expert group consisting of 55 members assessed and approved the statement.[237]

The existing problems in the reservoir area are reflected in a number of research reports, including the official *Statement*/EIA. The worst problems are considered to be erosion and bad water quality:

1. The erosion problem in the Three Gorges reservoir area is serious and is mainly caused by two factors: agricultural reclamation and tree felling.[238] The vicious circle of opening up new land and destroying the forest for agricultural purposes has resulted in serious ecological degradation. The reclamation of land only increases the problems in the area, and the quality of the reclaimed land is often poor, which also makes it difficult to resettle the peasants.[239] Erosion covers 58.2 percent of the land in the reservoir area.[240] In Chongqing municipality it covers 59 percent.[241] Each year in the Chongqing area erosion amounts to 200 million tonnes. Of these, 140 million tonnes go into the Yangtze River.[242] According to the *Statement*/EIA, the main reason for the erosion problem is low forest coverage, which is only between 7.5-13.6 percent in the entire Three Gorges reservoir area. The forest coverage in the Three Gorges reservoir area has shrunk from 22 percent in the 1950s to 11.8 percent in the late 1990s.[243] The floods in 1998 were a result of the extensive tree felling in the upper reaches of the river, and a tree felling prohibition was immediately issued. Grazing is another reason for the erosion problems. The erosion results in sedimentation[244] of the river, and experts warn about the Yangtze River becoming the Yellow River. A report issued in February 2001 by Wuhan University states that already the sediment problem

---

237 Some of the experts did not sign the report, and listed 11 ecological and environmental reasons for this, which will be discussed in chapter 7. Interview no.13, 7 September 1999.

238 Deng (1997).

239 See Wu (2001), pp. 49-50.

240 Zhongguo Sanxia jianshe nianjian bianjibu (1996), p. 137. Some scholars state that more than 65 percent of the reservoir area is seriously eroded. See Deng (1997).

241 Interview no. 18, Chongqing September 1999, Wang Yunnong, Vice-Secretary General, Chongqing municipal people's government.

242 Li (30 October, 2000). The article states that 63 percent of the Chongqing area is eroded.

243 Deng (1997), p. 51.

244 The accumulation of sedimentation in the back end of the reservoir where Chongqing municipality is situated may affect the access by ships to the municipality. The sedimentation issue would deserve greater emphasis than is the case in this book. However, due to the nature and scope of the book this has been only one of several issues that are discussed in relation to the resettlement policy change.

is very serious in the Yangtze River, and unless immediate measures are carried out, the river bed will become higher than the land along the river.[245]

2. Water pollution is not paid enough attention to, and even though the water in the Yangtze River is considered relatively clean compared to other rivers in the country, the water quality in the river is very bad in some river sections.[246] For instance, one-third of the water discharge in Chongqing does not satisfy the national standards.[247] Chongqing discharges 1.18 billion tonnes of wastewater each year. Of this, the treatment of 940 million tonnes industrial waste water and 245 million tonnes household waste water reach 27.8 percent and 8.27 percent respectively.[248] There are 22 major cities and 394 outlets for discharge of waste water along the river down to Shanghai. Of these outlets, 30 percent have not reached the national discharge standards, and 20 billion tonnes of waste water are discharged through these outlets annually.[249] When the natural river flow slows down in the reservoir, there is concern among researchers and officials that the reservoir will become a cesspool, as most waste water in the area is discharged into the river without proper cleansing.

Water pollution problems and the erosion problems are considered to be the most serious issues in relation to the resettlement issues. Other issues that are mentioned for instance in the *Statement*/EIA are logging and the frequent occurrence of natural disasters, such as flooding and land slides. Also, there is great concern about the project's impact on biodiversity, such as the extinction of animals and fish species, which could have an impact on people's livelihoods in the area.

The above issues are discussed in the numerous reports and articles about the resettlement and environmental capacity. From literature used for this book, it is clear that the environmental capacity problem is one of the most important issues of the Three Gorges resettlement discussion. This is due to the fact that the Three Gorges reservoir area has a population of 19.461 million people, of which 14.736 million belong to the agricultural population. The rural population constitutes approximately 50 percent of the total population to be resettled in the Three Gorges Project. Developing agriculture is the key to success in the Three Gorges rural resettlement according to authorities and scholars alike. The reservoir area is densely populated, with between 250-296 persons per km^2 on average, while 130

245 *Wenhuibao* (12 February, 2001).

246 For instance in the north of China the rivers are badly polluted; this is described in chapter 6. The majority of the water in the Yangtze River ranges from grade 2 to 4 on the national scale from 1 to 5; a very small portion of the river is 5. Grades 1, 2, and 3 permit direct human contact and use as raw water for potable water systems. Grade 4 is restricted to industrial use and recreational use other than swimming. Grade 5 is restricted to irrigation. World Bank (1997), pp. 91-2.

247 Chen, Xu and Du (1995), p. 9.

248 Li (30 October, 2000).

249 Ibid.

people per km^2 is the national average.[250] However, below 300 metres in the Three Gorges Project reservoir area a density of 1000-1200 people per km^2 is common.[251] The average farmland in the area is approximately 0.98 *mu* per capita, although some sources state only 0.81 *mu*,[252] while the national farmland average in China is 1.3 mu. This would be 0.49 *mu* less than the national avarage. These figures illustrate that farmland in the reservoir area is limited. The farmland in the mountainous reservoir area constitutes 16.7 percent of the total reservoir area.[253] Due to the shortage of farmland, farming on steep hills is common in the reservoir area. This is one of the main threats to the land, as farming land with more than 25 degrees slope is to a great extent the reason for erosion in the area.[254] The problems with farming on land on gradient slopes above 25 degrees was also stressed by Zhu in his May-99 speech, where he demanded that existing farming on land above 25 degrees must be stopped, and that future farming at this level must be prohibited. 17.6 percent of the farmland in the reservoir area is land being farmed on hills with more than 25 degrees slope.[255] In Wanzhou (former Wanxian), 30 percent of the farmed land is above 25 degrees. Deforestation in the upper reaches of the river is blamed for the flooding due to serious erosion problems. A policy consequence of the floods was to prohibit all tree felling. However, some scholars believe that disaster was related to the main focus being put on the construction of water conservancy projects instead of at the same time focusing on soil conservation measures.[256] When resettlement is carried out, the people carry out reclamation and cultivation of steep hills, resulting in increased soil erosion.[257]

The Chinese authorities would originally have liked to resettle people in the neighbourhoods by pushing them up the hills, in order to keep the rural population within the Three Gorges reservoir boundaries. It has become apparent that the area cannot accommodate such a large number of people whose main occupation would

---

250 There are a number of different figures for the population density in the Three Gorges area. Interview no. 29, Beijing, May 2000 with CAS academics, who state 296 per km^2. In 1997 the population density in the area was 282 people per km^2, while the national average was 122 people per km^2, see Gu and Huang (1999), p. 353. Some state 255 people per km^2, see Wu (2001), p. 48.

251 For instance, 1060 people per km^2 as stated by Deng (1997), p. 51.

252 SEPA (1999), p. 5; and Wu, Chen, and Yu (March 2000), p. 67, respectively. According to Wu et al, hillside farmland constitutes 95.3 percent of farmland in the reservoir area.

253 The reservoir area is 57.941 km^2, of which 9.683 km^2 is farmland. SEPA (1999), p. 5.

254 Ibid., p. 13.

255 Zhongguo Sanxia jianshe nianjian bianjibu (1996), p. 137. Wu (2001), suggests 50 percent of the land of the Three Gorges reservoir area. The reason for this large discrepancy may be that there is a difference in the figures for land availability in China. Wu's figure may be in accordance with the Land Resources Survey (i.e. actual farmland) that was carried out in the early 1990s.

256 Chen (1998).

257 Erosion in Sichuan amounts to more than 40 percent of the entire province, nearly 200,000 km^2. Ibid., p. 3.

be to cultivate the land. Some officials still claim that it would be possible to resettle the entire population within the reservoir area. The reasons for policy change, they say, must be seen in light of the erosion problems and the 1998 floods in order to reduce the pressure on the environment. Nevertheless, actual knowledge of how much land is available in the area may also be an important explanation, which will be discussed later in this chapter. CAS and the MWR held different opinions regarding the environmental impact of the project and the environmental capacity in the reservoir area, which was reflected in their final *Statement*/EIA. This was the first time that such a disagreement was reflected in a report.[258] The reason for disagreement about the environmental capacity in the area, was that the MWR at the time of the feasibility study for the project used remote sensing and infrared aerial photography to get an overview of the area, while CAS used more concrete methods to measure the land capacity.[259] As one interviewee put it in 1999: ‘First it is necessary to understand the environmental situation before talking about resettling people’,[260] implying that the environmental situation and the land availability still was not clear for this project. The impression from interviews with scholars is that officials now acknowledge that the remote sensing method was not satisfactory, although it is not discussed in public. Thus, increased information regarding the actual land capacity may also have been an important factor in the decision-making.

The above illustrates that the immediate problem in the reservoir area in relation to erosion and capacity seems to be the steep-hill farming. It will be difficult to implement the prohibition of farming on slopes above 25 degrees unless there are other measures that will compensate for the loss of land. If the prohibition is implemented, there will be even less land to farm. The lack of fertile farmland is the reason why farmers are farming in the steep hills. Farming of steep hills is also one of the main reasons for the increasing soil erosion in the area, as the trees are cut down. Thus, this vicious circle is hard to break. Despite the hopeless situation, Chinese scientists have tried to come up with remedial measures. Scientists at CAS for instance, suggest agroforestry as one possible measure that could enable the farmers to continue to farm hill slopes as well as practice soil conservation. Trees would be planted in the hills where farming at more than 25 degrees slope now is taking place, and grains would be planted between the rows of trees. The erosion problems will thus be reduced. Another measure suggested by CAS researchers is the import of grain into the reservoir area in order to ease the pressure on the farmland. However, they point out that the transportation problem would be an important hinder for this policy. Thus, the majority (at least 70-80 percent) of the grain must be grown in the reservoir area or the close vicinity. The extent to which these policy measures have been taken into consideration by Chinese authorities is not clear.

258 Interview no. 4, Beijing, August 1999 with MWR official.

259 Interview no. 27, Beijing, September 1999, and Dai (1998a), p. 56.

260 Interview no. 5, Beijing, September 1999.

To sum up, the environmental capacity was given as the direct cause for the Three Gorges resettlement policy change. Erosion and water pollution are the main concerns in the reservoir area. Erosion is due to tree felling and farming on hill slopes, and the silt runs into the river. Water pollution is caused by industrial activities as well as houshold discharge of waste water. Much of the industrial and city waste water is discharged into the river without treatment. There is concern that when the river flow slows down, the reservoir will become a cesspool. The large population in the area and lack of farmland aggravate this situation. When the rural population is resettled, new land will be cultivated and new erosion problems will appear. Some measures have been suggested by CAS to relieve the pressure on the land, such as agroforestry and import of grain. The extent to which these measures will be included in the authorities' resettlement policy is unknown.

Problems Emerged in the Three Gorges Resettlement Process

An assumption in this book is that one explanatory factor for the resettlement policy change is the problems that have emerged from the resettlement process. The lack of environmental capacity in the area was the stated reason. However, there are indications that problems in the resettlement process may have influenced the resettlement policy change. These problems were not mentioned by Zhu in his announcement. Yet from the discussions surrounding the resettlement problems that have appeared in the process both in scientific journals and reports, it is likely that these have been important factors for the change. Even though the term *waiqian* (move out) was included in the regulations from 1993, it has been clear that moving people out has not been the first choice, due to the long history of unsuccessful resettlement.[261] Thus, the main purpose of the resettlement plan has been to resettle the rural population within the boundaries of the reservoir area. The rural population should continue to base their living on agriculture according to article 9 of the 1993 Resettlement Regulations. This would be done by developing arable land, and by soil improvement. However, years of resettlement experience in the Three Gorges dam project have shown Chinese authorities that it will be difficult to resettle all the people in this way within the reservoir area. The explanatory factors for the decision to move a fixed number out, and to encourage more people to find their own way through the help of relatives and friends, are twofold: it is based upon the actual problems that have emerged in the process, as well as the perceived potential problems that may appear as described and discussed in numerous articles about the resettlement process. These factors will be discussed below.

In order to structure the issues that have emerged in the Three Gorges resettlement process, I will draw on the impoverishment risks and the recon-

---

261 During the discussions in the beginning of the 1990s, there were rumours that the rural population was to be moved to Xinjiang, Shandong, Heilongjiang and Hainan, but these were denied by Chinese authorities. See Reuters (19 December, 1992).

struction model (IRR) that has been developed by Michael B. Cernea.[262] The IRR model consists of eight impoverishment and reconstruction points. In the first part of this chapter I will use the impoverishment risk points to discuss the problems in the Three Gorges resettlement process. In the latter part of the chapter, the reconstruction points will be discussed in relation to the Chinese resettlement plan. These dimensions of the impoverishment risk model are useful reference points for the discussion, as they are based on the experience of numbers of resettlement projects around the world. Thus, the issues being discussed in the Three Gorges resettlement are discussed against an international background and in relation to internationally used reference points. The model is meant to facilitate the organisation of information. However, some of the dimensions discussed in the model may not be as relevant in the Three Gorges resettlement and environment discussion. Thus, I will use the model pragmatically. This, as well as the relevance of the IRR model for the Three Gorges project, will be summed up at the end. It may also be possible to draw some conclusions on the model's relevance for dam projects in general in China. Furthermore, references will be made to the World Bank, as it has funded numerous infrastructure projects in China, including hydro-projects, where the population has been resettled.[263] Below is a brief introduction to the impoverishment risks and reconstruction model and the World Bank's view on China's resettlement practice, followed by a description of the issues that have emerged in the Three Gorges resettlement debate that may have been instrumental in influencing a change in the resettlement policy.

Impoverishment Risks and Reconstruction Model and the World Bank

In general, the social variables have been the least focused on in dam projects involving resettlement, and this has been criticised in literature about dam building and resettlement.[264] Forced population displacement has been downgraded in resettlement projects, and has been treated as a minor problem. In China as well, there is official acceptance that one of the main reasons for the problems in resettlement was the fact that construction of water resources projects often was separated from resettlement. According to Cernea,[265] these results (on a global basis) are due to several factors such as i) lack of official policies governing dislocation, ii) absence of perfunctory planning for resettlement, iii) little resource allocation for something seen unworthy of high expenditures, iv) disregard for the human and civil rights, including the customary rights of the people affected; and v) organisational responsibility passed down the line to weak, often incompetent and unequipped agencies. Furthermore, authorities often understate the number of

262 See Cernea (1997); and Cernea and McDowell (eds. 2000).

263 A conservative estimate of resettled population due to World Bank infrastructure projects in China is 750,000 people. Meikle and Zhu (2000), p. 128.

264 See Cernea (1997); Cernea and McDowell (eds. 2000); McCully (1996); Goldsmith and Hildyard (1984); and World Commission on Dams (2000).

265 Cernea (1990), p. 3.

people to be resettled, due either to lack of accurate knowledge or deliberately in order to obtain acceptance for a project which could have serious implications for the resettled population.

During the 1990s, Cernea formulated and developed a model with 'risks to be avoided' in displacement, described in *Poverty Risks from Population Displacement in Water Resources Development.*[266] Cernea states that:

> the analysis of vast evidence coming up from numerous displacement and resettlement processes reveals patterns of recurrent characteristics. While each one of these characteristics is distinct and irreducible to the others, they have a common denominator: all are dimensions of a multifaceted process of impoverishment. These characteristics may be linked into a coherent typology or descriptive model of impoverishment risks through displacement.

The advantage of using the model is that the risks of resettlement are identified in advance. Thus the model could be used as a tool in strategy formulation and a device for practical planning. The eight dimensions of the impoverishment risk model described in the above-mentioned texts are: a) landlessnes, b) joblessness, c) homelessness, d) marginalization, e) food insecurity, f) loss of access to common property, g) morbidity, and h) social disarticulation.

A preliminary version of the model was applied in the resettlement review of nearly 200 projects by the World Bank in 1996,[267] which was led by Cernea. The model's emphasis was on identifying the impoverishment risks. The model has been further developed since then, and a recent example of the usage of the model can be found in the book, Cernea, Michael M. and Chris McDowell (eds., 2000) *Risks and Reconstruction, Experiences of Resettlers and Refugees.* In addition to discussing the impoverishment risks from resettlement, the model is turned on its head, giving a model of strategies to be adopted in resettlement:

> a) from landlessness to land-based resettlement,
> b) from joblessness to reemployment,
> c) from homelessness to house reconstruction
> d) from marginalization to social inclusion,
> e) from increased morbidity to improved health care,
> f) from food insecurity to adequate nutrition,
> g) from loss of access to restoration of community assets and services,
> h) from social disarticulation to networks and community rebuilding.

Cernea states[268] that 'these variables are interlinked and influence each other: some play a primary role and others a derivative role in either impoverishment or reconstruction (largely as a function of given circumstances)'. The model was

---

266 Ibid.

267 World Bank (1996) is a final version of the report World Bank (1994c). The impoverishment risks model is discussed in World Bank (1996), pp. 114-27 and in World Bank (1994c), pp. 4/8-4/15.

268 'Risks, safeguards, and reconstruction: a model for population displacement and resettlement', in Cernea and McDowell (eds. 2000), p. 19.

developed mainly in the 1990s, and Cernea states that it builds upon work that has been carried out in the resettlement field.[269]

According to Cernea, there are four distinct but interlinked functions that the risks and reconstruction model performs. These are:

1. A predictive (warning and planning) function: this results from the knowledge of past processes; these are predictions of likely problems in the resettlement process, manifested in the eight impoverishment risks. They are early warnings that can be issued in advance of the decision to displace people. The model thus equips planners with knowledge to anticipate what is essential in the planning process in order to avoid or reduce the risks. It enables them to search for alternatives.
2. A diagnostic (explanatory and assessment) function: refers to the models' capacity to explain and assess. The model functions as a tool for guiding assessment field work, it gives a prognosis of the situation, and reveals in advance information and recommendations for the project preparation and planning of counter-risk measures.
3. A problem-resolution function, in guiding and measuring resettlers' re-establishment: moves from prediction and diagnosis to prescription for action. The model becomes a compass for strategies to reconstruct resettlers' livelihoods, by setting the model on its head.
4. A research function for social researchers, in formulating hypotheses and conduction of theory-led field investigations: the model stimulates the generation of hypotheses about relations between key variables in both displacement and relocation, and facilitates the exploration of mutual linkages. By using the model, empirical findings are structured along the key variables. The key variables also make comparisons possible across cultures, countries and time periods.

Relating the above to China and the Three Gorges project's new developmental resettlement scheme, it seems at first sight that the points in the model should not be that relevant today, as China in many ways has succeeded in pre-empting the potential risks of the model. As described earlier, the resettlement performance in China (based upon China's own assessment as well as the World Bank's) was not very successful until the 1980s; it was frequently associated with impoverishment, and resistance to resettlement was common. However, according to the World Bank,[270] resettlement in China improved substantially from the 1980s, and is now

---

269 Some examples are: Chambers, R. (1969), *Settlement schemes in Tropical Africa*, London: Routledge; Nelson, M. (1973), *Development of Tropical Lands: Policy Issue in Latin America,* Baltimore, Maryland: Johns Hopkins University Press; Scudder, T. and Colson, E. (1982), 'From Welfare to Development: A Conceptual Framework for the Analysis of Dislocated People', in A. Hansen and A. Oliver-Smith, (eds.) *Involuntary Migration and Resettlement: the Problems and Responses of Dislocated People,* Boulder, Colorado: Westview Press. See also, Scudder, (1985, 1991).

270 World Bank (1996).

generally considered to work well (compared to other developing countries). World Bank resettlement experts consider reservoir resettlement performance in China among the best when it comes to the standard of maintaining real income.[271] The World Bank lists resettlement policy and several incentives initiated by the authorities as reasons for this,[272] which are as follows: i) the relocatees in urban areas receive one to three times their current living space; ii) host communities in rural areas receive compensation payments to receive resettlers; iii) decentralisation is another important factor, all responsibility for resettlement is given to local or city governments, and resettlement solutions are developed locally; iv) China has developed binding policies and procedures for each of the three major sectors involved in resettlement: agriculture, energy, and urban development; and v) many resettlers use the opportunity to launch new enterprises that they previously could not afford, often adding their own resources to improve their housing and economy.[273] This positive picture is further emphasised in a recent publication by the World Bank, *Recent Experience with Involuntary Resettlement, China—Shuikou (and Yantan)*,[274] which is considered to be a very successful project that resettled 67,200 people in Shuikou in Fujian province. The report praises the Chinese authorities for approaching resettlement as a development challenge, where the resettlement funds were used to open up new land for cultivation, improve existing land, and establish new enterprises.[275] It states that China in this way differs from most other developing countries that look upon resettlement as a burden.

These points may all be correct, and the optimistic picture drawn by the 1996 World Bank report on resettlement performance in China after the 1980s[276] could indicate success for the Three Gorges project resettlement programme. Nevertheless, theory and practice may differ, and from research about the resettlement process for this book it seems as if the promising picture drawn by the World Bank of resettlement in China may not be completely accurate in the case of the Three Gorges project. There could be a number of reasons for this, such as no World Bank involvement, as well as the unprecedented, large-scale Three Gorges reservoir resettlement. Moreover, resettling 1.13 million (or more) people may be more challenging than projects involving less than 100,000 people. As we will see later in this chapter, it is particularly difficult in an area with a large rural population without much formal education, and where there exists little room for

271 World Bank (1993), p. x. Where the standard of maintaining real income is followed, as it is in Bank funded projects.

272 See the World Bank (1998); and World Bank (1996), p. 122, box 4.7 for additional details on the reasons for success in resettlement performance in China.

273 During a visit to Wujiaxiang in Yichang in 1997, the author met with a resettled family that had built a house worth RMB 40,000. RMB 20,000 came from the authorities, and the remaining 20,000 from other sources.

274 World Bank (1998).

275 Ibid., p. 15.

276 World Bank (1996).

further development of land for cultivation. These are reasons why critics of the project, both inside and outside China, argue that resettling such a large population cannot be carried out easily, and view the potential for success as bleak.

Despite China's improved resettlement performance, problems exist in the Three Gorges project. Below I will present some of the issues that have appeared in relation to the resettlement of the Three Gorges project that give an impression of a difficult resettlement process.

Potential and Concrete Problems in the Three Gorges Resettlement

In the following, I will be using several of the points under the Impoverishment Risks and Reconstruction (IRR) model to structure the issues that have emerged in the Three Gorges resettlement process. The points appear in the order that I found natural, and not necessarily according to the listed points in the previous section. Some of the points have been merged into one section, as the issues discussed under these points are interlinked to such an extent that this seemed natural and more practical. To shed light on different aspects of the resettlement process, some of the arguments used may overlap somewhat thematically. Some points, such as morbidity and health, have not been the focus of the study, nor have these areas been the focus for the debate that I have followed regarding the resettlement and the environmental issues. Thus, I will not go into the potential health risks of the project.[277] On the other hand, there are a few issues that are relevant for the Three Gorges project that are not covered by the model. I have therefore divided the following section into two parts: the points that can be included under the Impoverishment Risks and Reconstruction model, and the additional points that have relevance for the Three Gorges project.

Impoverishment Risks in the Three Gorges Project

Landlessness,[278] *lack of available land and homelessness.* Lack of available farmland in the Three Gorges area is one of the biggest challenges for the successful rural resettlement in the reservoir area. The rural population is large in the reservoir area; 87.3 percent belong to the peasants category as is shown in Table 5.4. In the reservoir area, 8 counties in Chongqing municipality have a rural

277 The issues of morbidity and health have been discussed and research has been carried out in relation to the Three Gorges resettlement. However, they do not appear to be big and controversial issues in the Chinese debate about the resettlement process and the environmental impact. Chinese officials have stressed that for instance schistosomiasis does not pose a great risk to the resettled people in the Three Gorges area. However, the EIAD/CAS and RIPYWR (1995), pp. 23-4, states that schistosomiasis occurs in areas to be inundated in Chongqing municipality and Hubei province.

278 Landlessness does not necessarily mean that peasants will lose all land. However, with little average land at present, loss of any land due to reduction of size may have a serious impact on people's lives.

population of 90 percent or above, while there is one in Hubei province. The rural population is engaged in agricultural activities of various kind, and may be divided into several categories as shown in Table 5.5.

There are different figures regarding the amount of land to be inundated by the Three Gorges dam, which may be confusing. The discrepancy in the figures may be due to different definitions or terminology used for the land to be inundated, which includes or excludes some land types. According to some sources (see Table 5.6), the reservoir will inundate 513,000 *mu* (34,217 ha) of different types of land. This table includes everything from cultivated land, flood land, garden plots, forest land, fish ponds and firewood hills. In Table 5.7, a different terminology is applied, and paddy fields, non-irrigated land and citrus land, amount to 431,300 *mu* (28,768 ha). Finally, some official sources state that 390,000 *mu* (26,013 ha) of cultivated land and garden plots will be inundated.[279] The figures for the amount of farmland that will be available for the resettled peasants also varies. According to Li Peng, one way for Chongqing to obtain more farmland to settle the peasants on would be the *di er ci tudi chengbao* (the second round of contract), which would give 110,000 *mu* (7,337 ha) of farmland.[280] Gan Yuping states that Chongqing, with the largest burden of population to resettle, may obtain 190,000 *mu* (12,673 ha) of farmland from not renewing the lease for the rural population who no longer farm the land, and have found work in cities and towns.[281] Most land lease contracts for the peasants expired in the late 1990s,[282] and the decision not to renew these contracts provides the Chinese authorities with an opportunity to adjust the land lease contracts. Retrieving this land will, according to Gan Yuping, enable the resettlement of the remaining rural population in Chongqing. In addition, new farmland will be reclaimed in the hillsides.[283] Below are two examples of figures for land to be inundated by the reservoir.

279 Statement by Gan Yuping, the TGPCC vice-chairman and deputy secrtary of Chongqing CCP. Xinhua she (7 March 2001).

280 Li (11 January, 2000).

281 These people have according to Gan Yuping left their farmland idle in order to work in the towns.

282 See Ho (2001b), p. 395.

283 Taking these figures as basis, after having moved out 100,000 from the rural population into other areas, 201,265 remain in the reservoir area of Chongqing municipality. Taking 190,000 mu as the basis, it is possible to obtain 0.94 mu of farmland per head, which is close to the average for farmland in the Three Gorges area. This calculation is not accurate, as it is based on the *jingtai* figure for resettlement that does not include the population increase.

Table 5.4 Active Working Population in the Three Gorges Reservoir Area (%)

	Techno-logical personnel	Cadres	Office workers	Commerce, trade	Service	Pea-sants	Workers
Reservoir area	2.9	0.7	0.8	1.4	1.1	87.3	5.8
Sichuan province (now Chongqing Municipality)							
Changshou	3.9	0.7	1.3	1.6	1.9	78.7	11.9
Baxian	3.0	0.7	1.0	1.6	1.4	83.8	8.5
Jiangbei	2.6	0.6	0.7	1.5	1.2	86.9	6.4
Wanxian city	9.6	2.9	3.8	6.5	4.9	44.0	28.0
Wanxian	1.9	0.3	0.6	0.8	0.6	**92.5**	3.3
Kaixian	2.2	0.4	0.5	1.2	0.6	**92.0**	3.2
Zhongxian	2.3	0.7	0.6	1.0	0.8	89.6	4.9
Yunyang	2.1	0.4	0.5	1.0	0.6	**92.7**	2.7
Fengjie	2.5	0.5	0.7	1.3	0.7	**90.6**	3.6
Wushan	2.2	0.5	0.7	1.1	0.7	**91.7**	3.2
Wuxi	2.5	0.5	0.7	1.0	0.7	**91.0**	3.6
Fuling town	3.6	1.4	1.3	2.0	1.7	81.8	8.2
Fengdu	2.4	0.7	0.6	1.3	0.9	**90.6**	3.6
Wulong	2.5	1.0	0.6	1.0	0.8	89.5	4.6
Shizhu	2.5	0.6	0.9	1.2	0.9	**90.2**	3.6
Hubei province							
Yichang	4.2	1.2	1.1	1.7	1.8	80.8	9.2
Xingshan	5.0	1.6	1.4	1.8	1.8	80.0	8.5
Zigui	3.3	1.1	0.8	1.2	1.0	88.1	4.4
Badong	3.2	0.8	0.9	1.0	0.9	**90.0**	3.2

Source: Zhu and Zhao (eds. 1996), p. 120.[284]

284 Zhu et Zhao's source is the population census from 1990 in Sichuan and Hubei provinces.

Table 5.5 Rural Population to be Resettled

Resettlers who live on the outskirts of towns (*Chengjiaoxing yimin*)	10 %	Grow melons, vegetables, and work in processing, and may engage in small businesses
Resettlers with special skills (*Techangxing yimin*)	30 %	Grow fruits (citrus), tea, silk worms
Ordinary resettlers (*Putongxing yimin*)	60 %	Cultivate the land in general (no special skills)

Source: Qiu, Wu and Du (2000), p. 2.

Table 5.6 Land to be Inundated (*mu*)

	Reservoir area	**Sichuan (now Chongqing municipality)**	**Hubei province**
Cultivated land (gengdi)*	257,000 (17,142 ha)	229,000 (15,274 ha)	28,000 (1,868 ha)
Flood land/riverside land (hetandi)	58,000 (3,868 ha)	57,000 (3,802 ha)	1,000 (68 ha)
Garden plot (yuandi)**	110,000 (7,337 ha)	74,000 (4,936 ha)	36,000 (2,401 ha)
Forest land (lindi)	49,000 (3,268 ha)	37,000 (2,468 ha)	12,000 (800 ha)
Fish ponds (yutang)	5,000 (0,333 ha)	5,000 (333 ha)	0
Firewood hills (chaicao shan)	34,000 (2,268 ha)	19,000 (1,267 ha)	15,000 (1,000 ha)
Total	0 (34,217 ha)	0 (28,081 ha)	0 (6,136 ha)

* This figure includes non-irrigated land (109,000 *mu*/7,270 ha), paddy fields (126.000 *mu*/8,404 ha) and vegetable plots (23,000 *mu*/1,534 ha).[285]

** This figure includes citrus land (96,000 *mu*/6,403 ha) and other (14,000 *mu*/934 ha).

Source: Zhu and Zhao (eds. 1996), p. 2.[286]

285 There may be a mistake in the figures in the table, as the sum of these three figures amounts to 258,000 *mu* instead of 257,000 mu.

286 Zhu and Zhao (eds. 1996), which have based their figures upon research by the Yangtze River Water Resources Commission in 1993 (Changjiang shuili weiyuanhui 1993).

Table 5.7 Land to be Inundated (*mu*)

Paddy fields (*shuitian*)	110,700	(7,384 ha)
Non-irrigated land (*handi*)	246,200	(16,421 ha)
Citrus land (*ganjudi*)	74,400	(4,962 ha)
Total	0	(28,768 ha)

Source: Li, Boning (1992), *Lun Sanxia gongcheng yu kaifaxing yimin* (On the Three Gorges project and development resettlement scheme), booklet published by the Ministry of Water Resources and Electric Power press from 1992.

By the end of 2002, 645,200 people had been resettled since 1993 (both urban and rural), and during this process the resettlement has run into some problems. The lack of farmland for the rural population is most likely one of them. One of the eight dimensions under Cernea's model is landlessness, and the lack of farmland to compensate the peasants in the area may lead to a state of landlessness. The peasants that are located in a reservoir area and need to be resettled are entitled to receive compensation. Cernea states that empirical data from many countries shows that the peasants remain landless forever, i.e. they do not receive the compensation they are entitled to.[287] In the resettlement plan for the rural population of the Three Gorges project, the peasants are entitled to 1 *mu* of land for citrus cultivation, as well as half a *mu* of fields with high and stable grain yield.[288] Whether or not the peasants will receive this amount of land depends upon several factors:

- Scarcity of farmland:
  The scarcity of farmland in the area is one of the most important factors that may cause problems in the process of resettling the rural population. Even if 125,000 rural residents are relocated out of the reservoir area, it is questionable whether there is sufficient farmland for the remaining relocatees. Most of the land in the area is mountainous (78 percent),[289] and about 40 percent is already under cultivation. Furthermore, a third of the land is on mountainsides with gradient slopes of 25 or greater, where development is prohibited according to China's Water and Soil Protection Law. To illustrate the problem in the area, the population density in the reservoir area is restated: 296 persons per km^2 on an average, while 130 people per km^2 is the national

287 Although this is not always the case in a number of countries, it is not uncommon. McCully (1996), p. 322 for instance, mentions several examples from India, Columbia, Vietnam, where the peasants did not receive any compensation at all, or where they received a symbolic sum only.

288 Li (1992a), p. 5. The land amount may differ depending upon land availability and quality. See also Gao (2000), p. 14, that refers to the Resettlement Outline (*Gangyao*) for Wanxian where the average farmland or fruit orchards/garden plots amount to 1.3-1.5 *mu*.

289 Dai (1998a), p. 65.

average, with 1000-1200 people per km^2 in the areas below 300 metres. Hubei province may be able to provide the rural population with the amount of land stated above, as the province is not as densely populated, and the number to be relocated is small. The per capita cultivated land in Hubei province is 1.47 *mu*, while it is 1.01 *mu* in Chongqing municipality, with a gross output value from farming amounting to RMB 68.80 billion and RMB 25.49 billion respectively.[290] The figures illustrate the different situation for Hubei province and Chongqing municipality. Moreover, populous Chongqing municipality, which shoulders 85 percent of the resettlement burden consisting largely of a rural population, will have serious difficulties in providing the amount set by the resettlement outline, or even provide land at all. According to researchers, the subsistence base line is between 0.6 mu (6 *fen*) and 0.8 mu (8 *fen*), which is the lowest possible area to farm on (grain ration field, i.e. the minimum field for obtaining sufficient grain ration, *kouliangtian*).[291] In Yunyang county in Chongqing municipality, which is a typical county in the Three Gorges area, the average amount of land per capita is 0.87 mu; one-fourth of the rural population there only have 0.5 mu to farm on.[292] In some places in the county, 'new' farmers (due to population growth) cannot obtain contracted land, as there is no excess land. Another example is Fengjie county, also in Chongqing municipality. A statistical sample survey from that county disclosed that a number of peasants currently lacked contracted farmland, and the number of people lacking farmland was increasing steadily,[293] which illustrates that even *before* resettling people this is a problem in the Chongqing area of the Three Gorges reservoir area. When the Three Gorges dam is constructed, even though the land to be inundated in the various counties does not constitute such a large amount of the area, it is nevertheless the most fertile, high-yielding land. For instance, the farmland to be inundated in Yunyang constitutes 4.8 percent of the total farmed land; nevertheless, it is the land along the river which has been developed and cultivated for centuries. The remaining land, which will be divided among the resettlers, is hilly, dry and the soil is thin. Even though Chinese officials continuously have stated that there is enough farmland in the area for the rural population, which includes soil improvement of land, it is becoming increasingly clear that this is not the case. Ripening of the soil, for instance, takes 3-5 years.[294] The policy change

---

290 Guojia tongjijiu (1999), pp. 390 and 382. This may not be an appropriate comparison, as Hubei is a province with a population of ca. 59 million people in 1998 on a land area of 1.85 million km^2, and where 3.3 million ha is under cultivation. Chongqing, with a population of more than 30 million and a land area of 824,999 km^2, has 2.5 million ha under cultivation. See Hubei sheng tongjiju (1999); and Xibu da kaifa zhinan bianji weiyuan hui (ed. 2000).

291 Interview no. 6, CAS, Beijing September 1999 and Interview no. 29, CAS, Beijing May 2000, respectively.

292 Wei (1999), p. 14.

293 Zhu and Zhao (eds. 1996), p. 146.

294 Interview no. 6, Beijing, August 1999.

in May 1999 is a clear indication of the land problem. Furthermore, in August 1999, the TGPCC issued a notice regarding the resettlement of the rural population that pointed to the necessity of ensuring that the peasants receive at least 0.8 *mu* of farmland, which may indicate that this may not be always be the case in the reservoir area.[295] As described in the environmental capacity section, uncertainty exists regarding the exact amount of farmland available for the rural population.[296] This may be related to the different methods used in measuring the available land in the area, where serious disagreement existed between CAS and the Ministry of Water Resources (MWR) at the time of compiling the environmental impact assessment for the dam project. Should this be the case, it may have serious implications for the rural population.

- Citrus[297] policy:
  The scarcity of high yielding farmland was to be solved by giving the peasants some land on which to grow fruits. The peasants would receive half a *mu* each of fields with high and stable grain yield. Each peasant would in addition be given one *mu* of land for cultivation of citrus fruits (tangerines), other fruits, herbs, tea and so on.[298] The resettlement officials have particularly stressed the development of citrus fruit orchards, as the area was especially suitable for growing citrus, and is famous for these fruits. Researchers at CAS have been engaged in research on how to improve the different types of tangerines so as to increase the market value. One of the main proponents of this policy was Li Boning, a former vice-minister of the MWR and the director of the former Three Gorges area Office of Economic Development, who aspired to become governor of the then planned Three Gorges province.[299] It was anticipated that the peasants would be able to increase their income several fold by growing citrus, as compared with traditional

---

295 Guanyu zuohao Sanxia kuqu tiaozheng tudi anzhi nongcun yimin youguan wenti de tongzhi, (Notice regarding the related issues of adjusting the land for resettlement of the rural population in the Three Gorges resevoir), 1999, No. 85 in Zhongguo Sanxia jianshe nianjian bianjibu (2000), p. 28.

296 One scholar put it this way: '*Guojia gaobu qingchu you duoshao di*', (The state doesn't know how much land there is), interview no. 29, Beijing May 2000.

297 The Chinese term for this is *ganju* which means tangerines and oranges, or citrus. The term citrus is used in this book to cover these citrus fruits that are grown in the Three Gorges area.

298 See Li (1992b), p. 24.

299 Li Boning was eager to become the Governor of the planned Three Gorges province. The establishment of the province was meant to make Sichuan more receptive of the dam project, as the resttlement burden was greatest there. However, the plans were eventually abolished, and Chongqing was instead raised to administrative status directly under Beijing in order to take on the responsibility for the majority of the resettlement.

agriculture.[300] However, this plan for citrus growing was greatly modified due to the lack of a market for an increase in citrus production. Poor transportation facilities in the area, which resulted in piles of citrus rotting away outside peasants' homes caused the authoritites to reconsider this plan. Consequently, the importance of this citrus policy has been toned down. Researchers interviewed by the author have criticised this policy as being unrealistic, and state that this illustrates the low level of such high leaders (*shuiping hen di*) involved in the decision-making for the Three Gorges project.[301]

- Number of people to relocate:
  A third issue to complicate matters is the uncertainty regarding the exact number of people to be resettled. The authorities differentiate between the *jingtai* (static, not including the population increase) and *dongtai* (dynamic, including the population increase) numbers; and these numbers vary in different reports and newspaper articles. The official *jingtai* (static number) is 846,200 people, of which 361,500 belong to the rural population and 484,700 to the non-agricultural population (towns and villages).[302] Following the policy change in May 1999, the static number to be resettled within the Three Gorges reservoir area is 721,200, as 125,000 people have been moved out of the area. The number of people to be resettled has been adjusted during the resettlement process; for instance a source from 1995 says the *jingtai* (static) is altogether 725,500 people, of which urban population constitutes 392,900 and the rural population 332,600.[303]

As for the *dongtai* number, the official figure is 1.13 million people, and this figure is most frequently used in the media when officials are cited. Nevertheless, some official sources state that altogether 1.2 million people will be resettled,[304] while Li Peng stated 1.10 million.[305] According to Zhu Rongji in May 1999, the second phase of the resettlement process needs to resettle 550,000, while Li Peng states 450,000 in the same period. It is unclear why the numbers vary to such a great extent, but it could be that the last figure is static, but this was not indicated in the text. One conclusion to be drawn from this is that the exact number of people to be resettled is not quite clear to Chinese authorities.

Furthermore, there are two categories of people who will be affected by the reservoir: those who live below the inundation line belong to the group that is

300 See Li (1992b), p. 24.

301 Interview no. 17, September 1999.

302 Yangtze Valley Water Resources Protection Bureau, MWR and NEPA (eds. 1999), p. 42. Another source that uses this figure is Zhu and Zhao (eds. 1996), p. 3.

303 Chen, Xu and Du (1995), pp. 3-4.

304 Xinhua she (7 March, 2001). 1.2 million people was used in March 2001, by the vice mayor of Chongqing municiality, Gan Yuping, also a vice-chairman of the TGPCC. Xinhua she (7 March, 2001).

305 Li (11 January, 2000).

directly affected by the inundation (*zhijie yanmo*) and indirect inundation (*jianjie yanmo*). 1.13 million people (846,200 people static figure) are directly affected by the inundation of the reservoir; it appears that only the people to be resettled (*dongqian renkou*) are included in the number, i.e. those people whose houses and land will be inundated by the reservoir.[306] Those people whose homes will not be inundated but who will somehow be affected by the reservoir are not included in the number of people to be resettled, and therefore belong to the group of people who are indirectly affected (*jianjie yanmo*) by the reservoir. These people do not live below the inundation line of the reservoir, but their source of subsistence will be inundated such as shops and small enterprises. Also, according to the 1993 resettlement regulations, article 24, people who have moved into the area during the resettlement process are not eligible for compensation. Consequently, the figure for the population which is indirectly affected does not seem to be clear. Most literature read by the author only discusses the people who are directly affected (whose homes and/or land disappear) by the reservoir inundation.

From the above it is clear that there is some uncertainty regarding the exact number of people to be moved, which may seriously affect the land compensation should the number exceed that which is anticipated. In addition comes the question regarding what will happen to the people who are not living below the inundation line, but whose livelihoods are threatened. To make matters worse, critics claim that resettlement officials are promoting a low, untruthful resettlement figure.[307] Historically, underreporting has been usual in dam projects in China (and elsewhere) for mainly two reasons. The authorities have possibly not known the exact number due to poor statistics, or the low figure has been given deliberately. In the Three Gorges case, critics state that the numbers have been kept deliberately low in order to obtain acceptance for the project. As mentioned in the introduction, during the debate for the Three Gorges project, opponents have come up with numbers between 1.4 to close to 2 million people. Some scholars interviewed by the author thought that critics exaggerated numbers, while the authorities tended to reduce the number of people to be moved. Thus, it is likely that the correct number of people to be relocated may lie between these two extremes. Should under-reporting be the case, it raises many significant questions. In such a case, the resettlement sum allocated for compensation would not be adequate, and there would be a shortage of farmland to divide between the relocatees.

In sum, the lack of available farmland, failed citrus policy and uncertainty regarding the number of people to be resettled may largely impact the resettlement compensation as well as the potential for successful restoration of livelihoods.

---

306 Li, Heming (2000), pp. 136-8.

307 See Dai (ed. 1989); and Qi (1998) in Dai (1998a), p. 55. Under-counting is a common fact in a number of dam projects around the world, see World Commission on Dams (2000), pp. 104-5.

Food insecurity[308] A natural consequence of loss of land, landlessness, is food insecurity. Undernourishment is a symptom and result of inadequate resettlement due to sudden drops in food production and loss of harvest. When the area of cultivated land decreases, local food production becomes insufficient. In the Three Gorges area the existing lack of farmland already poses a threat to the local food production. With the inundation of fertile farmland it is expected that the problem will aggravate. Let us look at some of the issues that have been discussed in relation to the Three Gorges resettlement.

The Three Gorges area is poor, and there is a limited amount of fertile farmland in the area. It is questionable if the area will be self-sufficient in grain production when farmers move to less fertile land from the river up in the mountainous areas. The agricultural emphasis in the Three Gorges reservoir area is on grain production; wheat constitutes 22.56 percent, rice 10.41 percent and corn constitutes 29.99 percent.[309] When the normal pool level in the reservoir reaches 175 metres, the most fertile farmland along the river will be inundated. The productivity of the allocated land in the hilly areas is often low due to poorer quality, since the highly fertile lands are situated along the river. More land would be needed in the hilly areas to achieve the same economic result as before. Farmers in the Three Gorges area state that it would be necessary to compensate 1 *mu* by the river with 5 *mu* of land in the mountainous areas in order to achieve the same results.[310] Even before inundation takes place, fertile farmland in the Three Gorges area has been reduced. Fengjie county for instance, is a typical county in the Three Gorges area in many respects, quite poor with little available farmland. Here, a reduction of arable land has taken place from 1949: 1.0252 million mu in 1949 to 877,000 mu in 1993.[311] Each person has an average of 1 mu of arable land, while 2.3 mu was the average in 1949. The scarce available farmland in Fengjie has brought about serious conflicts among people. Furthermore, following the May 1999 policy change announcement, it is prohibited to farm land steeper than 25 degrees in all the counties in the Three Gorges area. Existing farming on land sloped above 25 degrees must cease, with land being restored to forest or grassland (*tuigeng huanlin*). In the case of Fengjie, steep hills are being used for farming, which indicates that there is little or no extra farmland to be handed out to the relocatees for development. According to statistics from 1993, steep hills constitute ca. 64 percent of all land in Fengjie, of which ca. 25 percent is made up of hills steeper than 25 degrees. As mentioned before, an increasing number of peasants in Fengjie have not been able to obtain contracted farmland. To solve the present and future problems, the import of grain has been suggested. However, as has been pointed out earlier, transportation would be difficult in the area. Farmland must also be taken to construct new homes for the peasants, which will worsen the grave situation.

---

308 Some of the issues under this point also refer to landlessness and lack of farmland.

309 SEPA (1999).

310 Ding (1998), in Dai (1998a), p. 83.

311 Zhu and Zhao (eds. 1996), p. 146.

A second problem threatening the provision of food in the area for a number of peasants is a decrease in income due to peasants' farming on land that is not registered in their name. When the Agricultural Responsibility System was adopted in 1978,[312] farmers were able to convert barren land to farming. However, this converted land did not have to be registered with the village, nor was it taxed by the government. Thus, when the authorities calculate the land compensation for the farmers, it will actually be much less than present, as this unregistered land will not be included. The peasants themselves expect a reduction in income amounting to 50 percent.[313] This will affect the compensation of the peasants and the potential for restoring their livelihoods.[314]

In sum, little available farmland, the need to restore farmed land that is steeper than 25 degrees, as well as the loss of unregistered land may result in a drop in cultivated area for the resettled peasants, and therefore food insecurity.

Joblessness The last point above leads to the next issue: joblessness. Some peasants will have to leave the rural areas altogether and become town and township citizens without any land at all. As described in the Developmental Resettlement Scheme, a number of farmers will have to change vocation. It may be difficult to find work, and with no land at all to farm on, their daily subsistence is questionable.

The low educational level in the Three Gorges area could influence the ability of the former rural population to find work in non-agricultural employment. The educational backgrounds from counties in Chongqing municipality and Hubei province are given in Table 5.8. The population of the counties Wanxian, Kaixian, Zhongxian, Yunyang, Fengjie, Wushan, Fengdu in Chongqing, and Zigui and Badong in Hubei province constitutes 75 percent of the total population to be resettled, and their educational level is among the lowest in the reservoir area.

In Fengjie for instance, illiterates constitute 24.92 percent of the total population, undergraduates from university 0.05 percent, and 54.65 percent have primary school level knowledge.[315] Low educational level is common in the Three Gorges area, as can be seen in Table 5.8. As mentioned, 40 percent of the rural relocatees, 144,600 people (the *jingtai* figure, not including the population increase), will be transferred from the farming sector to the secondary and tertiary

---

312 The Agricultural Responsibility System did away with the collective agricultural system of the People's Communes and granted the farmers the right to lease land for fifteen years or longer. Ding (1998), in Dai (1998a), p. 82.

313 Ibid., p. 82.

314 The problem of unregistered land may be larger than described in this book. However, the materials used for this book have largely focused on the environmental issues. A discrepancy exists between figures of cultivated land. In the Agricultural Census, for Sichuan the adminstrative villages reported a farmland area of 9,169,000 ha, while the official figure in the Agricultural Yearbook for 1997 was 6,165,000 ha. Based on communication between the author and Eduard B. Vermeer regarding the contents of Vermeer (2000).

315 Zhu and Zhao (eds. 1996), p. 156.

industry sectors.[316] Chinese authorities state that vocational training for former farmers will be carried out. Their reason for optimism regarding finding employment for this group is that when the 1600 factories, according to plan, are relocated,[317] many of the factories will go through an expansion, creating the need for additional workers. However, these factories' ability to compete in a market economy is questioned. As some point out, some factories, in order to accommodate more former peasants as workers, purchase old-fashioned equipment, more labour intensive equipment.[318] Thus, it may be problematic for these enterprises to survive in a fiercely competitive market. Moreover, Li Heming also points to the fact that the prospects for the rural population to move into urban areas and engage in nonfarm jobs are expected to be poor because of overstaffing and current unemployment in urban industrial sectors.[319]

Through the policy of *duikou zhiyuan* (counterpart support), enterprises from other areas in China that settle in the Three Gorges area are expected to provide work places for many people, both urban and former rural population. This may ease the employment problem for part of the rural population deprived of farmland. Nevertheless, vocational training can provide skills but not necessarily jobs. Even with some vocational training, it might still be difficult for former farmers and unskilled population to find jobs in enterprises, as they are not particularly in demand. Regardless of whom I have spoken with about the rural people in the Three Gorges area, be it researchers or officials, they all express identical views: the rural population has a low educational level, i.e. *suzhi di* (which directly translates as 'low quality'). During trial resettlement, peasants who had received some training (unskilled labour) had difficulty in finding jobs. At the time, some factory directors were reluctant to recruit people from the resettled rural population.[320] A second factor that may influence the possibility for unskilled peasants to find employment is that a number of the enterprises in the Three Gorges area will not be able to continue to operate. They will close down either due to economic and/or environmental reasons. Polluting factories will not be allowed to continue their production unless the enterprise purchases modern equipment and reaches a certain standard. One example of unsuccessful settlement of the rural population from the Three Gorges reservoir area is Wantong electrical equipment industrial company (*Wantong dianzi qicai gongye gongsi*) that employed 208 rural relocatees from Zhongxian, Wanxian, Fengjie, Yunyang and Wuxi through the *duikou zhiyuan* (counterpart support) system. After a few years the company had to let the workers go, as continued operations were economically

---

316 Zhu and Zhao (eds. 1996), p. 243. Also Yangtze Valley Water Resources Protection Bureau, MWR and NEPA (eds., 1999), pp. 45-6.

317 According to Chinese authorities 657 factories will be inundated. Li (1992a), p. 2. 1600 will be relocated. These will be located above the inundation line. See Li (11 January, 2000).

318 Wei (1999), p. 15.

319 Li, Heming (2000), p. 175.

320 See Jiang (1992), pp. 62-3.

Table 5.8 The Three Gorges Reservoir Area Educational Level from Age 6 and Up (%)

	University undergraduate	University professional training	Polytechnic school	High school	Junior middle school	Primary school	Illiterate
Reservoir area	0.13	0.40	1.14	4.53	23.67	51.18	18.94
Sichuan province (now Chongqing Municipality)							
Changshou	0.29	0.61	1.50	6.22	31.94	44.30	15.14
Baxian	0.23	0.38	1.04	5.55	32.12	48.92	11.76
Jiangbei	0.15	0.33	0.89	4.52	28.76	52.35	12.99
Wanxian city	1.14	2.70	4.36	12.20	33.58	32.75	13.27
Wanxian	0.07	0.21	0.78	3.41	22.45	55.61	17.47
Kaixian	0.06	0.20	0.78	3.28	20.73	58.05	16.90
Zhongxian	0.06	0.21	0.93	3.45	22.99	51.25	21.10
Yunyang	0.05	0.18	0.80	2.80	16.58	55.73	23.86
Fengjie	0.05	0.23	0.83	2.69	16.63	54.65	24.92
Wushan	0.05	0.19	0.88	2.41	14.38	58.69	23.39
Wuxi	0.03	0.20	0.90	2.40	12.89	53.20	30.39
Fuling town	0.30	0.98	1.68	5.80	30.85	45.76	14.62
Fengdu	0.08	0.27	0.91	3.28	25.21	50.76	19.50
Wulong	0.04	0.26	1.04	3.47	20.91	49.19	25.10
Shizhu	0.08	0.26	1.12	3.46	21.12	48.35	25.61
Hubei province							
Yichang	0.16	0.55	1.58	6.81	27.08	47.27	16.55
Xingshan	0.09	0.64	1.75	8.04	25.16	45.46	18.86
Zigui	0.04	0.35	1.04	5.21	20.84	50.94	21.57
Badong	0.05	0.31	1.28	5.29	19.90	51.19	21.97

Source: Zhu and Zhao (eds. 1996), p. 110.[321]

unsound.[322] The number of enterprises to be shut down is unknown. However, according to Li Peng this policy will be strictly upheld.[323] This may impact on the rural population's potential for finding a job. Yet, the possible impact that closure

321 The figures in the book are based upon the Sichuan and Hubei provinces population census from 1990.

322 Xia (1999), p. 5.

323 Li (11 January, 2000).

of enterprises may have potential solutions was not mentioned by Li. During the trial resettlement period in the 1980s, examples exist of newly established enterprises in the Three Gorges area that intended to employ laid off peasants who were unskilled, where production has come to a stand still.[324] Even though this must be a big issue in the resettlement process, the problem of how the potential unemployment among the former peasant population will be solved is not reflected in the official state media. A survey by the Labour Department in Yunyang county showed that 60 percent of the peasants-turned-worker, did not have a position to go to in the factory where they were promised work.[325] The transition to a market economy may also prove a challenge to the resettlement process. The role of the government has diminished, and the reform of the State-Owned Enterprises (SOEs) has resulted in their losing benefits. Increased economic efficiency of SOEs and TVEs may also contribute to unemployment, as lack of skills among the rural population does not make them attractive as labour force. The Government's diminished responsibility for TVEs make it more difficult to guarantee employment for the resettled rural population.[326] As seen in Table 5.9, the rural population affected by the reservoir inundation is fairly young, with an average age of 27.4. This may be positive in the sense that young people may be capable of adapting to new situations and new challenges (such as new work tasks). On the other hand, the large number of young people indicate an active work force either in the agricultural sphere or in the secondary and tertiary industries.

Table 5.9 Characteristics of the Rural Population Affected by the Three Gorges Reservoir Inundation (%)

	Gender structure
Male	50.2
Female	49.8
	Age composition
0-14 years	26.3
15-24 years	25.4
25-34 years	16.3
35-44 years	14.2
45-54 years	7.1
55-64 years	5.4
65 years and above	5.2
Average age	27.4

Source: Zhu and Zhao (eds. 1996), p. 217.

324 See Heggelund (1994), p. 27, or Jiang (1992), pp. 52-64.

325 Wei (1999), p. 15.

326 For more details on implications of the economic transition for displacees, see Meikle and Zhu (2000), pp. 131-4.

In sum, the low educational level of the population in the Three Gorges reservoir area may influence the peasants' potential for finding employment when they cannot continue to farm the land. Experience from trial resettlement and the resettlement process illustrates that employment of former peasants is a problem.

Marginalization and social disarticulation Marginalization in different ways may occur should the Three Gorges reservoir relocatees be unable to fully restore economic strength. This may take place because the rural population cannot use their acquired skills and farm the land as they did earlier. Even if they do not become entirely landless, their piece of land may have shrunk so much that they might fall below the poverty line. 'Economic marginalization is often accompanied by social and psychological marginalization, expressed in a drop in social status, in resettlers' feeling of injustice, and deepened vulnerability'.[327] Issues that have appeared in the Three Gorges case that are related to marginalisation and social disarticulation are:

- Economic weakness
  The economic strength of the farmers in the new areas is questionable. A phenomenon called *erci yimin* ('secondary migrants/relocatees') highlights the potential economic weakness among relocatees and is mentioned by several scholars as a serious problem in the Three Gorges area.[328] Secondary migrants are peasants who need to move to make way for the reconstruction of towns that will be inundated by the reservoir, and thus lose their homes and farmland. These people are not among the people who need to move because they live below the inundation line, and are therefore called secondary migrants. There are basically three ways to settle these peasants: they are given work in a factory, or they arrange their own position, as well as receive funds for living expenses. The former peasants become non-agricultural population and have to change their vocation, and it may be difficult to keep up their economic strength when they move into an urban setting. These people are marginalised as they have lost their land and house due to city and town construction. In addition to farming the land, many of the peasants carried out sideline occupations (selling vegetables from local stalls, etc.), which they cannot continue to do when the land is occupied. They have moved into storied buildings constructed for relocatees, and have not found a new and steady occupation. They live on the funds that the authorities give each month, as well as on ad hoc construction work in the area. Their average income amounts to RMB 30-40 each month,[329] which is half of what is regarded as a standard basic living cost, not including fees for school, doctor,

327 Cernea (2000), p. 26.

328 See Gu and Huang (1999), pp. 356-7, and Wei (1999), pp. 14-15.

329 RMB 80 per month is the lowest standard existence for a rural family that has lost its farmland; this does not include school fees or medical fees. Gu and Huang (1999), p. 357.

or other expenses.[330] The problem is that their income is based on a rural standard of living, while their living environment has become urban.

Loss of economic strength among former peasants is an ongoing trend in a number of places[331] in the reservoir area. They eventually move from the storied buildings into shacks, as they cannot afford the rent.[332] The standard compensation for a house is RMB 400/m^2, while the actual price for new housing is RMB 700-800/m^2, a difference of RMB 300-400.[333] The only solution is to borrow money to buy housing. Without work and steady income the expenses become impossible to handle.

The policy for resettling these people has been the three ways mentioned above: factory work arranged by the resettlement authoritites, finding work on their own intiative, and subsidies from the government. The growing tendency appears to be that the migrants receive subsistence allowance instead of finding work. In Yunyang county in Chongqing municipality for instance, which has to resettle 120,000 people, it has become increasingly difficult to settle migrants via factory work. This is due to the weak economy of the county, where there are few enterprises in the area. The existing middle and small-size enterprises and factories do not have the economic strength to take on more employees. Only a very small share of the migrants manage to find occupation on their own. Thus, subsidies are the main source of income for these people. These people are marginalised economically and may feel inferior as they have to live on subsidies from the government. The new ways of living in an apartment building without any farmland also differ greatly from their previous lives and may make them feel marginalised. It has been suggested by Chinese researchers to instead construct *yimincun* (resettlement towns),[334] and provide these rural relocatees with some farmland in order to ensure their daily subsistence.

Protests or conflicts are rarely (or never) reported in the Chinese media. However, Western media sometimes reports of protests in the area. In January 2001, Probe International reported that secondary migrants were beaten by the police and soldiers in Kai county, as they protested against their homes being demolished to relocate the county seat. According to the report, the people

330 Ibid., p. 357.

331 Examples sited are Yichang, Wushan, Wanxian, Kaixian, Yunyang, Fuling. Ibid., p. 357.

332 The phenomenon is described by this saying '*xinggao cailie banjinqu, busheng buxiang banchulai, loufang bu zhu, zhu wopeng*' (Moving into [the relocatee building] in high spirits, silently moving out [from the relocatee] building, [we] go to live in a shack instead of living in the storied building), Ibid., p. 357.

333 As mentioned in Chapter 4, compensation varies according to the standard of the house and the size. Thus, these figures may not be representative for the entire reservoir area.

334 *Yimincun* is defined as an area in the new town chosen by the government where the relocatees can construct homes according to their preferred style. Gu and Huang (1999), p. 358.

were forcibly moved out of their homes and had to share the remaining homes without any compensation of land or shelter.[335]

- High expectations not met

  The rural and urban population in the Three Gorges area expect that the resettlement funds will radically improve their lives, and have high expectations for the Three Gorges project. According to Meikle and Zhu, peasants are no longer satisfied by re-registration from rural to urban status when resettled.[336] They also expect to obtain well-paid jobs. Moreover, as can be seen in Table 5.10 from Fengjie county in Chongqing, which is based upon a survey carried out by Zhu and Zhao (eds. 1996), the willingness to move may depend upon where the peasants live and their income. They divided the peasants into two groups, the *chengjiao nongcun* (outskirts of towns) and *shanqu nongcun* (mountain areas).[337] From their study one can see that the majority of the rural population from the mountainous areas are positive towards resettling to other places, and the majority have expectations that the Three Gorges project may be an opportunity to improve their living standards. As grain growers in a mountainous area their per capita incomes are not high (RMB 892 in 1991). On the other hand, the rural population who live on the outskirts of towns are not very positive to moving. This may have to do with their present standard of living, with a stable income and a more urban lifestyle. Also, the per capita income is higher than the rural population in the mountainous area (RMB 1650 in 1991) from growing vegetables, which constitutes 94.3 percent of their cultivated land. The tendency of richer peasants being less willing to move away is also confirmed by Li Heming, who has carried out surveys among the rural population to be relocated in the Three Gorges area.[338]

  The high expecations are due to the propaganda, rhetoric and slogans used by the authorities in order to create enthusiasm and to convince the public of the gains from resettlement, creating an optimistic image of the future for the relocatees. The social stability in the area could be threatened if the promises are not kept and expectations not met. One indicator that the authorities have realised that it will take longer than anticipated to become more affluent or raise their living standards is one of the slogans used in relation to reservoir resettlement in general to convince the relocatees of the positive effects a project will have on their lives. *Ban de chu, wen de zhu, zhubu neng zhifu* (move out, settle down to a stable life, gradually become rich). The character

335 Probe International, (15 January, 2001).

336 Meikle and Zhu (2000), p. 134.

337 See Zhu and Zhao (eds. 1996), pp. 222-6.

338 The surveys are from Zigui, Yunyang and Kaixian counties. See Li, Heming (2000), pp. 163-4.

zhubu[339] (gradually) indicates that people cannot have such high expectations. Li Peng also stressed the term 'gradually', saying that the change (from 9 to 11 characters, i.e. added the gradually) has to do with the fact that it is not possible to become rich right away, but it is important to first create conditions for the relocatees.[340]

Table 5.10 Attitudes towards Resettlement in Fengjie County's Rural Population (%)

	Rural areas on the outskirts of towns	Mountainous rural area
Do you want to move?		
Positive	15.8	76.8
Negative	84.2	23.2
How will life become after resettling?		
Better than present	17.6	37.5
Worse than present	52.6	19.6
Don't know	29.8	42.9
What about production conditions after resettlement?		
Better than present	8.8	26.3
Worse than present	56.1	28.6
Don't know	35.1	45.1

Source: Zhu and Zhao (eds. 1996), p. 223.

- Conflicts between relocatees and host population and cultural identity
  Cernea points to China as being unique in fostering community solidarity such as sharing of losses (particularly land) and redistribution of non-affected village lands between the non-displaced farmers and their community neighbours.[341] Despite this, conflicts between the host population and the relocatees are common in reservoir resettlement in China. In the case of the Three Gorges project, the resettled population will receive new farmland that is taken from the peasants who formerly lived there, i.e. the land is divided between the host population and new population. When the migrants are resettled into host areas, an increase of pressure on resources and social services takes place, which may result in economic losses for the host population. This may create hostility between the two groups.[342] Furthermore,

339 The term *zhubu* (gradually), was added to the reservoir resettlement regulations in 1991, which is regarded as a relaxation compared to the 1985 requirement of maintaining real income in reservoir resettlement. World Bank (1993), pp. 11 and 37.

340 Li (11 January, 2000).

341 Cernea (1997), p. 1582.

342 See for instance Qiu, Wu and Du (2000), p. 3.

the relocatees receive favourable treatment, such as lower income tax and living subsidies for the first few years after resettlement. This favourable policy does not include the host population, which may influence their attitudes towards the relocatees, and conflicts may arise. One measure suggested by Chinese scholars to avoid conflicts between the host and the guest population is to establish a favourable policy for the host areas, and not merely focus on the reservoir area and the resettled population.[343]

Moving out from the reservoir area is negative for the relocatees in several ways, as networks are disintegrated. According to Zhu's recommendations in May 1999, one should try to relocate people in groups and social units. The number of people who may stay together or separate is not known. When the interaction between families is reduced, resettlers' obligations towards non-displaced kinsmen are eroded. In Chinese culture family ties and community are important. When people live among strangers, communication is difficult, favours are not returned, conflicts arise easily, etc. It is thus likely that the resettled peasants may be discriminated against and may even be bullied in the host areas. This is due to the fact that Chinese villagers very much stress being in the same clan (*shi*), bearing identical surnames. The relocatees are thus regarded as strangers, and the resettled peasants being looked upon as strangers could last one to two generations.[344] This will not improve the lives of the relocatees.

In sum, the economic weakness, high expectations that are not met, and conflicts in the host areas may contribute to the marginalisation of the resettled population.

Factors not Covered by the Impoverishment Risks and Reconstruction Model

In addition to the selected points of the IRR model discussed above, a couple of issues that are not part of the model need to be mentioned. These issues may have had a certain impact on the Three Gorges resettlement policy change, as they are regarded as problematic in the resettlement process.

Loss of education Education loss[345] among displaced children may be another result of the Three Gorges dam, which could have negative consequences in an area that is poor and where the educational level is low. A number of schools need to be relocated, and the resettlement funding as well as the *duikou zhiyuan* (counterpart support) are intended to cover this.[346] Nevertheless, there are signs that the relocation of schools may be an issue in the discussions about the project

343 Ibid., p. 3.

344 Interview no. 6, Beijing 31 August 1999.

345 Loss of education is also added as a ninth point by Mahapatra (1999).

346 From 1992-1998, altogether 364 schools were constructed in the reservoir area with support from this system; which means 50-60 schools per year. In 1999, 36 schools were constructed. Zhongguo Sanxia jianshe nianjian bianjibu (1999), p. 199.

resettlement, and that funding may be insufficient. Li Heming concludes in his dissertation that loss of education for children is a problem, and in his analysis he found that basic education was ignored, and school children were forced to interrupt their schooling as a result of migration.[347] Moreover, some coverage in the Chinese media about this topic underlines the potential problem, as described below.

Altogether 527 primary and middle schools will have to be resettled in counties, districts and towns under Chongqing municipality, involving more than 250,000 students and teachers.[348] Each year Chongqing must move between 50 to 60 schools in order to complete the relocation on time. It is estimated that RMB 15.3 billion will be necessary to resettle these schools. Chinese authorities will pay RMB 3.5 billion; the remaining RMB 11.8 billion must come from external aid as well as contributions from other places, such as counterpart support. Thus, the funding for relocation and reconstruction appears to be far from secured as yet. Wanzhou district (under Chongqing municipality) for instance, has to resettle more than 75 percent of the schools. It is one of the poorer districts in the area, and the article cited here gives a clear impression of a district that will have difficulties in relying on their own efforts to move the schools. The funding gap is also due to the population increase; from 1999 to 2002 middle school students in Wanzhou will increase from 180,000 to 380,000.

In sum, taking the low educational level into consideration, the relocation and construction of schools would be important in securing education for children affected by the reservoir. The development of human resources would also be important for the future development of the Three Gorges reservoir area. However, a shortage of funding as reported in the Chinese media may indicate that the relocation and construction of schools may not go as smoothly as indicated by official yearbooks.

Corruption, mismanagement of resettlement funds and false numbers

Embezzlement of resettlement funds has emerged as one of the main threats to the implementation of the resettlement policy, as it reduces the amount of money for resettlement. In January 2000, the National Audit Office publicly announced the numbers of the corruption that had been uncovered in 1999. RMB 500 million had been embezzled in 1999 by 14 individuals.[349] However, not long after this the media reported that some of the funding had been retrieved.[350] At a meeting in March, Gan Yuping, deputy Secretary General of the Chongqing CCP, deputy

---

347 Child relocatees may suffer due to schooling being interrupted. There is also a lack of funding to construct new schools in for instance Hubei province. Children being unble to continue their schooling was also one of the concerns of the migrants in the Three Gorges area. See Li, Heming (2000), pp. 253-5.

348 *Guangzhou ribao* (1 February, 2000).

349 Cong (24 January, 2000). This was referred to in a number of Chinese newspapers such as Huashengbao, Tianjin Daily, People's Daily, etc.

350 Wang (10 March, 2000).

mayor of Chongqing and vice-chairman of the TGPCC, stated that the embezzlement amounted to 350 million yuan instead of 500 million.

Some of the worst embezzlement cases in Chongqing municipality were pointed out in March 2000, when Gan Yuping attended a press conference in Beijing (during the NPC). One case occurred in Fengdu county, where the bureau chief of the *Guotuju* (Land Administration Bureau)[351] had been sentenced to death in February due to embezzling more than 12 million yuan of funds earmarked for resettlement. The second case involved a resettlement officer of the Wanzhou resettlement bureau who had used more than 1 million yuan of the resettlement funds and received a life sentence. In other cases even governments have misused earmarked funding, such as the government of Taipingxi in Hubei that used rural resettlement funds (approximately RMB 7.07 million) to construct office buildings, living quarters and hotels. The resettlement office in Fuling district of Chongqing used 3.8 million to establish a company. Zigui Education department in Hubei used RMB 361,000 to construct an office and housing for the department.[352] These are only a few examples of the mismanagement and embezzlement of the resettlement funds for the Three Gorges project.

In addition, falsification of relocatees figures is also common among local officials in order to obtain more funding for the locality, as well as get more into their own pockets. This is related to the short-term of the local officials, as some only stay in the position for a few years, and most do not remain in their positions until the end of the term.[353] Also, false settlements are common, where enterprises that need to relocate due to the construction of the dam receive resettlement funding in order to employ relocatees. However, after having moved, the resettled people still do not get work, and the settlement funding then becomes an extra income for the enterprise.

In summing up this section about the problems that have appered in the resettlement process, one may see that a number of problems exist in the Three Gorges resettlement process. This questions the positive picture drawn by the World Bank regarding the resettlement experience in China. It may be that the problems portrayed above are special for the Three Gorges project due to its size. However, it is also plausible that some of the problems may be found in other, smaller projects. Cernea's IRR model is useful for structuring the discussion about the project's resettlement. The problems which emerged in the resettlement process are structured according to selected points from the eight impoverishment risks and reconstruction points. The problems that have appeared in the process are: lack of available farmland, shortage of farmland in the area; peasants not receiving land at

351 The Ministry of Land and Natural Resources was elevated from bureau status (*ban buji*) in 1998 during the restructuring of the Chinese bureaucray. It was formerly called *Guojia tudi guanli ju* (State Land Administration Bureau, shortened to *Guotuju*). On the provincial level *Guotuju* is often still used. For more informaion about the task of the Ministry see Guowuyuan bangongting mishuju (ed. 1998).

352 *Renmin ribao haiwai ban* (31 January, 2000).

353 See Wei (1999), pp. 18-19.

all or receiving low yield farmland, failed policy (citrus policy), uncertainty regarding the number of people to be moved, decrease in income, food insecurity, joblessness due to vocational change (from peasant to factory or construction worker), economic weakness, expectations for a better life not met and conflicts between the host and the resettled populations. In addition to the IRR model, loss of education and corruption and mismanagement of resettlement funds are problems in the resettlement process that need to be mentioned.

The Chinese Authorities' Response to the Problems in the Resettlement Process

In light of the numerous problems portrayed above, it is natural to wonder about the reactions of the Chinese authorities to these issues. From the media coverage, where the majority of the articles are propaganda for the project, it is clear that the project is given high priority by the Chinese government. Attention is in particular being paid to the developments in the Three Gorges resettlement. As problems have emerged in the resettlement process, there has been a response from Chinese authorities to these problems, and measures have been introduced. Chinese authorities have reacted to the problems that have appeared in the resettlement process by initiating two steps: moving 125,000 people out of the reservoir area and issuing new resettlement regulations. The initiatives introduced by the government are described and discussed below, and the reasoning behind these decisions will be looked into, together with the potential benefits or disadvantages of these measures.

Out-moving (waiqian)

The decision to move 125,000 of the rural population out of the reservoir area must be regarded as a response to the lack of environmental capacity and lack of farmland in the Three Gorges area. The increased focus on these issues must be seen in relation to the floods in 1998 that made Chinese authorities aware of the serious erosion problems along the Yangtze River. Thus, the reason for moving the 125,000 out of the area is twofold. It is based on an urgent need for environmental protection as well as on the need to reconstruct livelihoods for the resettled population. Less pressure on the environment will improve their chances of recovering or improving their living standards. During the years since construction began on the project, there have been numerous meetings and conferences, and reports have been compiled. It is likely that during this process the problems that occurred during the first and second phases have been reflected, resulting in a shift in the original resettlement policy. Concrete experience and research about the resettlement have become increasingly important for decision-making in the project. (See Chapter 8 regarding the changing shape of decison-making and leadership change for further details on this issue). Thus, there is a certain amount of feedback in the Chinese system, in the sense that there is willingness to attempt to solve problems, which indicates a dynamic decision-making process.

The resettlement policy change was an attempt to solve the resettlement and environmental problems. However, even though the environment may improve, a solution to the recovery of livelihood problems may not be the case, as it is unknown how the distant relocatees will live in the new areas. The expenses for relocating out of the reservoir area are higher than resettlement in the reservoir area, and the need to increase the resettlement budget is emphasised by scholars.[354] Different views on this topic exist in China among Chinese scholars. Some believe that moving people out of the reservoir area may be an improvement. From interviews regarding the 846,200 (static figure) to be resettled, scholars seemed quite optimistic about the possibility of restoring livelihoods for the 125,000 people who will move out to ten other provinces and one city. With regard to the 721,000 people remaining in the reservoir area, scholars seriously doubted that they can be resettled properly due to the scarcity of farmland and the limited potential for making a living. Some scholars, such as Wei Yi, find that out-moving is only a choice between two evils, where the other is to stay in the over-populated and environmentally pressed Three Gorges area. His conclusion is that the problems will not be solved by moving out, as distant removal is one of the factors that made resettlement in early dam projects unsuccessful, with the relocatees continuing to be impoverished in their new home areas.[355] Taking experience from early dam projects into consideration, it is interesting that some scholars find out-moving an advantage. One reason why optimism exists for the Three Gorges out-moving, is the fact that the majority of these people will be moved to provinces that are situated in the eastern coastal provinces or along or close to the Yangtze River (in the Jianghan plains).[356] Many of these provinces are also beneficiaries to the Three Gorges project (electricity production or flood control). Out-moving to these provinces is expected to increase the chances of successful resettlement, as the farming methods would be similar, which would simplify recovery of livelihoods. Also, the cultural differences would not be as big, and the distance from original home areas would be shorter, making it possible to go home to visit relatives and friends. Furthermore, according to a survey, the resettlers are also more willing to be resettled in the flatlands of the Yangtze and Han rivers,[357] which may also indicate an increased emphasis on the opinions of the relocatees. These provinces are more similar to the Three Gorges area and the culture and customs would be more similar than for instance Xinjiang or Hainan Island, which have been tried out as possible locations earlier. Thus, two conclusions may be drawn from this: choosing a similar area to relocate the people could indicate that the authorities have given increased attention to the livelihood issue, as well as placing

354 Qiu, Wu and Du (2000), p. 3.

355 Wei (1999), p. 19.

356 These are the plains surrounding the Changjiang (Yangtze River) and Hanshui (Han River), one of the major tributaries to the Yangtze River.

357 Xia (1999), p. 6.

more emphasis on the wishes of the resettled rural population. However, the resettlement funding needs will increase due to the out-moving, and it is uncertain whether or not the the resettlement budget will increase.[358]

On the other hand, moving people out to other places away from their safe environments will always be difficult, as communities are dispersed and networks are broken. Reports already exist of unsuccessful resettlement in the provinces mentioned above.[359] It is difficult to move into a new village, and it is difficult to rebuild the lost networks. Several generations will often pass before the relocatees become assimilated into their new communities. Loss of cultural identity, social problems, etc. may be the result of such a move. One option that has been mentioned during interviews of scholars is the reduction of the dam height, which would reduce the resettlement figure. However, this has not been officially discussed by the Chinese authorities. A reduction in dam height in order to reduce the resettlement figure is also accentuated in the World Bank guidelines Operational Policies (OP) 4.12 on Involuntary Resettlement,[360] which stresses that 'involuntary resettlement should be avoided where feasible, or minimised, exploring all viable alternative project designs'. Also, '...a reduction in dam height may reduce the resettlement needs'. For instance, according to one scholar, if the normal pool level of the Three Gorges project were reduced to 160 metres (i.e., reduced by 15 metres from 175 metres), the number of relocatees would be reduced by more than 500,000.[361] Reducing the height would seem to be an easy option that would solve some of the resettlement problems. However, reducing the height will influence the planned benefits of the dam. The electricity output of the dam will not reach the estimated goal;[362] the navigational potential is reduced, affecting the shipping industry on the river; a lower dam means a reservoir with less storage capacity, which again reduces the flood protection potential. Usually in such cases, a comparison is made between the displacement costs and the project benefits. Cernea states that in many cases cost and benefit analysis (CBA) is inadequate, as it does not fully take social costs into account.[363] The purpose of a CBA is to find out whether or not a project's benefits outweigh a project's costs. Should the benefits of the dam outweigh the social costs, i.e. if for instance full compensation for displaced families is less than the value of the electricity, it is beneficial to maintain the dam height and displace the families. 'If the

358 Some scholars point to out-moving as a factor that will increase resettlement expenditures due to new surveys and inspections that have to be made, moving and transport costs, delay in the resettlement process, etc. Qiu, Wu and Du (2000), p. 3.

359 Reports state that promises are not fulfilled for peasants who have moved for instance to Jiangsu province. See Macleod, Calum and Lijia (1 July, 2001).

360 Page 1, (a) and BP 4.12 page 1, (b), footnote 4. These were the former OD 4.30 on involuntary resettlement.

361 Wei (1999), p. 19.

362 David Pearce states that the lowering of dam heights need not sacrifice electricity output, but may do so. Pearce (1999), p. 56.

363 Cernea (1997), p. 1578.

compensation costs are higher than the added value of electricity, then the designers should lower the dam.'[364] However, to make a qualified choice is not as simple as it seems. One problem is the lack of full and proper compensation for social costs, which according to Pearce include 'the loss of nonpriced environmental and cultural assets, the loss of social cohesion, the loss of market access, and psychological damage from dislocation'.[365] Furthermore, certain costs, such as psychological traumas from broken networks, may not be easily quantified, but the issues are real for the people in question.[366] If these assets were taken into account, the social costs may in the end outweigh the economic benefits. This view is further emphasised by the World Commission on Dams' report, which reproaches the approach taken by public authorities in general: '[t]he narrow nature of technical and economic analyses undertaken does not necessarily mean that public authorities that chose dams as a development option were unaware of the social and environmental costs. Rather, within the context of knowledge available and the value system of those making decisions at any given time, the sacrifices were judged to be worth the benefits of pushing ahead with the project. This approach to decision-making continues largely intact today'.[367] The report goes on to say that poor accounting in economic terms for the social and environmental costs and benefits of large dams has rendered the true economic efficiency of and profitability of these schemes largely unknown.

Not surprisingly, in the case of the Three Gorges project this would also seem to be the case. In China, resettlement due to reservoir construction is still regarded as a collective duty and a contribution to the country, since dam construction is viewed as an important part of China's ongoing modernisation process. As one of the solutions to the problems in the resettlement process, the Chinese authorities have chosen to move one-third (125,000) of the rural population out of the reservoir area to other provinces in the country. Unfortunately, there is not much material available to indicate whether the Chinese authorities have considered the option of a reduction in dam height. However, it is my impression, based on interviews and discussions with scholars, that the Chinese government regards reducing the dam height as a higher cost compared to the cost of moving the people out (as in many other countries). One may therefore conclude that the social costs related to reservoir resettlement are still regarded as a lesser cost than the perceived benefits of a project, and it is unlikely that the full social costs of resettling at least 1.13 million people have been sufficiently considered. The benefits of the Three Gorges project are power production, flood control and

---

364 Pearce (1999), p. 56.

365 Ibid., pp. 52-3.

366 Other costs may be cultural heritage and loss of cultural resources (such as temples, shrines, burial sites) and loss of cultural and archeological resources. In addition, environmental quality may be affected, which may impact on agricultural productivity, fisheries, etc. These items are usually not part of the planning process for dams. See World Commission on Dams (2000), p. 117.

367 Ibid., p. 120.

improved navigation on the river. However, the stated benefits from the dam have been questioned recently, such as the project's potential for flood control and power production. Experts doubt that the dam will be able to hold large floods as promised, which has been a concern since the beginning of the debate about the project. The accumulation of sediment is predicted to occupy space in the reservoir and hinder full utilisation of flood control capacity. Furthermore, the power production of the dam as well as the income from power production is questionable, as recently constructed dams have been unable to sell their power according to plan. One example is the Ertan dam in Sichuan province that was constructed with World Bank funding and completed in 1998. Ertan is presently operating at less than half speed[368] due partly to a monopolistic power market system that makes it difficult to sell power to other provinces. Experts such as Hu Angang, an economist at the Chinese Academy of Sciences, warn that this may be the case for the Three Gorges dam power production unless the power industry's monopoly is changed.[369] The Three Gorges project has boasted of its hydro capacity, but should the estimated power sales be reduced, one must conclude that the dam has not achieved one of its stated goals.

In sum, the out-moving is a reaction by Chinese authorities to the problems that have come forth in the resettlement process for the Three Gorges project, and signifies that the resettlement decision-making process is dynamic. Lowering of the dam height, which would reduce the resettlement figure but may impact on benefits such as flood control and power production, has not been an official alternative for the authorities.

New Version of the Resettlement Regulations in 2001

A second initiative, which attempts to solve the problems in the resettlement process, is the new and edited version of the resettlement regulations. The regulations were approved by the State Council on 15 February, 2001 and began implementation on 1 March, 2001.[370] The new regulations replace the regulations of 1993. In connection with the publication of the regulations, the State Council stated that it was necessary to issue new and revised regulations in order to ensure the successful implementation of the Three Gorges resettlement and the social and economic development of the area. The new regulations were issued due to the fact that '...new conditions and new problems have appeared in the resettlement

---

368 Since operation began at the Ertan dam, its electric power production could have reached 32 billion kWh. However, it has only reached 13.7 billion kWh, less than half of its capacity. Zhongxin Sichuanwang (24 August, 2001). See also McCormack (2001), pp. 22 and 24, about the Ertan dam.

369 In Sichuan, during the construction of the Ertan dam, several thermal power plants were constructed, thus reducing the need for power. See Zhongxin Sichuanwang (24 August, 2001). Chinese authorities are discussing establishment of a nationwide power transmission grid which may reduce the power monopoly by the provincial power industries. For future plans see *China Daily* (16 July, 2000).

370 Zhonghua renmin gongheguo guowuyuan ling (2001).

work'.[371] The new version of the resettlement regulations is a longer and more concrete and detailed version. For instance, the 1993 regulations consist of six sections and 43 articles, while the 2001 regulations have seven sections and 64 articles. I will not make a comprehensive comparison between the two sets of regulations. However, it is important to point out a few changes that reflect some of the resettlement problems that have been described in the above sections in this chapter, which have likely been the main factors in the policy change for the project. The following changes have been made:

Agriculture One major difference between the two sets of regulations is that the principle of taking agriculture as a basis (*yi nongye wei jichu*) has been deleted in the 2001 regulations. Thus, one may conclude from this that less emphasis is put on agriculture as a way of settling the rural population. In the 1993 version, article 3 states clearly that agriculture is to be the foundation for the rural population. Furthermore, article 9 in the 1993 version elaborates further on this issue, and goes into more detail on how the rural development should take place, for instance by developing new land, from low-yield to high-yield, etc. All in all, numerous references are made to the rural population in the 1993 regulations, and agriculture was to play a major role in the future lives of the rural population. In the 2001 resettlement regulations, since the principle of taking agriculture as a basis has been erased, the importance of agriculture as a means of making a living has been greatly diminished. Emphasis is now put on high-yielding agriculture as stated in article 13 of the 2001 version. Moreover, the regulations also recommend relying on the second and third industries for the rural population. The reason for this shift in focus must be seen in relation to the lack of available and arable farmland in the Three Gorges area.[372]

Environment A second major change in the resettlement regulations is the increased emphasis on environmental protection. All sections have articles (such as 3 and 4) that include instructions regarding the rational use of natural resources, environmental protection, and water and soil conservation. Article 13 stresses the need for ecological agriculture, while article 21 focuses on the need to close down polluting enterprises in the reservoir area. Furthermore, an article (62) under the penalty section has been added regarding penalty for the destruction of the environment. It states that whoever causes destruction of vegetation and the ecological environment as well as erosion, will be punished according to the environmental law and the water and soil erosion law. The serious environmental

371 Xinhua wang (25 February, 2001).

372 Furthermore, the decreased profitability of farming may have been a factor that influenced the change that is not discussed in this book. In 1998, 1999 and 2000, per capita net income from crop cultivation declined by 16, 45 and 98 yuan respectively. Prices have fallen since 1997 and continue to fall. Information is taken from State Statistical Bureau Bulletin (February 28, 2001) on China's Economic and Social Development. I am grateful to Dr. Eduard Vermeer, Leiden University for this information.

situation in the area is reflected in the regulation regarding the prohibition to farm land sloped above 25 degrees (article 26 under relocation and resettlement), which also stresses the need to make terraced fields of farmed hill slopes below 25 degrees. This was not mentioned in the 1993 regulations. The regulations also give directions with regard to tree felling (article 27), which is related to the serious erosion problems in the area. The increased emphasis on the environment must be seen in relation to the environmental problems and the erosion problems in the area, as well as the floods in 1998.

Supervision and management A third major change in the 2001 regulations is a new section of 11 articles regarding the supervision and management of the usage of the resettlement funds (*yimin zijin shiyong de guanli he jiandu*). Some of the articles in this section were also in the 1993 version. However, making it a separate section stresses the increased importance paid to solving this problem. This section is very concrete, and stresses that the authorities will not increase the resettlement funding. It lists six points on which the resettlement funding should be spent: i) compensation for rural resettlement, ii) moving and reconstruction of towns and cities, iii) compensation for moving and reconstruction of industrial enterprises, iv) reconstruction of infrastructure projects, v) environmental protection and vi) other resettlement projects that are initiated by the responsible resettlement management organisations under the TGPCC. In comparison, the 1993 regulations (article 21) only state that the funds must be spent on specific purposes, and that these specific purposes will be stipulated in other regulations. Thus, the problem with mismanagement of resettlement funding is one of the likely causes for issuing new resettlement regulations.

Penalty A fourth major revision is the penalty section, which in the 1993 version was quite brief and named 'award and penalty' (*jiangli yu chufa*). Of the four articles, one was related to awards that would be given to outstanding performance in the resettlement process either by units or individuals. In the 2001 version, this section has been expanded to nine articles, and made more concrete. The title is now only 'regulations on punishment' (*fa ze*), and the award part has been deleted. This could indicate that there is no 'outstanding' performance in the resettlement process, and that it has become more problematic than was perceived at the time. Furthermore, the regulations on punishment are more concrete than the 1993 version. Article 58 lists several points that one may receive punishment for: i) refusal to move or delaying resettlement, ii) migrants returning to home areas after having received compensation, iii) attempts to obtain compensation a second time. It is probable that these issues are problems encountered in the resettlement process. They may indicate that the resettlement process does not go very smoothly, and that people do not agree to moving away or out of the reservoir area. Furthermore, it also indicates that some people may try to obtain extra money. Article 59 lists four points in relation to mismanagement and embezzlement of resettlement funds that state that resettlement funds should not be spent on i) non-resettlement projects, ii) investment projects, iii) the purchase of bonds and stocks,

iv) other methods to divert resettlement funds. These regulations indicate that there are big problems in the management of the resettlement funding, which is further underlined by the great media coverage given to corruption and mismanagement of resettlement funding.

One may conclude that the learning process for the Chinese authorities during these years of resettlement has resulted in new regulations that in many ways are more realistic than the previous regulations. The above examples show that the shift in focus on agriculture reflects the situation in the area where there is little or no available farmland and the peasants must find alternative ways to make a living. Stricter management of resettlement funding reflects the problems of mismanagement and corruption related to the implementation of the project. The strengthened penalty section illustrates the problems that exist with actually getting the peasants to move. As a result of the corruption scandals in relation to the resettlement, Chongqing Municipal Party Committee and the city government have had discussions with the institutions involved in the resettlement, such as the Resettlement Bureau and the Land Administration Bureau in Chongqing.[373] In addition, Chongqing municipality has established an auditing network consisting of a three-step control system called *shiqian, shizhong, shihou* (which literally means before, during and after the event is implemented),[374] with new ways to manage the resettlement funds, which includes increased responsibility for the people in charge as well as improved supervision of funding allotments. Chongqing city Resettlement Bureau will have meetings with all bureau chiefs of the local resettlement offices in order to increase management control over the resettlement funds.

It is nevertheless difficult to predict how the new regulations will affect the dam resettlement, as no set of regulations will solve all problems. The shift in agricultural resettlement for the rural population may create new problems, such as difficulty in finding jobs. As mentioned earlier in this chapter, many enterprises will be closed in the area because of environmental pollution or if they are economically unsustainable. Moreover, the new regulations emphasise that the expenses in relation with an expansion of enterprises must be carried by the local governments or units,[375] which may impact on the number of jobs available for the resettled population. It is also difficult to see that punishment for point i), refusal to move, and ii) migrants returning home, will improve the situation very much, as concrete problems for the resettled people often are the reason for resistance. This leads on to an important issue that is not treated at all in the new regulations (nor in the previous ones): the legal aspect, as well as the rights and interests of the resettled population. Even though the regulations state (in article 43) that public notices should be put up with information about the resettlement spending, and urges the acceptance of public supervision, arrests have taken place of rural

---

373 Xinhuashe (22 February, 2000).

374 Liu (9 March, 2000).

375 Article 21 emphasises the need to shut down non-profitable and polluting enterprises; article 23 states that local governments or other units must bear the expenses relating to expansion of enterprises that surpass the allocated resettlement funding.

citizens who have pointed to mismanagement of the resettlement funding.[376] Without a strengthening of rule of law and public supervision of the project, it is difficult to see that the regulations will solve these problems.[377]

In sum, the new resettlement regulations issued in 2001 are a reaction to the problems that have appeared in the process, and an attempt to solve them. They are an improvement of the 1993 regulations, but fail to provide any legal aid and suport to relocatees.

The IRR Model and the Three Gorges Resettlement Process

As we have seen in the previous sections, the problems that have appeared in the resettlement process are numerous. These problems are related to the land capacity in the area, environmental condition, the number of people that need to be moved, and the backward economic situation in the area. The problems have received increased focus in the state media, which indicates that Chinese authorities are concerned about the resettlement process. We have also seen that Chinese authorities have attempted to resolve some of the issues in the resettlement process by moving people out of the reservoir area and introducing a new version of the resettlement regulations. Now let us turn to the Impoverishment Risks and Reconstruction model again and look at the reconstruction aspect of the Three Gorges resettlement policy. The following two issue areas will be discussed below: Does a risk and reconstruction perception exist in the Chinese resettlement planning process? The IRR model is a general model that has been developed based upon resettlement practice in a number of countries. How relevant is the IRR model for Three Gorges resettlement, and in China in general? Does China need the IRR model?

Risks and Reconstruction Perception

The Three Gorges project has been planned for many years, and the development type resettlement (*kaifaxing yimin*) was formed during the many years of discussion about the project's feasibility as described in Chapter 4. The resettlement plan was developed based on the resettlement experiences in China when the displaced population had difficulty in reconstructing their livelihoods. A development aspect was therefore included in the plan. The IRR model has existed

---

376 See Probe International (5 November, 2001).

377 Wei Yi, the author of the article in *Strategy and Management* believes that the regulations fail to solve some important problems such as that of fake migrants, who obtain migration status on false premises, which may create tension between real and fake migrants. See Probe International (29 March, 2001).

in a Chinese language version since 1998.[378] The extent to which the Chinese authorities are aware of the existence of such a model is unknown. However, both the World Bank and the Asian Development Bank (ADB) now require risk analysis and apply the IRR model as the tool to carry out this analysis in projects that involve resettlement.[379] This would have implications for China, as the country is a major borrower from both the World Bank and the ADB. Thus, one may expect that the model will be known to Chinese authorities, and will be used in projects funded by the World Bank and the ADB. For Chinese funded projects such as the Three Gorges, the application of the IRR model is less certain. Even if they were aware of it, they may not be interested in using it as a model to work with, as they may feel confident about their own methods. The challenges in resettling such a large population have inspired the Chinese government to initiate measures that would avoid some of the risks from resettlement that were experienced in early dam building. The measures are similar to many of the points in the IRR model. This assessment will not make a point by point comparison with the eight risks and reconstruction points as there is no such Chinese model to compare with. Instead, it will be an overall assessment of the policy measures initiated, seen in relation to the Impoverishment Risks and Reconstruction model. Cernea states that by turning the risk model on its head, one not only obtains predictions about which risks to avoid in resettlement, but one also obtains a guide towards counteracting the risks.[380] He also states that risk recognition is crucial for sound planning. The general risk pattern can be controlled through a policy response that mandates and finances integrated problem resolution, and a strategy response with specific resettlement programmes. Cernea states that '...a risk prediction model becomes maximally useful *not* when it is confirmed by adverse events, but rather when as a result of its warnings being taken seriously and acted upon, the risks are prevented from becoming reality, or are minimised, and the consequences predicted by the model do not occur'.[381]

With regard to the reconstruction perception, one issue in the discussions about resettlement is how resettlement is perceived in China, and whether or not resettlement policy/practice includes a reconstruction part as well. We have seen from early resettlement in China that only one-third of the displaced people in hydro projects were successfully resettled, while two-thirds were less successful or unsuccessful. The offical reason for the unsuccessful resettlement history was the 'lump sum' type compensation given, while their future livelihoods and social

---

378 Cernea (1998). An earlier Chinese version of Cernea's work is found in Cernea (1996). I am grateful to Michael M. Cernea for sending me the Chinese versions of the IRR model.

379 The ADB has recently adopted operationally the Impoverishment Risks and Reconstruction Model, which means that this directive will have to be applied by China in the projects that ADB will finance in China. Personal correspondence with Michael M. Cernea, March 2001.

380 Cernea (2000), p. 33.

381 Ibid., pp. 33-4.

factors were not taken into account. The experience has created an awareness among Chinese authorities for the need to reconstruct relocatees' livelihoods. This is also reflected in the Chinese term used for resettlement: *yimin anzhi* (which may mean resettlement and placement, settlement, find jobs for). Also, *anzhi hao yimin* (to settle the displaced people well) is frequently used. Thus, one can conclude that the Chinese authorities have an intention to restore people's livelihoods, and a will to create a favourable environment for the resettled people, to assist them in finding new occupations and to develop the areas in which the people are resettled. Moreover, as stressed by the World Bank, China regards resettlement as a development opportunity, as opposed to a number of other developing countries.[382] One could therefore say that China has experienced a shift in resettlement thinking and approach since the 1980s. It has attempted to move away from past failures, where resettlement was looked upon as a burden, to a new approach where resettlement is perceived as a development opportunity. Nevertheless, the issues that have appeared in the Three Gorges resettlement, indicate that the process is still rather problematic, even with a reconstruction perception present.

The Chinese authorities are already carrying out a number of the measures that are mentioned by Cernea in the *Risks and Reconstruction, Experiences of Resettlers and Refugees.*[383] Setting people back on cultivable land, converting unproductive hills and steep uplands into flat terraces or into forested areas, and training rural resettlers in new skills for non-farm employment are some of the measures that are carried out by Chinese authorites. One may therefore conclude that a reconstruction awareness exists, and that China's approach to resettlement is 'risk-conscious'.[384] Below I will briefly present some of the measures that have been initiated by the Chinese government in the planning process that may be perceived as reconstruction activities. These measures may fit under the points in the IRR model that were discussed in the previous section: from landlessness to land-based reestablishment, from food insecurity to adequate nutrition, from joblessness to reemployment, from homelessness to house reconstruction; from marginalisation to social inclusion. The examples of Chinese reconstruction measures are listed together, and not discussed according to a specific topic of the IRR model. The list of governmental measures below may not be complete. Some of the policy measures that may be included in a risks and reconstruction model are described below, including comments regarding the potential for success.

Trial resettlement In the planning process for the dam project, trial resettlement (*shidian yimin*) was initiated in 1985, several years before the project actually was approved (1992). Trial resettlement included preparation of new land and planting of orchards in advance of resettling the rural population. Even though this trial resettlement was on a small scale, it may have provided important information about the farming possibilities in the area. For instance, during the 1980s the citrus

382 World Bank (1998), p. 15.

383 Cernea and McDowell (eds. 2000).

384 Personal communication with Michael M. Cernea, January 2002.

fruit policy was regarded as one of the main ways to restore livelihoods for the resettled peasants. Chinese scientists were involved in research projects to improve the different sorts of citrus in the area in order to enable the relocatees to make a living from selling the fruits.[385] The trial period resulted in an awareness of the lack of market potential, as fruits were rotting outside of peasants' homes. Moreover, CAS was against this policy of only focusing on citrus. Eventually this policy was scaled down. Furthermore, trial resettlement for outmoving to Xinjiang and Hainan may have provided valuable information for the authoritites and may have resulted in the decision to move 125,000 persons to the flatlands along the Yangtze and Han Rivers. Thus, one could say that the reconstruction process was initiated before the resettlement started.

Training of peasants There are two categories of training for peasants. The first category concerns the training of peasants who continue to farm the land. This group may be trained by researchers from for instance the Chinese Academy of Sciences at training/research stations that are situated in the vicinity of both Wanxian in Chongqing and Zigui in Hubei. The training involves learning about new and more efficient agricultural methods. The peasants come to the stations to observe the methods that are being used and the results, which may provide an incentive to use these methods. According to interviews with CAS, letting the peasants themselves participate in such activities increased the chances for the peasants' applying new methods, as they often would prefer traditional methods. Agroforestry is one method mentioned, that involves trees being planted on hillsides where farming at 25 degrees is now taking place. Grains will be planted in between the rows of trees, which will reduce the erosion problems. The second category concerns training the peasants whose land is lost and who become non-agricultural population, in skills which will enable them to work in secondary and tertiary occupations. They will be provided with work in factories and receive training. For instance, the *duikou zhiyuan* (counterpart support) policy includes training of former peasants by enterprises. The training of resettlers in new skills may improve their potential market value. However, due to the low educational level in the area (as described in Table 5.8), finding and holding a job has proved problematic.

Duikou zhiyuan (counterpart support) This is described in the resettlement history in Chapter 4. However, it is mentioned here as it is also part of the reconstruction aspect. This additional source of funding to the resettlement funding allocated by the government is particularly important with regard to providing the former peasants with new job opportunities. For instance, Wanxian shi (now called Wanzhou) mentions 50 favourable policies to attract enterprises to establish branches in the Three Gorges area.[386] However, the success of this policy is still uncertain, and needs further study. Problems such as greediness and demands from

---

385 The information is based on interviews with CAS researchers involved in this work.

386 Wanxianshi Sanxia gongcheng yimin bangongshi (1996), p. 8.

local enterprises or units have scared off potential counterparts, and an overlap of types of enterprises exists in the area, which threatens the economic viability of these enterprises.

Later Stage Support Fund (Yimin houqi fuchi jijin) In 1996, the State Planning Commission (now State Development and Reform Commission) issued a circular that all projects must establish a Later Stage Support Fund, which was based on the post resettlement fund set up in the 1980s. 'Later stage' refers to the period after the resettlement budget has been spent and resettlement work is completed for the project.[387] The Fund will collect annual revenues from power generation that will be turned over to provincial governments for 'operation, maintenance, and further development of resettlement schemes behind large- and medium-size hydroelectric dams'.[388] Article 45 in the 2001 regulations states that a later stage support fund is to be established for the Three Gorges project. The fund will be divided between Chongqing municipality, Hubei province and the provinces and cities that receive the out-moving rural population. The Ministry of Finance and the relevant ministries and organs will come up with the concrete measures to be approved by the State Council. The Fund will be collected for 10 years. The provinces and municipal authorities will make the final decisions regarding maintenance of resettlement infrastructure and further development work. A conclusion about this Fund is that it is praiseworthy and illustrates the rehabilitation perception that exists in China. Nevertheless, the largest threat to this initiative is the existing corruption on all levels, which if allowed to continue unchecked will impact on the funding available for concrete measures.

Decentralised implementation Decentralisation increases the chance of success for resettlement, as local governments are more aware than the central authorities of local conditions and what is needed in their respective areas. The basic principle of the resettlement regulations from 1993 and the edited version of 2001 is that resettlement policy is carried out according to the central authorities' unified leadership, responsibility according to province, and takes counties as a basis (*zhongyang tongyi lingdao, fensheng fuze, xian wei jichu*).[389] This implies that much of the implementing responsibility lies on the provincial, county and local levels, while the funding for resettlement is allocated by the central authorities. In the 1996 World Bank report, China is praised for decentralising the implementation of the resettlement policy, where all responsibility for resettlement is given to local or city governments, and resettlement solutions are developed

---

387 I am grateful for this information regarding the 1980s post resettlement fund and the Later Stage Support Fund provided by Dr. Zhu Youxuan, resettlement expert and World Bank consultant, April 2001.

388 The rate is 0.005 per kWh. In the Shuikou project the World Bank expected an income of about US$ 2.5 million annually. World Bank (1998), p. 19.

389 Zhonghua renmin gongheguo guowuyuan ling (1993), article 6; and Zhonghua renmin gongheguo guowuyuan ling (2001), article 8.

locally. The disbursement of relocation compensation comes from the National Resettlement Bureau to the Bank of Construction, which passes on the funding to the Chongqing municipality and Hubei province. The municipality and province then divide the funding to cities, then from cities to counties, from counties to townships, from townships to villages, from villages to the relocatees.[390] This system ensures responsibility on each of the levels. However, it also provides an opportunity for local officials to engage in mismanagement and corruption of resettlement funds, as has been described earlier in this chapter.

Dialogue and popular participation Cernea points out that dialogue between the resettlers and officials is decided by law in China, both as individuals and as community groups.[391] The World Bank also states examples of popular participation and dialogue betweeen resettlers and the local authorities in Chinese dam projects, which has been beneficial for the outcome.[392] Some dialogue seems to exist between the authorities and the resettlers for the Three Gorges project, but this is mainly on the village level.[393] At the village level the heads of villages and resettled population may participate in 'the selection of resettlement sites, reallocation of land resources and the settlement of disputes between migrants and their hosts'.[394] Public participation in the early stages of the process, i.e. in the decision-making process is next to nothing, and at the provincial level the possibility for relocatee participation is low. The new resettlement regulations do not mention dialogue between authorities and the population to be resettled. However, they imply that some dialogue must take place, such as in article 12 that states that the individuals must agree to become *feinongye hukou* (non-agricultural household). The impression from interviews is that meetings may take place in villages where resettlement officials inform the relocatees about decisions already made. The intention is not to engage in discussions about important decisions, and the resettlers are not involved in the process from the very beginning. However, it may be that these meetings provide an opportunity for the relocatees to express their opinions. It is however, uncertain to what extent the relocatees' opinions are taken into consideration. From conversation with both Chinese local resettlement officials and researchers, it is clear that one of the major obstacles to the resettlement work is the peasants' feelings towards their land, roots and ancestors. They do not wish to leave the land even if they are promised good conditions in the

390 See an overview of the disbursement system in Li, Heming (2000), p. 193.

391 Cernea (2000), p. 41.

392 World Bank (1993), p. 38, point 3.44, which points to the Daguangba multipurpose project where one of the counties encouraged public participation with regard to village layout, alteration of house designs, etc.

393 See Li, Heming (2000), pp. 263-9, and p. 306. Little or no public participation has been common in the majority of dam projects involving reservoir resettlement. Lack of participation in the decision-making process has resulted in social instability, complaints and collective actions carried out by the relocatees in order to obtain better conditions following resettlement.

394 Ibid., p. 267.

new areas. According to resettlement officials both in Yichang and Wanxian, the majority of their work is trying to convince the peasants to move (*zuo sixiang gongzuo*). One survey showed that 90 percent of the rural resettlers in the reservoir area did not wish to lose their land.[395] Other surveys have shown that the willingness to move away may depend upon which category of peasant one belongs to, peasants on the outskirts of town (*chengjiao nongcun*) or peasants from the mountain areas (*shanqu*). One can imagine that 'difficult' peasants who are unwilling to move may be able to negotiate,[396] as the resettlement officials would be eager to settle the people before the deadline. According to officials, emphasis is put on obtaining the co-operation of the resettlers, and the resettlement officials have to go back to the same families several times to convince them of the advantages of moving to another place. Nevertheless, recent reports exist of rural population being detained, or having disappeared, after collecting peasants' protests and complaints about bureaucratic corruption.[397] One may conlude that some dialogue exists, albeit on the lowest level and in a very late stage of the process, which means that the resettled population has very little say about their future. Chinese scholars also point to the issue of participation during the decision-making process and the implementation process as imperative for successful resettlement, and some are critical towards the lack of dialogue.[398]

From the above we can see that a reconstruction aspect is present in the Chinese resettlement planning, and that Chinese authorities try to use the dam building as a development opportunity, as is suggested by the World Bank. The debate about the dam project under the CCP has been going on since the 1950s. Thus, the area had to await a decision on the Three Gorges project for forty years. Even though the inland provinces in general still lag behind the development level of the coastal provinces, it is generally acknowledged in China that the uncertainty surrounding the Three Gorges project has been one of the factors for the slow economic development in that specific area. Economic growth is expected in the Three Gorges area because of the dam project, in particular since many enterprises are expected to settle in the area. Nevertheless, uncertainties about economic development exist, which are related to the area being poor and the average educational level being low. Furthermore, in the new regulations, article 23 stresses that in relocating enterprises, any excess costs to enlarge the enterprise and raise standards for instance, must come from the local governments and not from the resettlement funding. Thus, the potential expansion of enterprises which would allow employment of former peasants, may seem difficult. The development potential in the Three Gorges seems to be disadvantageous compared with for instance the Shuikou project, portrayed in the World Bank report, which is situated in flourishing Fujian province on the east coast. One policy that has been introduced recently which may impact positively on the area, is the *Xibu da kaifa*

395 Qiu, Wu and Du (2000), p. 2.
396 For instance to select a good site to be resettled to.
397 Becker (21 March, 2001).
398 See Wei (1999), p. 16.

(development of the Western region, also called 'go west' policy).[399] This policy is intended to bridge the gap between the inland provinces (in the north-west and the south-west) and the more developed eastern provinces. The Three Gorges area is also a part of this plan as Chongqing is one of the targeted areas. It remains to be seen whether the Three Gorges project and the go-west policy will bring prosperity to the Three Gorges area.

In sum, the reconstruction aspect is present in the Chinese resettlement thinking, which is reflected in the Three Gorges project. Measures have been initiated which indicate this, such as trial resettlement, counterpart support, training of peasants, decentralised implementation, and the later stage support fund. Nevertheless, despite the existing reconstruction perception, the potential for success is questionable. One important reason for this is that the relocatees have to a very limited extent been involved in the decision-making process regarding their own future.

The IRR Model and The Three Gorges Project

The IRR model is intended as a tool for decision-makers to anticipate risks and as a guide to reconstruction of livelihoods for the resettled people. It is 'first a tool for generating and organising knowledge, but also a tool for guiding action, usable for policy and planning purposes'.[400] Cernea states that all the points may not be equally relevant in one project. The model captures the basic and most frequent risks in resettlement projects, and each combination of risks will appear different in different contexts.[401] One advantage to having such a model is its generality; it is in principal applicable in any project in the world. Nevertheless, since political, economic and social situations vary in different countries, the national condition will always play an important role for the application of such a model. In China for instance, even though there is an awareness regarding the resettled people's needs and where resettlement is looked upon as a development opportunity, certain problems are typical in the Chinese society that the model does not cover. The following factors create problems for the IRR model in China.

The corruption problem The growing corruption problem in China has also had implications for the Three Gorges project.[402] Funding that is earmarked for resettlement and reconstruction of infrastructure has been embezzled and spent elsewhere. This may be special for the Three Gorges project due to the large scale

---

399 The Western region is comprised of the following places: the south-west—Chongqing, Sichuan, Guizhou, Yunnan and Tibet; and the north-west—Shaanxi, Gansu, Qinghai, Ningxia and Xinjiang. For an introduction to these areas and the policy, see Xibu da kaifa zhinan bianji weiyuanhui (2000).

400 Cernea (1997), p. 1571.

401 Personal communication with Michael M. Cernea, January 2002.

402 Corruption scandals in China involving high-level officials are common. See Reuters (2 March, 2001), or Conachy (1 February, 2000).

of the project. It not only involves the construction of a dam, but it also includes the construction of towns, roads and other infrastructure projects. This may make it easier to divert funding into one's own pockets or for other projects. There is for instance no reference to corruption in the report about the Shuikou and Yantan projects. This is perhaps due to the World Bank involvement in the project, the smaller scale of the project and/or that corruption was less common during the 1990s. (The Shuikou I and II projects were completed in 1993 and 1997). Corruption in general in China is rampant, and even though the central authorities try to fight the corruption, apparently no improvement has taken place. Corruption is a phenomenon that is eroding the policymaking in relation to the Three Gorges project resettlement to the extent that the implementation of the resettlement is imperiled. The resettlement funding that is mismanaged or pocketed by local officials will result in a reduction of the compensation to the relocatees as well as for infrastructure construction. It goes without saying that this may have grave consequences, and may lead to social instability as in earlier dam projects.

Rule of law Strengthening of the Chinese legal system is gradually taking place, and new laws appear constantly. The Chinese state that *renzhi* (rule by man) often still prevails over *fazhi* (rule by law),[403] and one may say that rule by law and the comprehension of the existing laws in China is still in its nascent stages. Legal issues are important in the Three Gorges project, as they relate to freedom of speech, a more open and free press, and supervision by the public. In relation to the dam project there are several examples where rule by man has predominated. Research and the resettlement for the project are two areas of great relevance. With regard to research on the resettlement and environmental issues, scholars have been criticised for their 'negative' atttudes, i.e. research results, which have had serious impact on these scholars' professional and personal lives. Being passed over for promotion and early 'retirement' are measures being used by the authoritites to restrain and punish 'unruly' scholars. As mentioned above, limited dialogue in the resettlement process takes place. However, the resettlers' influence and participation in the planning process is low. A recent example of lack of legal protection for resettlers is from Yunyang county in Chongqing municipality, where a few rural resettlers have attempted to take petitions from the rural population in Yunyang to Beijing to the central authorities in order to focus on mismanagement of the resettlement funding. Local authorities blame the rural resettlers for creating disturbances in the area and for hindering the resettlement process.[404] Nevertheless, the rights and interests of the resettled people are receiving some focus. During the NPC in March 2001, a deputy suggested that 'China is badly in need of a law on protection of the rights and interests of people displaced by water control projects'.[405] The resettlement regulations (from both 1993 and 2001) state that

403 See for instance Dai (ed. 1989), p. 64.

404 Probe International (23 March, 2001). See also Probe International (20 April, 2001).

405 *China Daily* (17 March, 2001). The issue was raised by NPC deputy Jin Xing from Henan province.

responsibility for the resettled people lies with the host government, i.e. the local governments of Chongqing and Hubei in the reservoir area. Relocatees who have been resettled to other counties either within their old province or to another province, must approach the host government in cases where problems arise. A weakness in the new resettlement regulations is its failure to address these legal and rights problems. The establishment of an institution where the resettled population could obtain legal assistance would have been a great improvement for the relocatees in the Three Gorges project.

The socio-economic environment China is now in a transitional economy; it is in a transition from a planned economy to a socialist market economy. One area where this complicates matters in resettlement is when peasants must change vocation. Due to the shortage of land, a number of the rural population become non-farm workers and must find work in enterprises. With the market economy, organisational structures in relation to providing work for the former peasants may have become weaker. For instance in urban resettlement in China, at present the labour bureau no longer has sole responsibility to provide work for the resettled peasants.[406] In the rural areas no such function existed. The local resettlement project staff appears to be responsible for finding work for them, and in many cases the resettlement officials must negotiate with the enterprises on a case-by-case basis to secure employment for the resettled population. The project officials are individuals who may not have the network needed, and the process is bound to be time-consuming, as former peasants are not particularly sought-after labour force. Zhu and Meikle also state that these negotiations are 'partiluarly difficult to resolve in the absence of any clear operational guidelines regarding enterprise compensation'.[407] Also, a second point which complicates matters is the reform of the State-Owned Enterprises (SOEs).[408] Many of the SOEs have either been closed down or sold. Thus, former SOEs experience a new demand for efficiency, and are not interested in taking on former peasants without skills. The authorities may not have the same leverage vis-à-vis these enterprises any longer, as opposed to when they belonged to the State.

Natural resources and environmental pollution Population pressure, diminishing natural resources and environmental pollution are factors that must be taken into consideration when resettlement is planned. The scarce natural resources in China become even more evident under resettlement circumstances. 'The risk of landlessness is prevented through landbased relocation strategies' states Cernea.[409] This may be true, but in the case of the Three Gorges project there is little available land left to satisfy the needs of the rural population. The population density has

406 See Meikle and Zhu (2000), p. 137.

407 Ibid., p. 139.

408 On the reform of the SOEs in China see for instance *People's Daily* (4 August, 1999); or Dorn (21 September, 2000).

409 Cernea (1997), p 1578.

already exceeded the limits, the land is eroded, and the water pollution is increasing. The result is that many of the peasants who will not receive land will have to change their vocation, and may or may not be able to find a job. In short, the meagre natural resources and the increasing pollution problems are putting restraints on the potential for successful resettlement, not only in the Three Gorges project, but in China in general. In the future, environmental migrants may be the result both in the Three Gorges area and in other places in China.

Central versus local interests As discussed above, the decentralisation of authority is viewed as positive, and increases the chance for success. Nevertheless, decentralised implementation may also be problematic, as it would be difficult to control all actions by the local governments. One common feature of the Three Gorges project is the frequent mismanagement of resettlement funding. According to the Director of the Resettlement bureau, Qi Lin, three basic concepts exist in connection with the misuse of resettlement funds: i) *weigui* (break regulations/ rules) which points to spending of resettlement funds on unintended projects (*yong da jiangyou de qian mai le cu*; use the soya sauce money to buy vinegar); ii) *weiji* (break discipline) which is the use of resettlement funds for non-resettlement projects; and iii) *weifa* (break the law) embezzle the money for personal use.[410] The mismanagement of funding (mainly the first two concepts) may be due to central versus local interests. Local authorities prefer to invest in areas other than those they were instructed to, without heeding the potential problems this may create for the resettled population by leaving less funding for reconstruction. Thus, the different priorities may have a serious impact on the implementation of the resettlement policies set by the central government. In the Three Gorges project a large portion of the resettlement funding has been spent on other items than intended, which may be related to a conflict of interest between the central and local governments. One may therefore conclude that for this specific dam project of such a large size, the decentralised implementation may also pose a threat to successful implementation of the resettlement policy.

The above issues may influence the efficacy of the IRR model in Chinese resettlement. We have seen that Chinese authorities are aware of the need to reduce the impoverishment risks in resettlement, and risk-reducing measures have been initiated. Thus, one can conclude that the reconstruction aspect definitely is present, and an intention and awareness exist regarding the resettled peoples' restored livelihoods or improved living standards. It is nevertheless important to emphasise that the IRR model is relevant for China. First of all it could be a useful reference list for the authorities to cross off in resettlement projects, in order to ensure that potential risks in the resettlement process are paid attention to. All officials on higher levels should be made aware of the model, and officials and

410 According to Qi Lin, the first two categories are the most serious (made up 8.8 percent of the resettlement funding in 1998), while the third category is very small (0.14 percent of resettlement funding), amounting to 31.15 million yuan from 1993 until 2000. See Xu (10 December, 2000).

cadres at lower levels should receive training in the potential risks and reconstruction points. As mentioned, the IRR model is available in Chinese and is also referred to by Chinese scholars.[411] Secondly, the Chinese authorities need to pay particular attention to certain points in the IRR model such as 'from marginalization to social inclusion, and from social disarticulation to networks and community rebuilding'. The model stresses the need to facilitate the integration of resettlers within the host populations by for instance establishing group structures, both informal and formal institutions. The sociocultural aspect of resettlement appears less (or not at all) emphasised in the Three Gorges policy. Focus is mainly placed on the rehabilitation aspect, such as compensation of housing, land and finding occupations for the peasants who cease to work the land. Even though the 2001 resettlement regulations (article 14) stress both settlement in groups (*jizhong anzhi*) and scattered resettlement (*fensan anzhi)* (in this order), there is little focus on the *social* importance for groups being relocated together. In cases where the migrants will be split up, 'fill-in operations'[412] will be practised, where resettlers groups will be inserted and scattered within pre-existing communities. Moving in groups would help the resettlers to feel more at home, being with people from their own areas, since part of the old network will still be present. However, a number of relocatees will also be moved in a dispersed manner, meaning that families lose contact with their relatives, friends, and neighbours. As has been mentioned earlier in this chapter, the importance of belonging to a clan (*shi*) in rural areas in China, makes the resettlement to a strange place more difficult, as many generations may pass before a migrant family is accepted by the host population. The 'heaviest costs of all are the severing of personal ties in familiar surroundings, to face new economic and social uncertainties in a strange land'.[413] This cost seems to receive little emphasis in the Three Gorges official policy context. However, the social issue of resettlement is being discussed by scholars in articles about the project.[414] Thus, one may hope that this issue will receive increased focus in the years to come. This is particularly important since China is planning several hydro projects where reservoir resettlement will be necessary.[415] The limited focus on social costs and social issues is a general impression from Chinese society, which is also related to the above discussion of cost-benefit and common-good aspect. One sign of increasing official focus on the situation for the Three Gorges relocatees is the recent establishment of a group that will carry out a survey about the resettled

---

411 Gu and Zhong (1998).

412 See Cernea (1997), p. 1582.

413 Sowell, T. (1996), *Migration and Cultures: A World View*, Basic Books, New York; quoted from Cernea (1997), p. 1576.

414 Articles that discuss or touch upon social aspects of resettlement are Gu and Huang (1999); Han, Wu, Yan and Chen (2001); Wang, Huang, and Ding (1999).

415 Several dams are going to be constructed on the Yangtze River, such as the Xiluodu dam upstream from the Three Gorges dam. For more details about planned dams on China's rivers see McCormack (2001).

population's conditions and human rights, which will result in a report.[416] Expectations towards the report should perhaps be moderate; however, the establishment may be regarded as an intiative that may eventually have some influence on the Three Gorges resettlement. The Chinese Academy of Social Sciences (CASS) and the Resettlement Bureau under the TGPCC will be responsible for the study, with participants from universities and local resettlement officials.

Summing up

The number of people to be moved in the Three Gorges project is unprecedented in reservoir resettlement in China. In May 1999 Premier Zhu Rongji announced an adjustment in the resettlement plan for the Three Gorges project, which entailed moving (*waiqian*) a large number (125,000) of the rural population from the reservoir aera to other provinces. The stated reason for this adjustment is the lack of environmental capacity in the Three Gorges area, where erosion and pollution problems are serious. The lack of farmland has led to farming on hill slopes (above 25 degrees), which has increased the erosion problem in the area. The 1998 floods were an eye-opener for the Chinese authorities with regard to the erosion problems along the Yangtze, and a tree-felling prohibition was issued. In addition to the stated environmental reasons for resettlement policy change, one assumption in this book concerns the potential impact the problems in the implementation process have had on the resettlement policy change. In addition come the potential problems as portrayed in academic journals. Despite the resettlement policy improvements that have taken place since the 1980s, as described in the previous chapter, a number of issues have appeared in the resettlement process. The issues are structured according to selected points in the Impoverishment Risks and Reconstruction model (IRR, by Michael M. Cernea). The problems are: lack of available farmland on which to settle the rural population; failed official policy (citrus policy); uncertainty about the exact number of people to be relocated; food insecurity and allocation of low-yield farmland to the peasants; problems of finding or switching vocation for former peasants; economic weakness and the phenomenon of secondary resettlers (*er ci yimin*) who have lost farmland to infrastructure construction; expectations not met and conflicts between the resettled and host populations. These issues are all related to the environmental capacity of the Three Gorges area. Two issues are added that are not covered by the IRR model, and that are perceived as problematic in the resettlement process:

---

416 The study to be initiated in January 2002 is *Zhongguo Sanxia gongcheng yimin renquan baozhang shizheng yanjiu* (A study to gather substantial evidence that human rights are ensured in the Three Gorges project resettlement), which will produce the following report, *Zhongguo Sanxia gongcheng yimin renquan baogao* (The Report about the human rights in the China Three Gorges project resettlement.) See Fang (13 September, 2001).

loss of education due to lack of funding for relocation and construction of schools, and corruption and mismanagement of resettlement funding.

The Chinese authorities' response to the problems has been twofold: moving people away from the reservoir area to other provinces in China, and issuing a new version of the resettlement regulations in March 2001. Moving people away has been controversial and not a favoured alernative for the Three Gorges resettlement, as this has been an unsuccessful solution in early dam projects. Out-moving may also create new problems for the relocatees. However, emphasis has been put on moving the relocatees to provinces in the plains of the Yangtze and Han Rivers, of which many are beneficiaries of the project. The new regulations issued in 2001 focus on solving many of the problems in the resettlement process that are described above. The main changes from the 93-version are: less focus on agricultural resettlement and greater focus on settling relocatees in the second and third industries; increased emphasis on environmental protection; increased emphasis on supervision and management of resettlement funding, as well as an expanded section on penalty measures. The resettlement policy change illustrates Chinese authorities' responsiveness to the resettlement problems; capability and flexibility to adjust strategies exist. One conclusion from dam building in China, and elsewhere, is that the full costs of dam construction and reservoir resettlement most often are not taken into account. The social costs, such as loss of environmental and cultural assets, are difficult to price and are not included. One alternative to reduce the resettlement problems and the environmental pressure in the Three Gorges area that has not been officially discussed is to lower the dam height. This would reduce the number of people to be resettled, but also may reduce the promised flood control and hydroproduction benefits of the dam. From the discussion above on the cost-benefit aspects of the Three Gorges project, it appears that the social costs due to resettling people seem to be regarded as a lesser cost than the perceived benefits (flood protection, power production and navigational improvements) of this project.

One conclusion to be drawn regarding the relevance of the IRR model in China is that Chinese authorities have an awareness of the need to reconstruct relocatees' lives. This is reflected in the language describing resettlement in China, as well as in the resettlement policy for this particular project. This awareness is based upon many years of resettlement experience in China as well as on interaction with multinational agencies such as the World Bank. The World Bank praises China and resettlement due to the thorough planning in resettlement projects, and for viewing resettlement as a development opportunity. China already carries out many of the suggested measures in the IRR model. Nevertheless, the IRR model could be a useful planning tool for Chinese authorities. The Chinese emphasis in resettlement is put on rebuilding relocatees' lives. It focuses less, or not at all, on the social aspects and the social trauma of broken networks, which is emphasised in the IRR model. Thus, one may conclude that from the points discussed in this book, the main relevance of the IRR model is to enduce learning more about the social costs of resettlement, and how they can be diminished or avoided. The IRR model now exists in a Chinese language version. Furthermore, both the Asian Development

Bank and the World Bank require risk analysis and apply the IRR as a tool to carry out this analysis. China, as a major borrower, will therefore be exposed to this model in projects funded by the development banks. Also, the generality of the IRR model requires China to further develop it according to their needs and the conditions of the country in order to provide the most efficient resettlement method. Some of the problems that must be taken into consideration when applying the IRR model are corruption; the need for a strengthening of the legal system; the need to increase supervision by the public and a more open press; central versus local authorities where instructions by the central government are ignored; the transitional economy where the state's responsibility for people's lives is diminishing; the lack of natural resources and increased pollution problems that make resettlement projects even more difficult. One sign that China may be beginning to take the plight of the resettled population seriously is the establishment of a group consisting of academics and officials that will review the human rights conditions for the Three Gorges relocatees. Although expectations towards the report should perhaps be moderate, the establishment may be regarded as an initiative that may have some influence on the Three Gorges resettlement.

Chapter 6

China's Environmental Policymaking: Trends and Developments 1972-2001

Introduction

The purpose of this chapter is to discuss trends in environmental policymaking in China that may have been important for the Three Gorges project environmental policy-making. China's environmental problems are serious, despite the country's efforts to stop the environmental degradation. Economic growth is partly to blame for the continuing degeneration of the country's environment and the depletion of natural resources. Although some problems, such as deforestation, may also be rooted in policies prior to the economic reform policy that was initiated by Deng Xiaoping in the late 1970s. The environmental problems in China are diversified, and differ according to regions. North China is arid, drought-ridden and lacks water resources, while in South China water is in abundance and frequent floods occur. Moreover, water pollution is threatening lives and contaminating drinking water in a number of places in China. Acid rain, which affects between 30 and 40 percent of China's territory, has become a major problem in central and south China, in particular for cities like Guizhou and Chongqing.[417] Acid rain is caused by the heavy reliance on coal, on which 67 percent of all energy is based.[418] Thus, air pollution is serious, and some of the most polluted cities in the world are found in China, including Beijing.[419] Health problems such as respiratory diseases and lung cancer are increasing rapidly. Land degradation is widespread, and soil and water erosion is the cause of flooding in some of China's major rivers. The biodiversity resources are under pressure and endangered species likewise. The technological gap between the developed eastern provinces and the poorer inland provinces is large, which has an impact on efficiency and pollution.[420] A large environmental apparatus has been developed in China, and environmental laws and regulations are frequently issued. Enforcement is lax, however, and in addition to the general low environmental awareness, it must be regarded as one major cause for the continued environmental problems in the country. Literature about China and its environment is plentiful, and it may be divided into four groups: i)

---

417 Vermeer (1998), p. 977.

418 China Statistical Yearbook (2001), Table 7-2 p. 229.

419 For more details about air pollution in China see World Bank (2001a), pp. 77-97.

420 For more on regional and sectoral differences in pollution problems see Vermeer (1998), pp. 982-3.

environmental policymaking, ii) implementation of environmental policy, iii) description of pollution problems and the effects of this on nature, society and the economy,[421] and iv) environmental regulation and enforcement.[422] This chapter draws on a number of these books and articles compiled about the environment in China.

The time period for this brief overview of the developments regarding environmental policymaking in China ranges from 1972 until 2001.[423] One assumption in this book is that the increased importance of environmental protection and policymaking has positively affected the resettlement and environmental policymaking for the Three Gorges dam project. This chapter will also provide a background for the environmental discussion of the Three Gorges dam project in Chapter 7. The chapter, although not a study of the implementation of China's environmental policy,[424] will nevertheless give some examples of impacts and effects of environmental pollution in China in order to illustrate the situation. The environmental review consists of selected events among the environmental developments in China. The account is organised into sections according to topics; within the sections the account is chronological. The events selected are considered by the author to be representative and important in the development of environmental policy-making.

Furthermore, this chapter will provide examples from the environmental policymaking process that are relevant in relation to the Three Gorges project. Examples are water pollution, floods, forests and erosion. The review intends to show a trend in the environmental developments in China that has been important for the resettlement and environmental decision-making in the Three Gorges project. In order to understand the environmental developments for the Three Gorges dam project, it is important to know about the major trends in China's environmental history. A number of books have been compiled about China's environmental developments which have been useful sources for the early period described in this chapter.[425] In addition, and in particular for the recent environmental developments, this chapter also relies on the *Zhongguo huanjing nianjian* (China Environmental Yearbook) from the period 1990-2000.

---

421 See for instance Smil and Mao (1998); World Bank (1997); World Bank (2001a); and Vermeer (1995).

422 Ma and Ortolano (2000).

423 For further details on China's environmental developments see Zheng and Qian (1998); Cannon and Jenkins (eds., 1990); Edmonds (1994); Ross (1988). Furthermore, the *China Quarterly*, December 1998, number 156, is a special issue about China's environment, with authors such as Vermeer (1998), Jahiel (1998), Ash and Edmonds (1998), and others. For specific issues such as water, see Ma (1999).

424 For materials about the implementation of environmental policy in China see Jahiel (1994); Sinkule (1993); and Sinkule and Ortolano (1995).

425 For details on environmental policy developments see Zhongguo Huanjing nianjian bianji weiyuanhui (1990), pp. 1-12; Jahiel (1994); Jahiel (1998), *China Quarterly*, December 1998, No. 156, pp. 757-87. Edmonds (1994); Ross (1988); Maa (1993).

Institutional Setting and Environmental Policymaking: A General Overview

China's participation in the United Nations Conference on the Human Environment in Stockholm in 1972 marks the beginning of environmental discourse at the policy level in China.[426] Before this time, China's isolation and the Cultural Revolution (1966-76) that created total chaos during its first years had hindered China from taking part in the environmental awakening that was taking place in the industrialised world. Following the first chaotic period of the Cultural Revolution, China could begin to pay attention to practical issues such as the environment. The UN Stockholm Conference as well as national pollution incidents had made China aware of the need to start paying attention to environmental problems. Two of these were water pollution incidents in Dalian and Beijing in 1972.[427] Consequently, the first national conference on the environment was convened in August 1973, which issued the '*Guanyu baohu he gaishan huanjing de ruogan guiding*' (Certain Regulations on Protecting and Improving the Environment). The conclusion from the meeting was to immediately begin to pay attention to environmental protection in order to avoid problems later.[428] The State Council later (in November) approved the Regulations from the meeting, which for the first time laid out the guidelines for environmental protection: overall planning, rational distribution, comprehensive utilisation, conversion of the harmful into the beneficial, reliance upon the masses, everyone taking part, protecting the environment, and benefiting the people.[429] These Regulations were the basis for the initial phase of environmental work in China that was begun following the conference.

As a preparation for the UN Conference in 1972, a leading group responsible for environmental protection was informally set up in 1971 under the State Council to supervise preparation of the Chinese delegation's position at the Stockholm Conference.[430] At the UN Conference, China acknowledged the seriousness of the environmental problems in the country, but placed the responsibility for global pollution on the industrialised countries. Following the First National Environmental Protection Conference in Beijing in 1973, environmental protection agencies were created at both central and local levels.

The State Council Environmental Protection Leadership Group was established in 1974, and its responsibility was to co-ordinate environmental work that had

426 According to Qu Geping, the former administrator of NEPA, the Conference was the turning point for China's environmental cause. Zhongguo Huanjing nianjian bianji weiyuanhui (1990), p. 4. See also Ross (1988).

427 At the same time, environmental pollution incidents within China such as the Guanting reservoir pollution incident, put the topic on the leaders' agenda. Zhongguo Huanjing nianjian bianji weiyuanhui (1990), p. 4.

428 Ibid., p. 4.

429 Maa (1993), p. 56.

430 Ross (1988), p. 137.

begun in the country. The Leadership Group only convened twice during a period of nine years. Therefore, much of this work was taken care of by the Environmental Protection Office, which had been established the same year as a subordinate organ to the Leadership group.[431] From that time on, each province, autonomous region and city, as well as ministries and organs under the State Council, gradually established environmental offices, and environmental research as well as monitoring was initiated.[432]

Despite these efforts, the environmental situation deteriorated in China in the period 1977-1982. Campaigns such as the Great Leap Forward and the Cultural Revolution had put stress on the environment. For instance, deforestation waves took place both in the 1950s and 1970s.[433] These campaigns were labelled the 'three great cuttings' (*san da fa*), as deforestation waves took place during both periods. The first was the Great Leap Forward, and the second and third the Cultural Revolution, with devastating impact on the environment such as erosion and sedimentation problems.[434] These deforestation problems are related to the present-day flood disasters along the Yangtze River for instance. This was according to Qu Geping due to the 'leftist' policy-making of a period where ideological campaigns were more important than paying attention to the environmental problems. Environmental policy existed at the time. However, it was more a matter of curing immediate problems that imposed health hazards on people.[435] In the over-all political picture of China, the environment was a minor issue.

Following the chaotic years of the Cultural Revolution, the environmental problems of the country exploded. And 'by the late 1970s the lack of significant attention to environmental protection and the related lack of institutional authority began to change'.[436] This is attributed partly to the change from Maoist policy to Deng policy, as well as a great increase in pollution problems due to the emphasis on heavy industry. Furthermore, the Deng administration was aware that rapid economic growth would impact seriously on the environment, and this was the reason for the change in attitude and change of values.[437] One example of attitude change took place during the National People's Congress in 1978, when a section

---

431 Zhongguo Huanjing nianjian bianji weiyuanhui (1990), p. 4.

432 In the following years, environmental research projects were carried out, such as the Beijing quality study in 1973 and the pollution source of Bohai and Yellow sea in 1977. Ibid., p. 5.

433 One of the environmental effects of the Great Leap was deforestation. In order to fuel the backyard steel furnaces, trees were cut down en masse, which resulted in serious deforestation and erosion. See Shapiro (2001), pp. 80-84 for more about the Great Leaps' impact on forests. The Great Leap is mentioned here even though it occurred prior to the period that is reviewed here, as it has influenced the deforestation and erosion problems in China for a long period of time.

434 Ibid., p. 80.

435 Jahiel (1994), p. 79.

436 Ibid., p. 80.

437 Ibid., pp. 80-86; and Ross (1988), p. 139.

on the protection of the environment and natural resources was included in China's Constitution for the first time.[438] Furthermore, the Ministry of Urban and Rural Construction and Environmental Protection was established, in which the Environmental Protection Bureau was placed. Due to the Bureau's location within the new ministry, it was not especially effective with regard to compelling other agencies to comply with its directives.[439] In addition, the Leading Group under the State Council was abolished; it had been established in 1971 to supervise the work for the UN conference in 1972 and had the authority to co-ordinate various bureaucracies.

The environment received a new boost when the Second National Environmental Protection Conference took place in December 1983 and January 1984. This was an important meeting for the environment in China and is regarded by many environmental officials as the critical point of change for environmental protection, as the leaders showed a new level of recognition of these problems.[440] The meeting explicitly stated that environmental protection is a strategic task in the modernisation of the country, that it is national policy, and should be an integral part of the economic construction of the country. The meeting also came up with an initial plan for environmental protection until the end of this century, as well as a plan to strengthen environmental management in order to very quickly solve some of the less complex environmental problems that did not need large investments.[441] Following this meeting, in May 1984, the Environmental Protection Commission under the State Council was established,[442] with the Environmental Protection Bureau serving as its secretariat. This was an important commission,[443] as its task was to co-ordinate all agencies involved in environmental protection work in China. The commission consisted of ministers and vice-ministers from the line-ministries and commissions. Furthermore, the State Council increased the authority of the Environmental Protection Bureau and changed its name to the National Environmental Protection Bureau (NEPB) in December that same year.[444]

---

438 Zhongguo Huanjing nianjian bianji weiyuanhui (1990), p. 6.

439 Ma and Ortolano (2000), p. 78; Ross (1988), p. 141. Ross states that environmental protection resembled an appendage rather than an integral part of the new ministry.

440 Zhongguo Huanjing nianjian bianji weiyuanhui (1990), pp. 7-8.; Jahiel (1994), p. 103. According to the Yearbook, four important results came out of the meeting: i) environmental protection became a state policy; ii) new strategy of developing the economy at the same time as protecting the environment, 'yu fang wei zhu' (focus on prevention); iii) including environmental protection in planning, both on central and provinicial levels as well as implementing environmental protection; and iv) environmental management.

441 Zhongguo Huanjing nianjian bianji weiyuanhui (1990), p. 8.

442 Vice-Premier Li Peng was appointed chairman of the commission. Ross (1988), p. 141.

443 The Environmental Protection Commission was a short-term commission, and its authority less significant than the long-term commissions (such as the SPC). Jahiel (1994), p. 107.

444 This doubled its personell allowance from 60 to 120 perople. Jahiel (1998), p. 769.

The name change altered the status of the organisation, as it now could issue orders directly to the provincial EPBs, and the documents issued were of equal status to those of the Ministry of Construction.

A second round of institutional reform took place in 1988 when the NEPB became the National Environmental Protection Agency (NEPA), and was granted independence from the Ministry of Urban and Rural Construction and Environmental Protection. A combination of factors made this structural change possible. One of the factors that influenced this outcome was a letter that Former Premier Li Peng received, who at the time was the head of the Environmental Protection Commission, from all the thirty provincial EPBs that argued for the independence of the NEPB.[445] In 1989, NEPA convened the Third National Environmental Protection Meeting. It was an important meeting in the sense that five environmental policy strategies were formally approved by the State Council and announced by Qu Geping at the meeting. The five environmental policy strategies were new management tools: the contract responsibility system with city government officials, the comprehensive clean-up system, the collective control system, the pollution permit system and the dead-line clean-up system. These five were added to the three existing management tools that had been introduced at the Second National Environmental Protection Conference in 1983: *santongshi* ('Three Simultaneouses') policy, the discharge fee system, and the environmental impact assessment system.[446]

In 1992, China participated in the United Nations Conference on Environment and Development (UNCED), a conference that pushed the country's environmental protection work further forward.[447] One institutional outcome from this process was the Administrative Centre for China's Agenda 21 (ACCA 21) that was established in 1993 following the UNCED to implement China's Agenda 21, which is attached to MOST.[448] Moreover, it was a way for China to be accepted by the international community again, as the country had become somewhat isolated following the crushing of the student demonstrations on Tian'anmen in 1989.[449] China therefore saw the UNCED in 1992 as a good opportunity, and was preparing actively for the conference. In 1991, the State Planning Commission (SPC), Ministry of Foreign Affairs, State Science and Technology Commission (SSCT) and NEPA compiled *China's Environmental and Developmental National Report*, as preparation for the UNCED. That same year, Beijing hosted a Ministerial

445 Jahiel (1994), p. 125.

446 Ibid., p. 132. Barbara Sinkule prefers to use the term three synchronisations to describe the policy, as she feels this better conveys the Chinese meaning of the *san tongshi* policy: three activities that have to be synchronised but not necessarily take place simultaneously. However, the direct translation of the term *tongshi* (simultaneously) will be used in this book. Sinkule (1993), p. 122.

447 Zheng and Qian (1998), p. 139.

448 See Buen (2001), pp. 79-83, for details on ACCA 21 and relationship with MOST.

449 Although according to Jahiel (1994), p. 113, the environmental protection exchanges were not cancelled following the 1989 crushing of demonstrations.

Conference of Developing Countries on Environment and Development, and thus played a leading role in voicing the concerns of the developing world in relation to global environmental issues. The result from the meeting was the Beijing Declaration on Environment and Development, which stated that poverty, under-development, and over-population are the causes of environmental degradation. It also stated that developing countries had the right to develop their economies, and that the industrial countries were responsible for the degradation of the global environment.[450] At the UNCED, as one of the first countries in the world, China ratified the Convention on Biological Diversity (CBD) and the Framework Convention on Climate Change (FCCC).[451] The Biodiversity Conservation Action Plan was issued in May 1995 funded by the Global Environment Facility (GEF) through the UNDP, with the World Bank as executing agency. NEPA was the leading agency for the publication, while members from various ministries and commissions as well as the scientific community (both national and international) participated in the compilation. Following the UNCED, China completed its White Paper, the National Agenda 21 (A21) as one of the first countries in the world, and it was approved by the State Council in March 1994.[452] An international roundtable conference was convened in July 1994, attended by Jiang Zemin, where a list of priority projects were presented to the international community to obtain funding.[453] In 1996, China's Agenda 21 was integrated into the Ninth National Five-year Plan (1996-2000) for Social and Economic Development and the Outline of Long-term Social and Economic Development Objectives for 2010, and adopted by the NPC.

An important institutional change took place in 1998, when NEPA was promoted to ministerial status and officially renamed the State Environmental Protection Administration.[454] The head of SEPA, Xie Zhenhua, was raised to ministerial rank. The elevation to ministerial status was part of a general

---

450 For more details on the process leading up to the UNCED see See Chayes and Kim (1998), pp. 514-15.

451 Ross (1998), p. 816. China has signed and ratified a number of international treaties, conventions, agreements and protocols in addition to the FCCC and the CBD. Among them are the Vienna Convention for the Protection of the Ozone Layer (1985), the Montreal Protocol on Substances that deplete the ozone layer (1987, 1990, 1992), the Basel Convention on Control of Transboundary Movements of Hazardeous Waste and their disposal (1989), the UN Convention on the Law of the Sea (1982), the Convention on the Prevention of Marine Pollution by Dumping of Waste and Other Matter (1972, London Convention), the Convention on Nuclear safety (1994), etc. For a comprehensive list see Wang, Xi (2000), pp. 16-17; and Bergesen, Parmann and Thommesen (eds. 1999/2000), pp. 200-201.

452 Zhongguo 21 Shiji Yicheng Bianzhi Lingdao Xiaozu (1994).

453 Zhongguo Huanjing nianjian bianji weiyuanhui (1995), pp. 1-2; and SPC and SSTC (1994).

454 Another sign of increased importance for NEPA in the process leading up to the elevation to ministerial status, was NEPA moving into a new building in September 1995.

government restructuring, where a number of ministries were abolished or merged into other ministries or commissions. This elevation in status is symbolically very important for environmental protection work in China, as it signified a new level of understanding among China's leaders with regard to the serious environmental situation in the country. The main purpose behind the elevation would clearly be to raise the status of the environmental administration in order to solve the long-lasting problem of NEPA being lower in the institutional hierarchy than the ministries it was set to monitor. The Environmental Protection Commission was abolished in the restructuring process, and SEPA took over its co-ordinating function. Furthermore, the abolishment of the State Environmental Protection Commission would also ensure that the proposals from SEPA could be turned into legislation more quickly, as they would not have to pass through the Commission. On the whole, one may say that SEPA came strengthened out of the restructuring process and received a large increase in its responsibilities.[455] As seen in Fig. 6.1, the Environmental Protection Bureaux (EPBs) are important players in the provinces, municipalities, districts and counties. Their main task is to implement environmental regulations. EPBs are subordinate both to SEPA and the provincial, municipal and district governments, and conflicts of interest may appear due to the dual relationship. The environmental policy instructions come from SEPA, while their main source of funding comes from polluting discharge fees as well as from the governments at their respective levels (provincial, municipal, district and county). For instance, EPBs are supposed to approve EIAs for any construction project before the project is constructed. Nevertheless, this is often disregarded by the local governments, and as the EPB is subordinate to and dependent upon local governments for funding, the EPB most often will have to let it pass in order not to jeopardise its own position.[456] The mayor may often be instrumental in selecting the EPB director, and also influences decisions regarding the size of the staff.[457] Their capacity to influence the implementation of environmental policy is therefore great. County-level EPBs found their status downgraded in 1993-1994, when a round of administrative reforms took place in order to reduce the bureaucracy. As a result, the county-level EPBs were reduced in status and staffing. At the Fourth National Environmental Conference in 1996, one of the many matters raised during the conference concerned the staffing of the EPBs, and many of the provincial representatives called for an increase of staff at the county level.[458]

455 SEPA took over responsibilities from the abolished Ministry of Forestry (which became the Forestry Administration) for biodiversity, nature reserve management, wetland conservation; received increased right to control marine pollution two miles from shore from the State Oceanographic Administration; as well as for effects from mining, and assumed a more important role in influencing nuclear energy developments. Jahiel (1998), p. 774.

456 See for instance the example from Chongqing in 1992, when the mayor instructed the EPB to give approval of the EIA without even having carried out one, in Ma and Ortolano (2000), p. 82.

457 Ibid., pp. 62-3.

458 Vermeer (1998), p. 962.

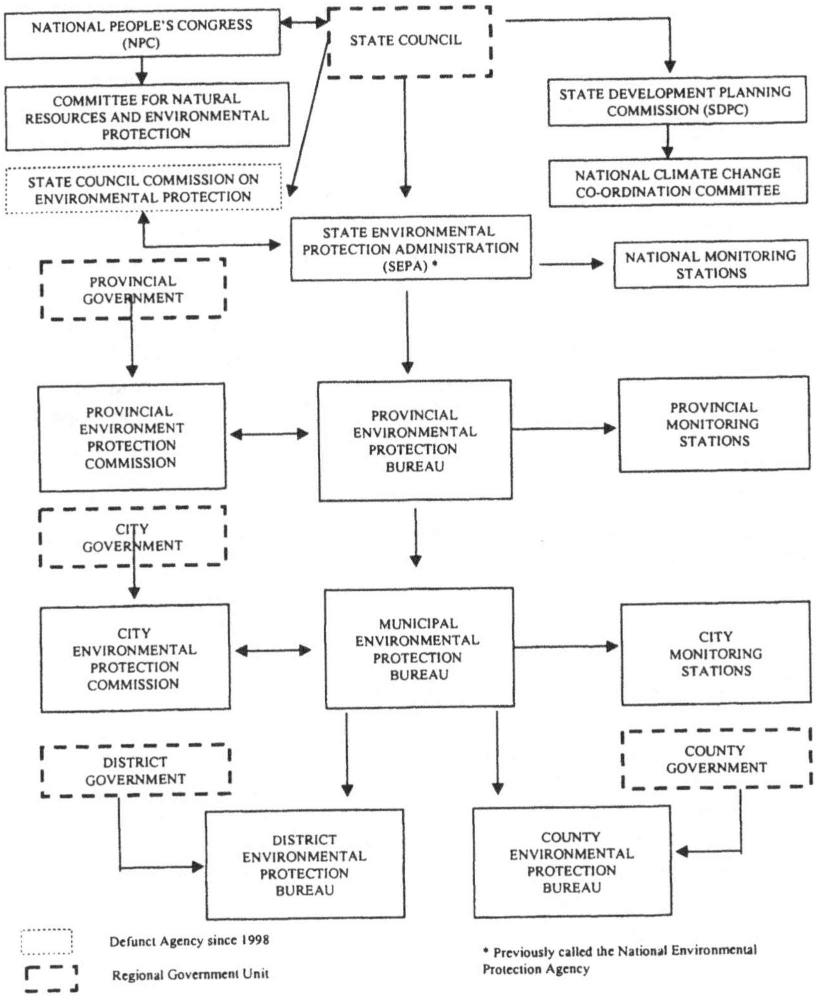

Figure 6.1 The Chinese Environmental Protection Apparatus[459]

In addition, an important restructuring took place in 1998, which placed the responsibility for co-ordinating the climate change work in China with the SDPC. The co-ordinating responsibility for this work in China formerly belonged to the

459 Adapted from Ma and Ortolano (2000), p. 56; Jahiel (1998), p. 760; and Tangen, Heggelund and Hu (2000), p. 6.

China Bureau of Meteorology (CBM).[460] The change in responsibility is a sign of increased importance put on the climate change issue in China by the leadership, as the SDPC is a powerful commission.[461] Furthermore, the change of responsibility is a sign that economic and energy issues have become increasingly important on the domestic agenda in China. SDPC oversees economic development issues in general, which is the reason why SDPC has been given the co-ordination responsibility for the climate change work in China.[462]

With regard to future developments for the environmental administration in China, there are at least two options that are being discussed. One recent suggestion from the State Council which is under discussion is to merge the Forestry Administration, Ministries of Water Resources, Land and Natural Resources and SEPA into one ministry.[463] The reasoning behind this would be to avoid overlapping among the various ministries, which would make policymaking more efficient, reduce the conflicts that need to be resolved by the State Council, and reduce competition for funding. The ministries may not be happy about such a development, as it would mean a reduction of staff. A second option being discussed is elevating the administration to become the Ministry of Environment. This is apparently the development hoped for by SEPA, and according to sources in the administration, rumours exist concerning the possibility of SEPA being elevated to a ministry in the next round of government restructuring.[464]

Environmental Laws, Regulations and Programmes

Laws and regulations are important measures in the environmental protection work in China. The environmental laws and administrative regulations that are being passed also reflect the consciousness on the part of Chinese authorities with regard to the environmental situation in the country. There are basically four important actors in the law-making arena in China: the SEPA, the National People's Congress (NPC), the State Council (SC) and the Chinese Communist Party (CCP). Their roles are as follows:[465]

460 For the institutional changes see Tangen, Heggelund and Hu (2000).

461 SDPC is a latecomer in the climate change policymaking process, however, has assumed an increasingly important role as the economic and energy issues have climbed on the domestic agenda.

462 Interview with SDPC official in 1999.

463 Interview no. 33, August 2000.

464 See also Ross (1999), p. 298. The restructuring during the NPC in March 2003 concerned the economy sector, such as the merger of the SETC and Ministry of Foreign Trade and Economic Co-operation (MOFTEC) into a new Ministry of Commerce. The restructuring of the environmental and natural resources sectors will most likely take place in the next NPC in 2004. Author's communication with SEPA, March 2003.

465 The following draws heavily on Ma and Ortolano (2000), pp. 13-18.

- SEPA: SEPA suggests environmental policies that are generally legitimised by being included into statutes or legally binding administrative regulations.
- NPC: The NPC, also called China's parliament, has approximately 3000 representatives from the entire country. It meets once a year, and in the meantime the NPC Standing Committee, consisting of more than 130 members, looks after the work. The NPC is responsible for approving projects and passing laws, One important decision made by the 14th National People's Congress in 1993, was to approve the establishment of the Committee for Natural Resources and Environmental Protection of the National People's Congress.[466] Qu Geping, who has been an important person in the development of China's environmental policymaking, has headed this committee since its establishment. The committee has become a visible player in Chinese environmental policymaking, and is responsible for revising and drafting new series of laws, as well as monitoring the implementation of environmental regulations.
- SC: The SC consists of representatives from all the ministries, commissions and administrative agencies. The SC's principal role is to implement the laws and policies of the NPC, however, the SC also drafts law proposals and passes them on to the NPC for approval. After the NPC Standing Committee passes a law, the State Council issues administrative edicts that provide specific details necessary for implementation.[467]
- CCP: The Party and the State in China are closely interlinked, and the CCP influences law-making by controlling the appointments to key legislative and administrative posts. The central committee of CCP also may issue resolutions that relate to environmental issues.[468]

After this brief introduction to the lawmaking institutions in China, let us look at some of the environmental laws that have been established. In the late 1970s, the attitude towards environmental protection began to change. The lack of attention to these issues and the related lack of institutional authority had become more obvious, and several national conferences in 1978 stressed the importance of environmental protection. One aspiration of China's leaders was to implement

---

466 It was originlly named the Environment Protection Committee, but the name was changed in 1994.

467 In addition, national departments can issue 'measures' and 'notifications' in order to clarify a law. Examples used to clarify environmental laws include: regulations (*fagui*), decrees (*guizhang*), orders (*mingling*), provisional measures (*zhanxing banfa*), measures (*banfa*), decisions (*jueding*), resolutions (*jueyi*), directives (*zhishi*), notifications (*tongzhi*), and administrative circulars (*tongbao*). Ma and Ortolano (2000), footnote 6, p. 31.

468 One example is the *Zhonggong zhongyang guanyu nongye he nongcun gongzuo ruogan zhongda wenti de jueding* (CCP Central Committee Resolution regarding some major problems in agriculture and rural work, article five), Issued 14 October, 1998. Zhongguo Huanjing nianjian bianji weiyuanhui (1999), pp. 11-12.

environmental policy through the use of law, which is illustrated by the number of laws and regulations that were initiated in the late 1970s and the following decades. In December 1978 the State Council issued Document # 79, also called the State Council Environmental Leadership Small Group's Ten-Year Plan.[469] Document # 79 acknowledged that environmental pollution had become a problem for citizens in certain areas in China. In the 1978 constitution, environmental protection and pollution control appeared for the first time as a key state responsibility. Article 11 of that constitution stipulated that 'the state protects the environment and natural resources and prevents and eliminates pollution and other hazards to the public'. In December 1978 important recognition of the environmental problems in China was given by the Communist Party, when the first directives came from the Chinese Communist Party Central Committee in relation to document # 79. The CCP stated that China should not take the road of first polluting, then cleaning up. Coming from the Party, this was significant as it gave signals to pay attention to the environment on all party levels, and the work to issue an environmental statute began. The environmental bureaux of seven northern and eastern provinces participated in drafting a statute with the help of experts and the SSTC, with references to experiences in foreign countries. Nearly one year later, in September 1979, the first law on environmental protection, Environmental Protection Law of the PRC (*shixing*, for trial implementation) was enacted.[470] The statute was approved in draft by the Standing Committee of the NPC in September 1979, and the law was a trial-use format. The purpose of the trial law was to be able to get laws into place quickly and to be able to eventually make changes as needed.[471] The 1979 passing of the environmental law in China is also regarded as a turning point in the country's environmental history. Before the law, it was necessary to rely on administrative methods to deal with environmental pollution problems, but now it was possible to depend on the law itself. On the whole, it signified an increase in the authority of the environmental protection apparatus. The trial implementation period ended in 1989, when the NPC Standing Committee passed the PRC Environmental Protection Law. The Environmental Protection Law is the basic law for environmental protection in China, provides basic guiding principles and a regulatory system for environmental protection. It was important, as it established the environmental administration system in China. According to Ma and Ortolano,[472] the Environmental Protection Law was important, as it established specific responsibilities for governments at the national, provincial, and county levels, and thus emphasises that it is not only the responsibility of the Environmental Protection Bureau. In addition to the Environmental Protection Law, a number of specialised environmental laws have been passed, such as:[473]

469 Ross (1988), p. 139; and Jahiel (1994), p. 80.

470 Zhongguo Huanjing nianjian bianji weiyuanhui (1990), p. 6.

471 Ross (1988), pp. 139-40.

472 Ma and Ortolano (2000), p. 16.

473 Wang, Xi (2000), pp. 9-32.

Pollution Control Laws

- Marine Environmental Protection Law (1982)
- Law on Prevention and Control of Water Pollution (1984, revised in 1996)
- Law on the Prevention and Control of Air Pollution (1987, revised in 1995 and 2000)[474]
- Law on Prevention and Control of Pollution Caused by Solid Wastes (1995)
- Law and Prevention and Control of Noise Pollution (1996)

Natural Resources Laws

- Forest law (1984, 1998)
- Grassland Law (1985)
- Fishery Law (1986)
- Mineral Resources Law (1986, 1996)
- Law on Coal Mining (1996)
- Land Management Law (1986, 1988, and 1998)
- Water Resources Law (1988)
- Water and Soil Conservation Law (1991)
- Wild Animal Protection Law (1988)
- Cultural Relics Protection Law (1982, 1991)[475]

In addition, China has issued administrative regulations to implement the laws. An example regarding environmental pollution is the The Provisional Regulations for Water Pollution Prevention and Control in the Huai River Basin 1995, and regarding natural resources, the Water and Soil Conservation Law Implementation Regulations 1993. There are also national environmental programmes to assist in attaining standards, for instance the environmental impact assessment, three synchronisations, and the pollution discharge fee system.[476] Finally, China has in addition to the national laws also signed a number of international treaties on the environment.

Let us now take a look at some areas of relevance for the Three Gorges project.

---

474 The Law on the prevention and Control of Air Pollution was amended in 2000, when the law was basically rewritten with concrete goals for reduction of air pollution. See U.S. Embassy (2000a).

475 In addition to these laws, China has a number of laws and regulations that are related to environmental protection such as the General Principles of Civil Law (1986), and Criminal Law (1979, 1997). For a comprehensive list of the laws, please refer to Wang, Xi (2000), pp. 13-17.

476 Ma and Ortolano, (2000), p. 20.

Water Pollution

The Law on Prevention and Control of Water Pollution that was revised in 1996 focuses on inland waters such as the rivers, lakes, reservoirs, canals.[477] This section focuses on rivers, as this is related to the Three Gorges dam and the Yangtze River. There are seven river basin commissions in China. One of them is the Yangtze River Water Resources Commission (YRWRC, previously called the YVPO as described in the history chapter).[478] These commissions, subordinate to the Ministry of Water Resources, are the main administrative body within a river basin. Under the Commissions are River Valley Water Resources Protection Bureaux that engage in monitoring and research. The river commissions focus on control of floods, sediment and drought, as well as monitoring the water quality in the river basins in co-operation with ministries such as SEPA. As has been described in the history chapter of this book, the Yangtze River Water Resources Commission has been an important actor in the long history of the Three Gorges project; it was for instance involved in project design.

The water quality in China's rivers varies. The Yangtze River Valley is considered among the less polluted rivers in China, with the majority of the river being grade 2 (30 percent), grade 3 (45 percent), grade 4 (20 percent) and grade 5 (5 percent).[479] One of the worst polluted rivers in China is the Huai River. The source of pollution is paper mills, dyeing factories, chemical dyeing, leather and liquor factories.[480] In 1994, the Huai River incident was an example of failed water protection, and an eye opener for the Chinese authorities with regard to the status of the water quality of China's rivers. In July 1994, 200 million m^3 of polluted water covering 70 kilometres[481] was discharged into the river, creating a pollution crisis in Anhui and Jiangsu provinces where for instance the drinking water was contaminated. The State Council and the State Council Environmental Protection Committee held a meeting in December that year to discuss the situation in the Huai river. The meeting came up with the draft *Huaihe liuyu shuiwuran fangzhi tiaoli* (The Regulations for water pollution prevention and control in the Huai River Basin) for comments, which was eventually issued in August 1995, *Huaihe liuyu shuiwuran fangzhi zanxing tiaoli* (The Provisional Regulations for water pollution prevention and control in the Huai River Basin).[482] These were the first

---

477 Zhongguo Huanjing nianjian bianji weiyuanhui (1996), pp. 45-9. It does not concern the oceans, article 2.

478 The other six are the Huai River Basin, Yellow River Basin, Hai River Basin, Pearl River Basin, Tai Lake basin, Song Hua and Liao rivers River Basin.

479 These are approximate percentages. See Chapter 5 regarding the categories for water quality standards in China.

480 Vermeer (1998), p. 971.

481 Zhongguo Huanjing nianjian bianji weiyuanhui (1995), p. 230.

482 Zhonghua renmin gongheguo guowuyuan ling, di 183 hao, 'Huaihe liuyu shuiwuran fangzhi zanxing tiaoli', Zhongguo Huanjing nianjian bianji weiyuanhui (1996), pp. 49-51, and p. 266.

regulations for water protection of the river basins, in addition to the introduction of plans to restore the river to normal. Consequently, a new model was set up to solve the pollution problem in the Huai River, which was mainly based upon the regulations above. The regulations set concrete goals such as timing for the clean-up of the Huai river basin, and enterprises were not to exceed the set discharge quotas after 1 January 1998.[483] Furthermore, they assigned tasks to for instance the Huai River Valley Water Resources Protection Bureau. The regulations were important for the protection of the other six river basins in China, as they served as an example of increased integration between different agencies in protection work. In addition, a leadership group was established headed by the administrator of NEPA (at the time) and the MWR vice-minister, consisting of representatives of the four basin provinces[484] as well as others. The situation in the Huai River is still serious after several years. The River is one of three rivers (in addition to Hai and Liao) that are focal areas for reduction of pollution, which was stressed at the Fourth National Protection Conference in 1996 by Li Peng.[485] The goals set for the Huai River following the pollution crisis have not been met. According to the *China Environmental Yearbook 1999*, 48 percent of the river section that was monitored had water quality below grade 5.[486] The World Bank also stated that the Huai River improved after 1996, but noted signs of deterioration of water quality again in 1998 and 1999.[487]

The Township and Village Enterprises (TVEs) [488] are among the main water polluting industries in China, and their environmental pollution has also been more difficult to control than the state-owned enterprises (SOEs). In order to control the TVE environmental pollution, NEPA and the Agricultural Bank of China issued in 1997 the *Jiaqiang xiangzhen qiye wuran fangzhi he baozheng daikuan anquan de tongzhi* (The Notice regarding strengthening of TVE's pollution prevention and bank guaranty security) in an attempt to strengthen enforcement of the regulations.[489] The TVEs in China have been important for the economic growth of

483 Article 18 of the *Huaihe liuyu shuiwuran fangzhi zanxing tiaoli,* (The Provisional Regulations for water pollution prevention and control in the Huai River Basin). Ibid., p. 50.

484 Jiangsu, Shandong, Anhui an Henan, See Ibid., p. 146 for work plans.

485 Ibid., p. 7.

486 The remaining were grade 2: 11 percent, grade 3: 17 percent, grade 4: 18 percent, grade 5: 6 percent, see '1998 nian Zhongguo huanjing zhuangkuang gongbao' (Report on China's State of the Environment 1998) Zhongguo Huanjing nianjian bianji weiyuanhui (1999), p. 118.

487 World Bank (2001a), p. 49, footnote 3.

488 Also called TVIEs Township- and Village Industrial Enterprises.

489 See Zhongguo Huanjing nianjian bianji weiyuanhui (1997), pp. 27-8. Another example of NEPA collaborating with other agencies and ministries is in prohibiting the import of hazardous wastes to China, where NEPA, MOFERT (Ministry of Foreign Economic Relations and Trade, now Ministry of Foreign Trade and Economic Co-operation, MOFTEC), the Customs Administration, the Industrial Commerce Administration, the Commodity Inspection Bureau, which jointly issued

the country, but they have also been a major source of pollution. The Notice states that no loan/credit should be given to enterprises that should be closed down according to the 'Resolution Regarding Some Environmental Protection Problems' by the State Council in 1996 (See section Increased Environmental Awareness in China for details about this resolution). Also that one should make sure before making payments to enterprises that the *san tongshi* ('Three Simultaneouses') policy is adhered to, which states that environmental aspects should be included at the design stage, construction stage and operation stage.

Project Construction

The Environmental Protection Law lays the basis for Environmental Impact Assessments (EIA) in China, while the *Management Guidelines on Environmental Protection of Construction Projects* issued in 1986 gave specific instructions regarding projects.[490] In November 1994, the MWR, SPC and NEPA jointly issued a circular, the *Kaifa jianshe xiangmu shuitu liushi baochi fang'an guanli banfa* (Water and Soil Conservation Management Measures for Development and Construction projects),[491] measures intended to clarify the *Zhongghua renmin gongheguo shuitu baochi fa* (the PRC Water and Soil Conservation Law) that was passed in 1991. The management measures stress the need for incorporation of water and soil conservation into the EIAs of projects such as water conservation projects (and railroad, public roads, mines, power enterprises, etc). In December that same year, NEPA issued regulations for environmental protection installation (facilities) in construction projects, *Jianshe xiangmu huanjing baohu sheshi jungong yanshou guanli guiding* (the management regulations for checking environmental protection facilities on project completion). These regulations specify that each project on completion will be checked to see whether the environmental protection installations are in accordance with the requirements of the *san tongshi* ('Three Simultaneouses' policy) before being accepted.[492] When the local or municipal environmental administrations accept, the constructed project can begin operation. Should it be detected during trial operation (usually no longer than 1 year) that the 'Three Simultaneouses' requirements are not fulfilled, the environmental administrative organs under the State Council can halt operation.

---

the *Feiwu jinkou huanjing baohu guanli zanxing guiding* ('Regulations for the environmental management of imported wastes'); (Guojia huanjing baohuju, Duiwai maoyi jingji hezuo bu, Haiguan zongshu, Guojia gongshangju, Guojia shangjianju guanyu lingbu feiwu jinkou huanjing baohu guanli zanxing guiding) Zhongguo Huanjing nianjian bianji weiyuanhui (1996), pp. 62-4. This was an important alliance; in 1996 alone, 200 European and American ships carrying solid waste were turned back from China. Jahiel (1998), p. 778.

490 Sinkule (1993), pp. 50-51.

491 Zhongguo Huanjing nianjian bianji weiyuanhui (1995), p. 30.

492 *Jianshe xiangmu huanjing baohu sheshi jungong yanshou guanli guiding* (Management regulation for checking and accepting environmental protection facilities of completed construction projects), Ibid., pp. 30-32.

Despite these regulations, many projects are accepted even if they do not operate according to the 'Three Simultaneouses' policy. This policy sets requirements that design, construction and operation of a new factory (or an existing factory that is expanding or changing the production processes) be synchronised with the design, construction and operation of appropriate waste treatment facilities.[493] The factory must address environmental concerns from design, through construction and into operation. The local EPB must approve the design for the environmental facility planned. If the requirements are not met, EPB theoretically has the power to hinder projects. This rarely happens due to the economic relationship it has with the local governments, which makes it difficult to oppose the government's wishes.[494] During the construction phase, the EPB can also inspect the facilities, and before operation finds place, the factory and the environmental facility must be inspected and approved. Since 1983, when the State Council issued a resolution, the 'Three Simultaneouses' concept has been applied throughout China. It is also important in the Three Gorges reservoir area for the construction of new enterprises and relocation of the old ones, as the water pollution problems may increase when the Yangtze river flow slows down and becomes a lake.

Criminal Law

The revised Criminal Law (*Zhonghua renmin gongheguo xingfa*) of March 1997,[495] *Pohuai huanjing ziyuan baohuzui* (The crime of destroying the environment and natural resources) specifies environmental crime as a new category of crime. The law includes areas such as illegal tree felling, protected animals, minerals, emissions, environmental accidents, etc. This was an important step for environmental protection and opens up for long prison sentences (above ten years for certain serious crimes) as well as fines, depending upon the crime. The first environmental crime court case was tried on September 12-17, 1998 in Shanxi province.[496] The case concerned a paper factory that according to the State Council *Resolution Regarding Some Environmental Protection Problems* that was issued in 1996, should be closed (as its production level was under the 5000 tonnes limit set by the Resolution). The factory had not been closed according to the instructions, and in October 1997 waste water from the factory flowed into a reservoir and drinking water system, and the water supply company had to cut off the water supply for 3 days. The factory manager was arrested on pollution charges, and was sentenced to 2 years in jail. Furthermore, the factory had to compensate the water supply company with RMB 358,815. The first environmental pollution crime case

493 Sinkule (1993), p. 122.

494 Ibid., p. 125.

495 Zhongguo Huanjing nianjian bianji weiyuanhui (1997), 7-8 *zhailu* (extracts), p. 7. Chapter 6, *Fanghai shehui guanli chixu zui* (The crime of jeopardising public management order).

496 Zhongguo Huanjing nianjian bianji weiyuanhui (1999), p. 159.

was followed closely by the media, and CCTV (China Central TV, national TV) had a special programme about the case. The use of the criminal code for environmental crimes may also be important in relation to the environmental crimes in the Three Gorges area, as a measure to control tree felling, erosion problems and water pollution problems.

Increased Environmental Awareness in China

In 1996, the Fourth National Environmental Protection Conference took place. For the first time, both the Secretary General of the CCP, Jiang Zemin, and then Premier Li Peng participated in the meeting, which signified a new level of attention paid to environmental protection by China's leaders. Their speeches reflected the need to further incorporate environmental protection into economic development.[497] One of the important goals for the Ninth Five-year plan (1996-2000) was to achieve control by the year 2000 of the environmental situation so that the environment would not further deteriorate. Particular challenges in this process that were mentioned by the leaders at the meeting were the specific pollution problems in the cities and the countryside, and the health problems that follow suit. Raising environmental awareness in particular among leaders and cadres was raised as an important issue, as well as their capability for environmental management. During a process of economic development, the number of projects carried out are enormous, and environmental impact assessments were stressed as crucial for all new projects, and for existing reconstruction projects.

Following this conference, one important resolution was issued by the State Council in August 1996 (as mentioned briefly on the previous pages): *Guowuyuan guanyu huanjing baohu ruoganwenti de jueding* (*Resolution Regarding Some Environmental Protection Problems*). The objective of the *Resolution* was to enable the realisation of the environmental goals of the Ninth Five-year plan, which had been approved earlier that year (in March) by the NPC (together with the Outline of Long-term Social and Economic Development Objectives for 2010). Sustainable development was an important strategy under the Five-year plan, and the *Resolution* gave instructions on how to reach these goals, such as requiring industrial pollution to reach national or local standards by 2000, and that EIAs must be carried out according to the environmental capacity when constructing projects.

---

497 Zhongguo Huanjing nianjian bianji weiyuanhui (1996), pp. 1-7. Other speeches where the environmental issue was stressed were when Jiang Zemin gave the speech, *Jingji jianshe he renkou, ziyuan, huanjing de guanxi* (The relationship between economic construction and population, natural resources and the environment) at a Central committee meeting held in 1996, where the focus was on sustainable development; and when Li Peng in his annual report to the NPC also emphasised the need to strengthen environmental protection when developing the economy.

One important requirement in the *Resolution* relates to small polluting enterprises, mainly TVEs, called the *15 xiao* (15 small). The *Resolution* points to 15 types of polluting enterprises that will be shut down if the enterprises annually produce less than a certain amount set by the authorities. Examples of such enterprises are paper mills producing less than 5000 tonnes per year, and factories with leather production of less than 30.000 sheets annually. These would have to cease production and be closed down.[498] Following this decision, which is related to the Huai river pollution crisis in 1995, NEPA (SEPA from 98) has closed down 66,000 small, polluting enterprises.[499]

NEPA introduced in 1996 the *Kuashiji lüse gongcheng jihua* (the Trans-century green engineering plan, hereafter the Green Engineering Plan), with a first phase covering the period of 1996-2000 (altogether three phases). The Green Engineering Plan is a part of the Ninth Five-year plan and the 2010 long-term objectives. Its goal is to induce the various ministries and local governments as well as enterprises to collaborate in focussing on some of the most serious environmental problems of the country. The Green Engineering Plan consists of concrete projects (altogether 3000 projects in the first phase), which focus on priority areas in China for environmental pollution clean up. One priority area, water pollution, has been given special attention due to the Huai River crisis, so the focus is on the *san he* (3 rivers, the Huai, Hai and Liao), *san hu* (3 lakes, Tai in Jiangsu and Zhejiang provinces, Chao in Anhui province, and Dianchi in Yunnan province). Another priority area is atmospheric pollution, acid rain and SO_2 problems in two regions (*suanyu kongzhiqu he eryang hualiu wuran kongzhiqu huafen fangan*, shortened to *liang kong qu*).[500]

1998 was an important year; one may call it a turning point in many ways for environmental policymaking in China. In addition to the above policy measures, NEPA's elevation to ministerial level and renaming as SEPA, was symbolically very important. Let us take a look below at some of the events that make this year stand out, as well as some future trends for environmental developments in China with regard to public participation:

The Floods in 1998

One may say that Chinese environmental policy making has in many cases developed in response to an environmental crisis, such as the Huai River pollution crisis mentioned above. In 1998 another crisis took place that made Chinese authorities realise the seriousness of the environmental situation along the Yangtze River. Large floods occurred in the summer months from June through August in

498 Ibid., p. 9.

499 Interview no. 33, August 2000.

500 Zhongguo Huanjing nianjian bianji weiyuanhui (1996), p. 18. Zhongguo Huanjing nianjian bianji weiyuanhui (1997), pp. 16-17. See Zhongguo Huanjing nianjian bianji weiyuanhui (1998) for an overview of the cities and provinces that are involved in the *liang qu* policy, pp. 34-5.

many of the largest rivers in China: the Yangtze river, the Songhua River, the Zhu Jiang (Pearl River), and Min Jiang (the Min River). As they were the largest floods for many years, they made Chinese authorities realise the seriousness of the tree-felling along the Yangtze River. As a consequence, Chinese authorities decided that from 1 September 1998 it was forbidden to cut natural forest along the middle and upper reaches of the Yangtze river and Yellow River, as well as in the Northeast.[501] The state decided to provide funding for afforestation and Sichuan, for instance, will receive RMB 480 million per year.[502] China's forests have been depleted since the 1950s, and tree felling is still a problem, although there are indications that the trend of steady deforestation is turning.[503] According to Smil and Mao, every year more than 5 billion tonnes of soil are eroded in China, and excessive deforestation may be the cause of half of the annual soil erosion. Deforestation reduces the water retention capacity, and increases the chances for floods. The Forest Law (1984, 1998) and the Water and Soil Conservation Law (1991) are intended to be important measures in the struggle against deforestation. Nevertheless, the floods in China in 1998 illustrate that this work has not been sufficient in the struggle to conserve China's forests. The Forest law was revised in 1998, only a few months before the floods.[504]

Accordingly, water was the theme for many meetings that year where the reasons for the serious floods were discussed, such as climate change, siltation, erosion, and tree felling.[505] In 1998, several decrees were issued by the Chinese authorities that may be seen in connection with the severe floods in several parts of China that year, in particular the Yangtze River floods. Criticism had been directed against the water conservancy administrations, in particular the Ministry of Water Resources, due to the lack of attention paid to the strengthening of dikes and other measures. In October that year, *Zhonggong zhongyang guanyu nongye he nongcun gongzuo ruogan zhongda wenti de jueding* (CCP Central committee Resolution regarding some major problems in agriculture and rural work),[506] article five, stresses the urgent need to strengthen the management of the Yangtze River and increase the flood retention capacity. It also emphasises the need to speed up the construction of the Three Gorges project in particular (in addition to the

501 Zhongguo Huanjing nianjian bianji weiyuanhui (1999), p. 6.

502 Chen (1998), p. 5.

503 Smil and Mao (1998), p. 37.

504 Zhongguo Huanjing nianjian bianji weiyuanhui (1998), pp. 16-22.

505 Water resources, or the lack of water resources was also a topic, as disparities in water resources are large: more than 82 percent of the water resources are situated in the Yangzte river basin and below, and only 36 percent of the agricultural activity takes place there, while the areas to the north of the Yangtze River have only 18 percent of the total water resources in China and cultivated land constitutes 64 percent of the total. Zhongguo Huanjing nianjian bianji weiyuanhui (1999), p. 8. The speech of Li Ruihuan *Guanyu woguo shui de wenti* (Regarding some water problems in our country), to the CPPCC on 23 october 1998.

506 Issued 14 October, 1998. Zhongguo Huanjing nianjian bianji weiyuanhui (1999) pp. 11-12, *zhailu* (excerpts).

Xiaoliangdi dam that has recently been completed on Yellow River with funding from the World Bank). The following month, the State Council issued *Jianshe xiangmu huanjing baohu guanli tiaoli* (Regulations on environmental protection management in construction projects)[507] for the purpose of preventing construction projects from increasing pollution or creating ecological and environmental problems. The Regulations include most construction projects that eventually pollute, or that may have some impact on the environment, such as water conservancy projects. These regulations would be important for the Three Gorges area as numerous new projects will be under construction. The Regulations give instructions on EIAs as well as the legal responsibility.

In November 1998 the State Council issued the notification *Quan guo shengtai huanjing jianshe guihua* (The programme for national ecological and environmental improvement),[508] which is the new catch phrase in environmental policy. The purpose of this programme is to, during the next 50 years, reverse the negative process that has been going on in many areas of China. The Programme sets concrete short-term, mid-term and long-term goals where the reduction of the erosion and increase of forest coverage are important goals. The short-term (1998-2010) goal is to control man-made erosion within 12 years, and before 2003 is the key period. During the mid-term period (2011-2030), emphasis will be put on continued water and soil erosion work, as well as reforestation with a goal of reaching 24 percent forest coverage on a national basis. During the long-term period (2031-2050), the work described above will continue and as a result reach a state of improved environment, with 26 percent forest coverage. Strengthening of legislation is also one of the goals of the programme. The upper and middle reaches of the Yangtze River is one of the priority areas due to the erosion problems,[509] and the goals are to curb deforestation and protect the forests as well as stop farming on land that is steeper than 25 degrees.

The floods on the Yangtze River that year were important for the Three Gorges project resettlement policy change, as they made the authorities aware of the fragile eco-system, leading them to decide to move parts of the rural population out of the reservoir area. Furthermore, the flood control capacity was questioned again, as some academics expressed doubts about the Three Gorges dam's ability to retain a flood of this type.[510] Furthermore, the poverty issue was also pointed out as being closely linked to the floods, since people reclaim land for agriculture and cut trees to make a living. Articles about the floods stated that the floods were man-made,

507 *Zhonghu renmin gongheguo guowuyuan ling, Di 253 hao* (Decree of the The PRC State Council no 253), issued on 29 November 1998, Ibid., pp. 12-15.

508 There are different ways to translate *shengtai jianshe*; some translate it as eco-construction, while others translate it as ecological improvement. Communication with CAS scholar.

509 Zhongguo Huanjing nianjian bianji weiyuanhui (1999), pp. 15-20.

510 Chen (1998), p. 4, states for instance that the dam can only have partial effect, only in the upper reaches, and will not be effective in the middle reaches from Yichang and down.

and that tree felling was the main cause.[511] Since the 1950s and the campaign to smelt iron and steel (*da lian gangtie*) during the Great Leap, tree felling and erosion problems have been serious along the Yangtze River.[512] In Sichuan for instance, erosion now covers more than 40 percent of the province.[513] The sediment discharge has increased from 520 million tonnes in the 1950s to 634 million tonnes in the 1980s.[514] The riverbanks of the Jingjiang dyke were 8 metres above the ground (during the floods). A second reason for the floods was the lack of ability to retain the water, as well as too many '*bingku*', i.e. reservoirs that do not function according to standard. The floods in 1998 were an environmental lesson for Chinese authorities, who introduced the policy of *tuigeng huanlin* (return farmed land to forestry), and *tuitian huanhu* (return the land to the lakes). As mentioned earlier in this section, the CCP Central Committee later that year passed (on 14 October 1998)[515] a Resolution regarding important issues for agricultural work in the countryside, where one part was directed specifically towards the strengthening of flood control and water resources, which must be regarded as a direct result from the floods that year.

The Media, Public Participation and NGO Development

The environmental developments in China have been gradual, and have often sprung out from crises that have emerged, as discussed in the previous two sections. The Tenth Five Year Plan (2001-2005) for economic and social development also reflects the increased importance placed on environmental protection. Seven hundred billion yuan (US$85 billion) will be invested in environmental protection during this period, which will account for about 1.3 per cent of the gross domestic product (GDP).[516] During the previous Five-Year Plan

---

511 Changjiang hongzai de ren wei yuanyin (The cause for the Yangzte floods is man) published in the 1998 May and August editions of the *Zhongguo lüse shibao* (The China Green Times) in Zheng and Qian (1998), p. 144.

512 In particular in the areas of the upper reaches of the Jinsha River, which is the name of the upper half of the Yangtze River on its course through Tibet, Yunnan and a small part of Sichuan. After Yibin (in Sichuan) the river becomes the Yangtze River. See Van Slyke (1988), pp. 15-20; and *Zhonghua renmin gongheguo fensheng dituji* (1990).

513 Chen (1998). The eroded area amounts to 199,800 square kilometres.

514 Published in the 1998 May and August editions of the *Zhongguo lüse shibao* (The China Green Times) in Zheng and Qian (1998), p. 144.

515 *Zhonggong zhongyang guanyu nongye he nongcun gongzuo ruogan zhongda wenti de jueding*, Zhongguo Huanjing nianjian bianji weiyuanhui (1999), pp. 13-14.

516 *China Daily*, (29 May, 2001). *China Daily Hong Kong Edition*, (13 November, 2001). This is below the stated wish of Minister Xie Zhenhua, who believes 1.4 percent of GDP would enable China to reach the environmental targets set in the Tenth Five Year Plan. See U.S. Embassy (2001b). This view was accentuated by SEPA Vice-minister Wang Yuqing who stated that it is still impossible to satisfy the demand for environmental protection. Interview, 23 April, 2002.

(1996-2000), the investment in environmental protection reached 360 billion yuan (US$43.5 billion), or about 0.9 per cent of GDP. One important factor determining whether China will be able to achieve its stated environmental goals will be the increased environmental awareness among leaders, officials and the Chinese public. Public participation in environmental protection work is still low in China. The main priority for Chinese authorities in the past few decades has been to alleviate poverty, develop the economy and raise living standards. With the environmental degradation in the country, China's leaders have become aware of the need to educate the public and raise environmental awareness. This was stressed by the minister of SEPA, Xie Zhenhua in the report to the Fourth National Environmental Conference in 1996.[517] Xie noted several ways to do this, such as using the television, radio, newspapers and periodicals; popularising knowledge about environmental science; educating school children and providing environmental training of officials on all levels. Moreover, Xie also emphasised that the role of the media as supervisor should be strengthened, so that lawbreakers and environmental accidents are reported. Due to the implementation of some of these activities as well as an environment that has continued to deteriorate, a growing awareness is emerging among the public, albeit only on a small scale. Parts of China's population are beginning to show concern for the environment, particularly in the urban areas, and wish to participate in environmental protection activities. This section intends to give a brief introduction to how the raising of public awareness is being carried out. In addition, a consequence of the increasing environmental consciousness is the emergence of social organisations[518] and non-governmental organisations (NGOs). The impact of these organisations on environmental policymaking is still unknown. However, as this is a growing trend in China, it may eventually have some impact on environmental policymaking in the country, and therefore deserves some attention.

Increasing environmental awareness The media is an important tool for the Chinese authorities through which information to the public about environmental protection is spread. Newspapers, television, radio, and periodicals are used in this effort. Newspapers such as the *People's Daily*, *Guangming Daily*, *Economic Daily*, etc, often publish articles about environmental issues in order to inform the public about certain campaigns or policies and to raise environmental awareness. Moreover, the *Zhongguo Huanjingbao* (China Environment News) is a newspaper

---

517 The title of Xie's report is: *Jianding xinxin, kaituo jinqu, wei shixian 'jiuwu' huanjing baohu mubiao er fendou* (strengthen confidence and continue to forge ahead in order to struggle to realise the environmental protection goals of the Ninth Five-Year-plan), Zhongguo Huanjing nianjian bianji weiyuanhui (1996), pp. 14-20.

518 The terms vary for these organisations: *shehui tuanti* (social organisations), Knup (1997), p. 9, *huanbao qunzhong tuanti* (environmental mass organisation), Zhongguo Huanjing nianjian bianji weiyuanhui (1996), p. 13, or *minjian zuzhi* (non-governemntal organisation or folk organisation). See for instance Zheng and Qian (1998), that uses *minjian zuzhi*, p. 318, *Fei zhengfu zuzhi* (non-governmental organisation) is also used.

entirely dedicated to environmental news published by SEPA. It was first published in 1984 and was originally published only once a week; the newspaper is now issued six days a week[519] and illustrates the increased focus on environmental issues. The above newspapers may reach only a limited number of people, as they may be either too political in their message or too specialised for the ordinary public. Furthermore, one may also find environmental articles in more popular newspapers such as the *Beijing Wanbao* (Beijing Evening News), the *Zhongguo qingnian bao* (China Youth Daily) and *Nanfang zhoumo* (Southern weekend). These may be less policy oriented and instead provide practical information on how people can make an effort in their own lives, such as how to recycle paper. Nevertheless, the most efficient way to reach the public is via television. Special environmental programmes are frequently shown on national TV.[520] One of them is 'Time for Environment' that is produced by an environmental NGO, Global Village of Beijing (GVB).[521]

In accordance with the Environmental Protection Law of China (1979, revised in 1989), Chinese authorities began publishing the *Zhongguo huanjing zhuangkuang gongbao* (China Environmental Situation Bulletin) in 1989. The following year came the first issue of the *Zhongguo Huanjing Nianjian* (China Environmental Yearbook), which carries an overview of all new laws and regulations, important speeches at meetings, the environmental work that has been carried out that year, environmental crisis, accidents, crimes, etc. The environmental yearbook is useful, as it gives some indication of the policy directions and of the environmental situation in China. However, it has been criticised for not realistically portraying the environmental situation in the country.[522] In 1997 the policy of environmental reporting to the public was implemented further, when 27 cities in China began publishing weekly air pollution reports. Xie Zhenhua stated

---

519 The author is grateful to Zhu Rongfa for the information about China Environment News. The *Zhongguo huanjing bao* (China Environment News) is published by *Zhongguo Huanjing baoshe* (the China Environmental News Press), which is an institution under SEPA.

520 There are several categories of environmental programmes in China: I) news related programmes: *Xinwen lianbo* (news), *Jiaodian Fangtan* (Focus interview), *Xinwen diaocha* (News Probe), *Dongfang Shikong* (Oriental Horizon); II) Knowledge based programmes: *Dongwu shijie* (Animal World), *Diqiu Gushi* (Earth Story), *Ren he ziran* (Man and Nature) and III) Special environmental programmes: *Zhonghua huanbao shijixing*. I am indebted to Zu Rongfa for this information about the different types of environmental programmes.

521 The full name in Chinese is *Beijing diqiucun huanjing wenhua zhongxin* (Global Village of Beijing environmental and cultural centre), and it is known as Global Village of Beijing. For more information on GVB activities see www.gvbchina.org/English/englishintro.html.

522 Scholars interviewed by the author view the situation as far worse than what has been reported in the Yearbook. Interview no. 32, Beijing, May 2000. A new yearbook has been introduced recently by the Chinese Academy of Social Sciences as a supplement to the SEPA yearbook: *Zhongguo huanjing yu fazhan pinglun* (China Environment and Development Review). See Zheng and Wang (eds. 2001).

at a press conference in June 2001 that an increase in public disclosure of environmental information would take place. Last year, China began broadcasting daily air-quality reports for 42 major cities, and information on 55 cities was available through local media. Xie also announced that air-quality forecasts would be provided for the same 42 cities. Moreover, he stated that a telephone hotline would be established for reporting environmental problems.[523]

In addition to the yearbook, critical reporting takes place by the Chinese media regarding environmental incidents around the country (such as the Huai River crisis in 1995). Such reporting is accepted and encouraged by the authorities. One such environmental publicity campaign was initiated in 1993, the *Zhonghua huanbao shijixing* (the China century information campaign on the environment)[524] by Qu Geping, the head of the NPC Environmental and Resources Protection Committee. A Resolution on this was issued in April 1994 by the NPC Environmental and Resources Protection Committee and NEPA as well as several ministries such as the Ministry of Broadcasting, Film and TV.[525] Each year a group of representatives from the relevant ministries, television and newspapers, travel around the country to inspect the environmental situation regarding a specific topic. Should the group discover that serious environmental pollution is allowed to take place (i.e. neglect by the local authorities) it will be reported in the national media. This campaign is carried out in order to force local authorities to adhere to the laws and regulations of the country, as well as to raise environmental awareness among the public.

The *Quan guo huanjing xuanchuan jiaoyu xingdong gangyao* (Outline for national environmental propaganda and education action 1996-2010) was issued in December 1996 by NEPA, the CCP Central Committee, Ministry of Propaganda, and the Education Commission. The purpose of the Outline was to activate the entire nation to protect the environment and natural resources in a period when economic development was taking place.[526] One part concerns environmental education that focuses on the training of for example teachers at all levels, as well as heads of EPAs (above county level, article 2.9.3). Another part concerns environmental propaganda, which aims at educating the public with regard to

---

523 U.S. Embassy (2001c).

524 The China century information campaign on the environment is based upon the *zhiliang wanlixing chanpin* which was a journey leaders undertook around China to check the quality of goods. Interview no. 33, August 2000.

525 The Resolution was issued by: *Quan guo renda huanjing yu ziyuan baohu weiyuanhui* (NPC Environment and Resources Committee), *Zhonggongzhongyang xuanchuanbu* (CCP Propaganda Department), *Guangbodianying dianshibu* (Ministry of Broadcasting, Film and TV), *Guojia huanjing baohuju* (NEPA), *Linyebu* (Ministry of Forestry), *Nongyebu*, (Ministry of Agriculture) *Shuilibu* (Ministry of Water Resources) *Gongqingtuan* (Communist Youth League), Zhongguo Huanjing nianjian bianji weiyuanhui (1995), pp. 21-3. The Ministry of Broadcasting, Film and TV is now defunct, and has become the State Administration of Radio, Film and Television.

526 Zhongguo Huanjing nianjian bianji weiyuanhui (1997), p. 20; the entire Outline goes from pp. 20-42.

environmental concern. Moreover, one part concerns establishing a framework for a national environmental education and information network, which would constitute a system whereby the public could participate in environmental protection as well as in environmental supervision. In sum, the Outline is a large-scale national plan to increase environmental awareness in China.[527] This work to increase environmental awareness may have had some effect, as we will see in the next section.

Public participation and social organisations Even though public participation in environmental protection work in China is limited and environmental organisations are still in their nascent stages, I believe that public participation is a growing trend. The combination of raised environmental awareness and raised living standards as well as increased environmental pollution, has prompted people to become more conscious of the environmental situation in the country. Moreover, the people who are eager to participate in the environmental movement genuinely care about the country's environmental problems and wish to participate on their own intitiative and not merely due to governmental environmental initiatives. Chinese leaders also emphasise the need to engage the public in environmental protection work and this was emphasised by Song Jian[528] in his speech at the Fourth National Environmental Protection Conference in 1996. Song also mentioned the NGO Friends of Nature (*Ziran zhi you*) as one organisation[529] that was carrying out good work, which signalled the importance of public participation in environmental protection and supervision in law enforcement.

The growing trend of environmental concern among Chinese citizens may be illustrated by several surveys during the past decade. One study, carried out from 1995 to 1996 by the People's University on this topic, concluded that people had superficial knowledge of environmental protection issues.[530] The majority of the respondents did not believe that there was any direct linkage between rapid economic growth and environmental deterioration. Nevertheless, the respondents stated water, air and solid waste pollution as the most urgent problems to be solved. Not surprisingly, respondents with higher education were more aware than respondents with lower education. A survey on environmental awareness was

527 Environmental education is ongoing in schools in China, and donors such as UNDP and WWF are co-operating with Chinese authorities in a comprehensive approach to teach the children about the environment (not only in the science class), in order to teach them that individual actions matter.

528 Speech at the Fourth National Environmental Protection Conference in 1996; Song Jian, was State Council Councillor and head of the now defunct Environmental Protection Commission at the time, and previously the president of the Academy of Engineering.

529 The term used by Song Jian was not NGO but *huanbao qunzhong quanti* (environmental mass organisation). Zhongguo Huanjing nianjian bianji weiyuanhui (1996), p. 13. For more information about Friends of Nature, see www.fon.org.cn/.

530 4,000 people were asked to participate, and 3,662 people responded to the study. Zhongguo Huanjing nianjian bianji weiyuanhui (1996), p. 248.

carried out by the State Forestry Administration (SFA) and the Horizon Market Research Group and released in March 1999.[531] One-third of the respondents chose clean air as the most indispensable thing for an ideal living environment. In 2000, a survey was made among 3000 urban citizens (Beijing, Shanghai and Guangzhou, altogether 10 cities) regarding their greatest concern, of which environmental protection was listed as number one.[532] Finally, a survey was carried out of 15,000 Chinese citizens in all 31 provinces, municipalities and autonomous regions in the country. 68 percent were willing to accept higher taxes for a cleaner environment, while 49 percent declared that environmental protection is China's greatest problem.[533]

Increasingly, the public has complained about environmental pollution to the environmental administrations in China. During the Eighth Five-year plan period (1991-1995), national and local environmental administrations received 283,224 letters from the public that complained about water pollution, air pollution, noise pollution, and solid waste.[534] Furthermore, 433,027 personally visited environmental offices to complain about environmental problems.[535] Moreover, a hotline for pollution problems was opened in Beijing in 1999 that offers advice and legal aid to pollution victims.[536] Although these numbers are small in a country of nearly 1.3 billion people, the public concern for the environment is a growing trend.[537] Raised environmental awareness and concern has resulted in increased participation in activities organised by social organisations. New environmental organisations in China are emerging rapidly, adding to the existing ones.[538]

---

531 Liang (12 March, 1999).

532 *Yangcheng Wanbao* (31 October, 2000). The others were unemployment, educating the children, public security, etc. In 1999, the top concern was an honest and clean government.

533 U.S. Embassy (2001d).

534 The 283.224 letters concerned: water pollution 20.5 percent, air pollution 36.1, noise pollution 33.3 percent, solid waste 2.8 percent, other 7.2 percent. Zhongguo Huanjing nianjian bianji weiyuanhui (1998), p. 149.

535 In December 1997, *Di yi ci quan guo huanjing baohu xinfang gongzuo huiyi* (the first National environmental working conference on public letters and calls of complaints) took place. Ibid., p. 149.

536 *China Daily* (22 November, 1999), p. 2.

537 Reported by Xinhua, 16 November, experts speaking at a forum on Chinese and international environmental non-government organizations, China now has more than 2,000 environmental NGOs of various types, ranging from government-sponsored NGOs to independent, grassroots groups. Over 90 percent of the NGOs are located in cities of the relatively well-developed eastern coastal region. See U.S. Embassy (2001e).

538 Presently 50 citizen environmental groups are registered with the government. In addition, a number of unregistered groups engage in environmental volunteer work through Internet groups, informal volunteer groups, clubs, etc. Government organised NGOs (GONGOs), organised by central or provincial government, number between 1000-2000. 184 university student groups, located at 176 universities in 26 provinces. See Turner, Jennifer L. (2003).

The environmental organisations may be divided into three groups:[539]

1. GONGOs—Government organised NGOs (GONGOs) have existed for some time, they tend to be associations or foundations with close ties to state agencies and consist mainly of scholars, policy-makers and government officials.[540] The activities include environmental awareness raising, policy advice, and assistance to Chinese authorities in mobilising public participation.[541] The GONGOs have traditionally received funding from the state, but they are becoming increasingly autonomous both in funding and work.
2. NGOs—Non-governmental organisations (NGOs), a fairly recent phenomenon in Chinese society, emerged in the 1990s. One major difference from the GONGOs is that NGOs are established on personal initiative, often with personal savings.[542] The emergence of these organisations signifies a new level of awareness among the public, in particular among urban Chinese.[543] They are active in environmental awareness raising, educating school children, and other environmental activities.

---

539 Some groups are more difficult to define, such as Hand in Hand (*Shou la shou di qiu cun*), which is an initiative in the field of environmental education, and is affiliated to more than 800 schools in China. I am grateful to Eva Sternfeld at the China Environment and Sustainable Development Reference and Research Centre for providing this information. Furthermore, a number of green students movements exist at the universities. Thus, the number of NGOs listed here is far from extensive.

540 The Center for Environmental Education and Communication (CEEC) under SEPA; funding is obtained by projects and not from SEPA. Information provided by Eva Sternfeld at the China Environment and Sustainable Development Reference and Research Centre. The National Natural Science Foundation under CAS is another GONGO. See Knup (1997), pp. 11-12, for more information about GONGOs. They are elitist in nature, and due to their close ties with state agencies they may be quite efficient. Another example of an environmental GONGOs is the China Society of Environmental Science (CSES) that was established in 1979 under NEPA. Members are scientists, teachers and environmental professionals throughout China.

541 Ibid., p. 11. In 1994-1995 the CSES established a Women and Environment subgroup in connection with the UN Fourth Conference on Women in 1995. In 1997 the Women and Environment group was allocated funding from UNDP for a women and environment project that included environmental training and education of women cadres. The author participated in the formulation process for this project at UNDP.

542 Liao Xiaoyi, Global Village Beijing, is such an example.

543 One of the first environmental NGOs in China was Friends of Nature (*Ziran zhi you*), which was established in 1994 and registered with the Academy of Chinese Culture. Liang Zongjie, the founder of FoN, is a professor and a former member of the CPPCC. Eventually more groups have emerged, such as Global Village Beijing (*Beijing diqiu cun*). Led by Liao Xiaoyi, who received the Sofie prize 2000, an environmental prize (of US$ 100.000) established by the Norwegian author Jostein Gaarder. Other NGOs are The Center for Legal Assistance, Mama Volunteers (Xi'an), Greenriver (Chengdu), Green Community, Green Plateau Institute (Yunnan), Green Friends Volunteers (Hebei).

3. Volunteers—Voluntary organisations are not registered with the Ministry of Civil Affairs, and do not have a formal organisational structure. One such example is the *Lüjiayuan* (Green Earth Volunteers) led by Wang Yongchen, a journalist of the Beijing People's Radio (*Beijing renmin guangbo diantai*). Individuals with common interests get together to watch birds, plant trees, attend classes on environmental issues in their spare time, and so on.[544]

The environmental organisations in China, be it GONGOs, NGOs, or volunteers, work closely with the media.[545] In fact several of the activists are journalists themselves. One example of how the organisations may use the media channel to mobilise people was a campaign to *lingyang* (adopt) trees, an initiative by Wang Yongchen and Green Earth Volunteers in 1996.[546] The initiative, which was approved by the Beijing Forest bureau, was broadcasted on the Beijing People's Radio station, and many listeners phoned in right after the programme to participate in the adoption of trees on the outskirts of Beijing. The importance of this activity is that it was based upon the people's own wish to participate and not by the authorities' initiative. People are willing to be responsible for the trees, as opposed to top-down tree planting initiatives where the majority of the trees did not survive.[547]

One issue that emerges in relation to NGOs in China is the degree of autonomy the NGO maintains from the state, as all NGOs had to be registered and regulated by the Ministry of Civil Affairs.[548] In order to be as efficient as possible within the set political frame, the NGOs work in accepted areas and their work is an important supplement to the environmental work of the authorities. Some NGOs, in order to escape the bureaucracy and long application process (see below), therefore register the organisation as a private enterprise. The future of environmental NGOs in China is difficult to predict. New regulations were issued in 1998 regarding registration for social organisations, including NGOs. Previous regulations were issued in 1989, which were provisional regulations). The 1998 regulations illustrate that China's leaders recognise the need for 'social intermediary organisations', as they are willing to take on responsibilities that the

---

544 One of the activities of Green Earth Volunteers in 1997 (that the author also participated in) was growing organic vegetables on rented plots outside Beijing.

545 As mentioned, GVB has environmental programmes on national TV. For more informtion on GVB activities see www.gvbchina.org/English/englishintro.html. 'It is effective to work with the media to make Chinese people more aware of environmental problems' says Liang Congjie. Chen (15 September, 2000).

546 See Zheng and Qian (1998), pp. 316-17.

547 '*Shangmian pai renwu, xiamian chou ren qu, zhong shu bantian hou, sihuo quan bu zhi*'. (The authorities assign tasks, subordinates whip people to go, after having planted trees for a while, whether they live or die we do not know), illustrate tree planting activities initiated by the authorities. Zheng and Qian (1998), p. 317. See also Ross (1988), about the forest problems.

548 Knup (1997), p. 12; Saich (2000), p. 129; and Ho (2001a), p. 903.

government does not have the capacity to carry out.[549] Nevertheless, the new regulations also attempt to control the activities of the social organisations as well as their numbers. The application process to register as an NGO (or social organisation) is a cumbersome process, and the two-step process may drag on for several years. It is first necessary to find a professional unit that will act as sponsor for the organisation (also referred to as *popo*, mother-in-law). With the new regulations, this may not be easy as the sponsor is held responsible for the organisation's actions.[550] This is the main difference from the previous regulations (issued in 1989), which did not specify the duties and details of the sponsor. The second step of the process is to apply to the Ministry of Civil Affairs, which may reject the registration of the organisation.[551] Moreover, the NGOs and GONGOs also compete for funding. Thus, the authorities may not welcome new groups. Saich notes that a decline in actual number of social organisations may in fact be the official intention of the 1998-regulations.[552] Also, fear of social unrest makes Chinese authorities suspicious of the organisations. In order to avoid the cumbersome process, some organisations have registered as a company, which is a simpler process.[553] This practice of registering the organisation under an identification other than their their true nature may result in the government's losing an overview of what organisations exist, as these will vanish from the government's view.[554] Thus, one may say that uncertainty exists regarding the process for establishing NGOs. However, environmental concern is growing and it is likely that the public will increase their participation in environmental activities and eventually become more vocal about these issues.

From the above, one may conclude that public awareness regarding environmental protection has developed rapidly during the 1990s. The environmental organisations described above all play important roles for increasing environmental awareness in China. They work together with the authorities and to some extent the authorities rely on them to carry out environmental awareness-raising. This relationship appears to be beneficial to China at this stage. At present, the number of people engaged in environmental protection is small, however, there is a growing trend of environmental concern in China which eventually may impact on environmental policymaking. The authorities' decision to include two environmental NGOs in the environmental preparations for Beijing Olympics may be a sign that they are taken seriously, although expectations with regard to the influence of the NGOs in this process should be moderate.[555] The organisations

---

549 The term 'social intermediary organisations' was used by Jiang Zemin in a speech held in 1997. See Saich (2000), p. 128.

550 Ibid., p. 130.

551 Ibid., p. 129 and Knup (1997), pp. 10-11.

552 Saich (2000), p. 131.

553 U.S. Embassy (2000b).

554 See Ho (2001a), p. 906.

555 In relation to Beijing's bid for hosting the 2008 Olympics, the 'Action Plan for a Green Olympic' has apparently been compiled with the help of Liao Xiaoyi, Global

have not been involved in work regarding the Three Gorges project, as this would be too controversial.[556] The exception is Green Earth Volunteers that were engaged in a dam related issue; a 1997 campaign was initiated to save the Yangtze dolphin, an endangered species that may disappear with the construction of the Three Gorges dam. Nevertheless, it was not a campaign aimed at the dam itself.

Summing up

This brief overview of China's environmental history from 1972-2001 illustrates that China's environmental development has moved gradually forward on all levels, in particular with regard to institution-building and legislation. The first period, designated as an awakening period, started with China's participation in the Stockholm Conference in 1972. National environmental pollution incidents that took place thereafter also made China aware of the need for environmental protection. Consequently, environmental institutions were set up in the 1970s. A few developments in that first decade of environmental protection stand out as particularly important, such as the passing of the provisional environmental protection law in 1979, which is the turning point for environmental protection in China. During the second period, in the 1980s, a number of laws were passed and it was a time of legislative strengthening. Perhaps the single most important event during that decade was when NEPA became an independent agency in 1988. The 1990s stand out as a time where China's leaders really have begun to understand the seriousness of the environmental situation in the country, and that environmental pollution may affect economic growth and human health. This is illustrated by leaders taking part in national environmental conferences such as in 1996, which has lifted environmental protection up to a new level. The major event in the 1990s is the elevation of the environmental administration SEPA in 1998 to ministerial level. The Tenth Five-year plan (2001-2005) for economic and social development also reflects the increased importance placed on environmental protection. An increase of spending on environmental protection will take place in this period, from 0.9 percent under the previous Five-year plan to 1.3 per cent of the gross domestic product (GDP) of the current plan.

Nevertheless, institution building and policy making are not identical to environmental action and enforcement of environmental legislation. Despite increased environmental awareness among China's leadership and environmental policymaking, the country still faces great challenges with regard to implementation of its environmental policy. As implementation goes beyond the intended scope of the chapter, with the exception of a few examples this has not been discussed. Suffice it to point to some of these challenges ahead: the ability to

---

Village Beijing, as well as Liang Congjie, Friends of Nature. See Yu (4 September, 2000).

556 Some individuals, such as Dai Qing, have been vocal in this process. However, no organisiation has participated in the debate about the dam.

implement central policy at the local level, which is related to behaviour change and increased awareness among lower level leaders; and to what extent SEPA will be able to carry out its role and shoulder the environmental responsibility.

Furthermore, the public's concern for these environmental issues has been manifested in increased participation in environmental activities and the emergence of environmental organisations such as GONGOs, NGOs and volunteer groups during the past decade. The influence of these environmental organisations is still limited, and they do not engage in controversial issues such as the Three Gorges project. Bureaucratic regulations make it a difficult and time consuming process to register new groups. Therefore, some groups instead register as enterprises. However, a growing concern in China may result in increased significance in environmental decision-making for these groups.

Chapter 7

The Three Gorges Project Environmental Developments—1972-2001: Linkage between Resettlement and Environment

This chapter will give an overview of the environmental developments for the Three Gorges project from 1972 until 2001. One assumption made in this book is that the increased environmental focus in China in general is one important factor that has allowed the change in the resettlement policy for the Three Gorges project to take place. This is related to the fact that China as an environmental actor is paying more attention to the environment, as described in Chapter 6. Furthermore, the fact that Premier Zhu Rongji stated environmental capacity as the reason for the policy change may indicate increased awareness with regard to the relationship between environment and resettlement. Resettling the population within the reservoir area as originally intended would increase the pressure on the environment, and existing environmental problems would aggravate. The policy change has indicated that Chinese decision-makers realise the close relationship between environment and resettlement; that they are mutually dependent.[557] By stating the environment as the cause for policy change, the resettlement and environment have been linked in a new way in the Three Gorges project, and the importance put on this relationship lifted the issues to an unprecedented political level.

In this chapter I will discuss the gradual development that eventually linked environment and resettlement in 1999, as well as look into the environmental developments of the dam project in general. It will look at the developments in the environmental area for the project by presenting and discussing some important incidents that may have been instrumental in effecting policy change. The chapter will be divided into four periods, and will introduce and discuss the trends of each period. The first period describes the developments in the 1970s when environmental considerations in the deliberations about the dam played a minor role. The second period concerns the heated discussion about the pros and cons of the dam. During this time, from 1980 until the project was approved in 1992,

557 This mutual dependency will not be stated explicitly in all sections in the text as it would be too repetitious, but should be kept in mind throughout the chapter as the background for the discussion.

scientific research for the dam project gradually paid increased attention to the environmental impact of the dam and resettlement of such a large population. The third period relates to the time following the project initiation in 1993 until 1997, when realities created a shift in scientific focus for the project. The fourth period begins in 1998 when Zhu Rongji became Premier and head of TGPCC, and when the awareness of the mutual dependence between resettlement and the environment seems to have reached a new level, and cumulates with the decision to *waiqian*, move some people out of the reservoir area. Emphasis will be on the last three periods, since these are more relevant for the recent development of the project.

The 1970s—the Buds of Environmental Consciousness

In the early years following the UN Conference in Stockholm in 1972 there seems to have been little environmental activity for the dam project. A Bureau for the Protection of the Environment Against Pollution at the Three Gorges was established in 1976,[558] which indicates that the environment had begun to receive more focus in the project policy as well as in national policy. Following this establishment, indicators exist of increased focus being paid to the environment.[559] In May 1979, for instance, a site selection meeting began, where the participants visited the two possible sites, and upon their return came together in Wuhan to discuss the results. The meeting included more than 200 participants with different backgrounds; professors, specialists and technicians involved in a number of areas, including the environment.[560] In a report from the meeting by Xinhua,[561] only a few references are made to the environmental issue. First, the report mentions in the

558 Lieberthal and Oksenberg (1988), p. 309. Silt problems were emphasised by Mao Zedong in 1963, ibid. p. 306. Zhongguo Sanxia jianshe nianjian bianjibu (1996) states that environmental issues were focussed upon already from the 1950s in reports by the YVPO, as well as CAS carrying out research on issues such as hydrology, climate, geography and resources, which would establish a basis for the environmental impact resarch, p. 133.

559 One study was initiated in the early 1970s by the Yangtze River Valley Water Resources Protection Bureau. In the overview of the research history for the project, one study is mentioned from the 1950s by the YVPO that included some environmental aspects (see footnote above). Then the overview skips the 1970s and continues from 1980. See Zhongguo Sanxia jianshe nianjian bianjibu (1996), p. 133.

560 Lieberthal and Oksenberg (1988), p. 313-14. Other areas were water resources, hydropower, geography and earthquakes, shipping and communications and machine manufacturing. The choice of site was between Taipingxi and Sandouping; Sandouping was eventually picked as the site for the dam.

561 Li (1985), pp. 106-13. The title of the report is: Chen Baokang, Guo Wanli (1979) 'Dui Sanxia shuili gongcheng nengfou hen kuai shang you bu tong yijian yiji shangma xuyaojiejue de yixie zhongyao wenti' (Regarding some different opinions on whether or not the Three Gorges project can be given an early launch, as well as some important issues that must be resolved).

introduction that environmental specialists participated in the meeting. Second, the report also mentions that criticism was raised against the organisers for not having invited experts on aquatic plants or fish, as the fish would need to migrate through the dam area. Third, the report raises the issues of silting up of the reservoir, mainly as a hinder to normal navigation on the river. Resettlement is mentioned mainly in relation to insufficient planning and underestimated costs. The environmental problem in relation to the resettlement of the population is not mentioned at all.

Thus, one can see from the above that the environmental issue in the Three Gorges project in the 1970s was nearly non-existent. The problems that were focused upon at the time were rather the height of the dam, the site, the electricity production and flood control as well as improving the shipping facilities. Environmental policymaking was in its initial phase in general in China, and in the discussion about the dam project this was not a hot issue.

1980-1992: The Beginning of the Feasibility Report Era and the Three Gorges Project Approval

The 1970s were a period when an awakening took place with regard to the environment in China. One could say that with the issuing of the environmental protection law in 1979, a new era begins. Also, following the political reforms in 1979 and the introduction of the Four Modernisations, scientific research was emphasised.[562] One important change in the 1980s was that science and research became increasingly important for the making and implementation of environmental protection policy. Furthermore, the environmental protection law introduced a system of regulations, monitoring activities and enforcement based upon two principles of 'pollution prevention' and 'polluter pays'. Regulations requiring environmental impact assessment were introduced in the beginning of the 1980s, and greater attention was paid to resettlement costs.[563] Initially, the environmental impact assessment process was confined to the post approval stage and usually had little if any influence on political decision-makers. Article 13 of the Environmental Protection Law now mandates that the process takes place before a project is approved.[564] Even if environmental impact statements were initiated for manufacturing and infrastructure projects in the early 1980s, according to some observers[565] it is doubtful that too much attention was paid to these

---

562 This will be further discussed in chapter 8. See Spence (1999), pp. 618-24 on the Four Modernisations, and Maa (1993), p. 58-9 about science and research.

563 Ross (1992), p. 636.

564 NEPA, Zhonghua renmin gongheguo huanjing baohufa (The Environmental Protection Law of the PRC, 1989, 26 December) in NEPA, (1993), *Zhonghua renmin gongheguo huanjing baohu fagui xuanbian* (A selection of environmental protection laws and regulations of the PRC), p. 3.

565 Edmonds (1994), p. 19.

statements. In a centrally planned economy and totalitarian system, pet projects would still be given the green light. However, the fact that these statements were written indicates that politicians were increasing their focus on environmental issues in the country's development plans.

This had implications for the Three Gorges project, which was still under discussion, as an environmental impact study would be required for the project. The 1980s were a period when numerous studies were carried out. Units under the project proponents carried out some studies in an effort to pave the way for the project. Other studies were carried out by more neutral units such as the Chinese Academy of Sciences (CAS), that apart from the research funding involved, did not have the same interest in the dam project as a proposing ministry. In 1980, one of the first impact assessments for the dam project was carried out by the *Changjiang liuyu shuiziyuan baohuju* (Yangtze Valley Water Resources Protection Bureau, YVWRPB) under the Ministry of Water Resources. The Bureau came up with an environmental impact assessment of a 200-metre dam scheme.[566] During the summer of 1981, the US Department of the Interior's Bureau of Reclamation (USBR) visited China and held talks with Chinese units who promoted the dam, the YVPO and the MWREP.[567] This USBR delegation adopted a generally positive attitude towards the dam project, but suggested to the Chinese that they undertake feasibility studies of a dam lower than 200 metres. Accordingly, soon afterwards the YVWRPB carried out an environmental feasibility study for a 150-metre dam, and in March 1983 compiled the report *Sanxia jianba dui huanjing de yingxiang* (The Three Gorges dam and its impact on the environment). The report was a result of the influence of the USBR delegation, and illustrates that the MWREP took the recommendations seriously.

In 1979, the YVPO, a subordinate organ to the Ministry of Water Resources and Electric Power, had compiled a report to the State Council in an attempt to launch the project. A discussion was held in the State Council with relevant provinces and ministries, where it was decided that the State Science and Technology Commission (SSTC) should organise an expert meeting to discuss the scientific aspects of the dam project. The meeting was organised in May 1983, when the State Planning Commission (instead of the SSTC) organised over 350 experts to examine the report on the feasibility study for the project in general submitted by the YVPO. This meeting was the first since the debate on the project began where such a large number of scientists and experts were present. The participants came from sixteen ministries, three provinces (Sichuan, Hubei and Hunan), 58 scientific research, design, and construction units and factories, and eleven colleges and universities. They were divided into seven specialised groups,

---

566 Zhongguo Sanxia jianshe nianjian bianjibu (1996), p. 133. The Yearbook 1995 mentions the work carried out in the 1950s in relation to the dam project's environmental issues, the project's impact on back water, economic effects of human activities, earth quakes, siltation, organisms, reservoir inundation and resettlement and diseases.

567 Lieberthal and Oksenberg (1988), p. 317.

including environmental protection.[568] The conclusions by the conference participants to the feasibility report of the YVPO were mixed, and found the study to be too superficial and too much focused on the project proper instead of focusing on the impacts the project would have on the area. It was weak on social and political impacts as well as environmental and ecological effects; in particular the pollution sources of Chongqing and investments to clean this up were mentioned.[569] One concrete suggestion was to include afforestation and soil conservation into the general project plans.[570] Lieberthal writes that the subgroups that discussed the environmental protection issue concluded that the resettlement problem had been seriously understated, as well as too little funding allocated for relocation due to population growth.

Even though a few feasibility studies had been carried out in the beginning of the 1980s as mentioned above, 1984 is regarded by both Chinese scientists and officials as the year when the government really began to pay attention to the environmental issues of the Three Gorges project. This may be related to the approval in principle of the dam by the State Council that year.[571] This new emphasis on the environment included extensive research about the relationship between the environment and the potential for resettling such a large population in the Three Gorges area. Academics state that research focusing on identifying problematic issue areas was first initiated during the 1980s, while the agenda was more or less set in earlier research.[572] This is illustrated by the large study that was initiated in 1984 with the help of the SSTC, CAS mobilised between six and seven hundred scientists from thirty-eight research institutions to study impacts of the project on ecology and the environment. During a period of three years, they collected over 20 million pieces of scientific data and produced 98 research reports on various aspects of the ecological and environmental impacts of the project. The findings were published in the *Changjiang Sanxia gongcheng dui shengtai yu huanjing yingxiang ji qi duice yanjiu lunwen ji* (A collection of papers regarding the research on the impact of the Yangtze Three Gorges project on ecology and the environment and the measures to deal with the issue).[573] The significance of this study is the different approach of the CAS compared to the research that had been

568 Li (1985), pp. 13-14. Yin (1996), p. 431.

569 Li (1985), p. 148. See also Lieberthal and Oksenberg (1988), p. 322.

570 A suggestion by the head of the agricultural committee of Hunan province, Shi Jiazhu. Li (1985), pp. 139-40.

571 In addition, the Second National Environmental Protection Conference took place, as well as the establishment of the Environmental Protection Commission under the State Council.

572 Interview no. 24, Nanjing 1999. See also Chen and Chen (1993), p. vi. See also Yin (1996), p. 489.

573 Edited by Zhongguo kexueyuan Sanxia gongcheng shengtai yu huanjing keyan xiangmu lingdao xiaozu (research project leading group for the Three Gorges Project ecology and environment, CAS), Beijing kexue chubanshe, 1987. Part of the collection was also the *Changjiang Sanxia shengtai yu huanjing dituji* (Atlas of the Yangtze Three Gorges Ecology and Environment (Beijing, Kexue chubanshe), 1989.

carried out earlier. The majority of earlier research had been carried out by institutions directly under the Ministry of Water Resources and Electric Power. These reports were not comprehensive in their approach, did not go deeply into the complexity of the environmental issues, did not come up with countermeasures, and the research agenda was set by the MWREP. Another sign that the environment was being taken seriously was the establishment in 1985 of the Ecology and Environmental Verification expert group (*Shengtai yu huanjing lunzheng zhuanjia zu*) by the SPC and SSTC on the initiative of the State Council. Its purpose was to investigate the potential environmental impact of the schemes with a normal water level ranging from 150-180 metres.[574]

The same year that the above mentioned research by CAS was published, in 1987, the CAS and SSTC included the Three Gorges project and the environment as a research topic into the Seventh Five-year plan for China (1986-1990), which lifted the environmental issues of the Three Gorges project up to a new level. Consequently, CAS then initiated a new study with the participation of 300 scientists. In 1991 the results of the study were published in the series *Sanxia gongcheng yu shengtai huanjing* (The Three Gorges Project and the Ecology and Environment), consisting of nine reports.[575] The conclusions from the study indicated that many of the researchers at CAS were sceptical towards the ecological and environmental effects of the Three Gorges project, and found that the potential negative environmental impacts of the project outweighed the potential benefits. In the two large research projects carried out by CAS, extensive research was conducted in a number of areas, and counter measures were proposed. The principal findings of the research were: i) the ecological and environmental situation in the Three Gorges area is fragile, and the construction of the dam will speed up the environmental and ecological deterioration in the area; ii) the main impact of the Three Gorges dam will be in the reservoir area. However, both the middle reaches, lake areas and the river mouth will be affected; iii) the largest impact on the environment will be made by the resettled population as their land is inundated, and erosion will increase; iv) funding for environmental measures is imperative, and to be started before constructions begins; v) there is need for further environmental research. In brief, one could say that the conclusion of the research concerned the complexity of the problems and the long-term adverse effect on the environment. It was fairly sceptical towards the potential negative environmental impacts of the dam, and recommended that it would be better to postpone construction until further research had been carried out.

At the same time as CAS was carrying out these environmental impact studies, other government units were also making assessments of the environmental impact of the project. The Ministry of Water Resources and Electric Power initiated one study in 1986. In accordance with a circular from the Central Committee and the State Council, a Three Gorges project ecological and environmental expert group

---

574 Zhongguo Sanxia jianshe nianjian bianjibu (1996), pp. 133 and 217.

575 Ibid., p. 133; and Chen and Chen (1993), p. vi. Another book in the series was Xu and Liu (1993).

was established, consisting of 55 experts. They studied and checked past results, and organised a study by topic carried out by the Yangtze Valley Water Resources Protection Bureau (YVWRPB) and the Environmental Impact Assessment Department of CAS. YVWRPB and CAS completed the *Changjiang Sanxia gongcheng shengtai yu huanjing yingxiang ji duice de lunzheng baogao* (The Yangtze River Three Gorges Project Verification Report for the Ecological and Environmental Impact and Countermeasures) in January 1988. In March 1991, a preliminary hearing was held where the ecology and environment expert group under the State Council Three Gorges project inspection committee put forward opinions about the report. In July the same year the inspection committee examined and approved the feasibility study phase of the verification report.[576]

Following this approval, based upon the regulations and the measures for environmental protection management in construction projects, as well as requirements from the State Council Three Gorges inspection committee, the EIA was compiled. As described in the section about the environmental capacity in Chapter 5, the environmental impact assessment for the project, the *Changjiang Sanxia shuili shuniu huanjing yingxiang baogaoshu* (*Yangtze River Three Gorges Project Water Conservancy Project Environmental Impact Statement*) was approved by NEPA on 17 February 1992. The *Statement*/EIA was compiled by the Environmental Impact Assessment Department, Chinese Academy of Sciences and the Research Institute for the Protection of the Yangtze Water Resources, Ministry of Water Resources. The *Statement*/EIA was based on research that CAS had carried out in the period before,[577] as well as on the report by the MWR. Before the approval of the project by NEPA, a preliminary hearing took place on 21-24 January, 1992 regarding the results from the Environmental Impact Statement. In addition to the research groups from the two institutes, the 55-member expert group and 88 representatives from the State Council and related cities and provinces also took part in the hearing.[578] At the meeting, the *Statement*/EIA was presented by the Environmental Assessment Department of CAS and the Yangtze River Valley Water Resources Protection Bureau, which had compiled and edited the report. After having given the presentations, 'serious' discussion took place.[579] The two institutes had concluded differently regarding the environmental capacity for resettling such a large rural population in the area, as well as the potential environmental impact by the dam. Several environmental scholars in the expert group of 55 agreed with the conclusions of the CAS, and did not give their consent to the *Statement*/EIA. Despite the opposition by several of the experts, the *Statement*/EIA was officially approved in February 1992 by NEPA. A couple of

576 Zhongguo Sanxia jianshe nianjian bianjibu (1996), p. 218.

577 Chen, Xu and Du (1995). This report was one of several reports that were carried out under the national research programme (identical title as the above report) of the 7th five-year plan.

578 Zhongguo Sanxia jianshe nianjian bianjibu (1996), p. 130.

579 Ibid., p. 130.

months later, the Three Gorges project was formally approved by the National People's Congress (NPC) in April 1992.

Additional Comments

From the process that is described above, five points stand out that need to be commented upon:

Background for EIA and the co-operation between the CAS and YVWRPB[580] In 1990-1991, during the debate about the environmental issues of the Three Gorges project, the former minister of MWR, Qian Zhengying, had stated that the EIA for the dam project already had been carried out by the Yangtze River Water Resources Commission (former YVPO). The conclusion of the ministry and of Qian at the time was that it would not be necessary to do another study, as the conclusion from the above study was clear: the project would have some negative impacts on the environment, but these could be solved by countermeasures. The project was therefore feasible with regard to the environment. At the time, the chairman of the NPC Environmental and Resources Committee, Qu Geping, pointed out to Qian Zhengying that if the quality of the study was not high enough, the project might not be approved when taken to the NPC. Consequently, Qian Zhengying decided that a new EIA would be carried out jointly by YVWRPB under the MWR and the Environmental Impact Assessment Department of the CAS. The conflicts were apparently numerous, due to different opinions regarding i) the resettlement issue, ii) increased water pollution and iii) influence on the river mouth. The final *Statement*/ EIA included opinions of both the MWR and CAS, which was the first time a report included differing opinions.[581] Nevertheless, the summary report, which was made public, the Environmental Impact Statement for the Yangtze Three Gorges Project, did not include the divergent conclusions. 'In reality, CAS participated only in name', states one interviewee.[582] This view implies that CAS was invited to participate in the process in order to make the EIA justifiable. The actual power was in the hands of Qian Zhengying and the MWR.

Loophole in the environmental policy The above point highlights a factor that influences the validity of EIAs in China. The construction or design units often compile the EIA themselves, i.e. assess themselves, which does not ensure an unbiased study. As for the Three Gorges project, this has been the case several times, since both the Yangtze River Water Resources Commission and the YVWRPB have been responsible or involved in the EIA for the project. These two are both subordinate to the Ministry of Water Resources, which has been the main

580 The background for the EIA was described to the author under several interviews in 1999. The interviewees were involved in the EIA research process for the project for a number of years.

581 Interview no. 4, Beijing, 30 August 1999, with MWR official.

582 Interview no. 17, September 1999.

proponent of the project for a number of years. The fact that units responsible for the design or construction of a project can carry out EIA for a project is perceived by scholars interviewed as a big problem for the environmental policy of the country.

The significance of the EIA in the environmental decision-making process[583] Two weeks before the approval of the *Statement*/EIA by NEPA, the executive meeting of the State Council agreed to construct a dam at the Three Gorges.[584] Thus, the decision to construct the dam was made before the *Statement*/EIA was approved. Moreover, NEPA did not even agree with the TGPCC regarding the *Statement*/EIA for the project.[585] This raises several questions: to what extent did scientific research about environmental impact have a politically justifying effect; did the EIA matter at all? From the debate about the project in general before the decision was made to construct the dam, it is evident that the dam project was a politically sensitive project. The proponents did what they could to have the project approved by publishing articles in the press that focused on the positive aspects of the project only. The critics did not have the same opportunity to air their doubts, but were able to publish some books in the late 1980s (as described in the history chapter). Following the crushing of the democracy movement in 1989 and the floods in 1991, it became increasingly difficult to come forward with criticism, be it environmental or otherwise, and the project was approved by the NPC in 1992. From the process above, one may question the sincerity of the Chinese government with regard to the environmental issue at the time, as it would seem natural to await the conclusions of the EIA before endorsing the project. At the same time, large funding had been handed out to CAS and other units to map and explore the possible environmental impact by the resettled population and the dam in the area, which illustrates a genuine concern about the project's impact. Nevertheless, having an EIA justified the decision, and it is possible that the proponents of the dam knew that the EIA would not stop the project. Furthermore, the project was approved by the State Council, while the EIA was approved by NEPA, which was only an agency at the time without political power to halt a project due to negative environmental impacts. One may conclude that the environment was not the decisive factor in the deliberations over whether or not to construct the dam. One interviewee put it this way: 'Number one is politics and two is science and technology. Science and technology are servants for politics. Politics take charge

583 The significance or lack of significance with regard to EIA in the decision-making process is not typical for China only.

584 Interview no. 3, August 1999, Beijing.

585 According to Dai Qing this was reported in the *Huanjing bao haiwaiban* (overseas edition of the environmental newspaper), and the journalist responsible was criticised for this.

of science and technology' (*zhengzhe lingdao kexue jishu*).[586] The political forces involved in the decision-making process for the dam were stronger than science.

Experts refused to sign the Statement/EIA The majority of the 55 experts who evaluated the *Statement*/EIA found it to be feasible and gave it their signatures. They also suggested a number of policy measures to be carried out in their reply comments regarding the *Statement*/EIA. In the letter, the experts pointed out the positive and negative environmental effects, and the negative effects outnumbered the positive effects.[587] Despite this addendum with the suggestions for measures, a few of the 55 experts who participated in the evaluation did not put their names on the consenting letter. They listed 11 problems which hindered them in endorsing the EIA. Some of the problems they sited were: resettlement, siltation, influence on the aquatic system (Chinese sturgeon), loss of natural fish pond, influence on the river mouth,[588] reduced water temperature in the lakes Dongting and Poyang (which would impact on the water birds in the area); geological disaster; earthquake; increased reservoir pollution; loss of scenery and destruction of cultural and historical sites. According to one interviewee, MWR officials tried to lobby the experts who refused to sign the report by phoning them in the evenings, as it would look better to have a report that was endorsed by all.[589] The disagreement among several experts regarding the environmental impact of the project is perhaps not significant, as a difference of opinion would be natural among experts. Nevertheless, knowing the history of this controversial dam project where it has been difficult to voice diverging views, one may conclude that the significance lies in the mere airing of these different beliefs, and refusal to sign. It is perhaps even more significant that the majority signed the EIA knowing that there still was so much uncertainty about the impact of the project, as reflected in the number of counter-measures that were suggested.

586 Interview no. 5, Beijing, August 1999. The scientist interviewed had been forced to an early retirement due to airing critical opinions towards the environmental impact of the dam.

587 The suggestions of measures to be carried out were the following: The need for a comprehensive plan in the area which takes into account both the resettlement, economic development and environmental protection; the industrial distribution and the selection of less polluting enterprises; the implementation of pollution control policy (in particular the *san tongshi*, the 'Three Simultaneouses' policy); carry out EIAs for infrastructure projects and the resettlement project during the process; establish an environmental monitoring network; further carry out research regarding the environmental impacts and potential preventive actions; further assess the cultural relics that need to be moved; include the environmental costs in the project's overall budget; and the establishment of an environmental fund etc. See Zhongguo Sanxia jianshe nianjian bianjibu (1996), pp. 129-30.

588 In winter the water level in the reservoir will be higher than at present; in summer the water level will be lower than at present. When it is lower it will influence the sea water flow (backwards). More sea water will flow into the river.

589 Interview no. 13, Beijing, 7 September, 1999.

Objective reporting The above points regarding the *Statement*/EIA for the project are related to an issue that has come to the fore in the discussions about the dam project, namely the difficulty in reporting objectively about the negative consequences that the project may have on the environment. Based on information from academics involved, it is apparent that the officials in charge of the project have not been interested in hearing about potential problems or negative impacts. When faced with such impacts, the reaction is often negative, as it is interpreted as criticism of the project, and thus criticism of the policymaking. One such example is the EIA appraisal meeting that was held at the Friendship Hotel in Beijing in January 1992. Top officials for the project participated in the meeting, such as Qian Zhengying and Li Boning, as well as the scientists who had taken part in the study and the expert groups. At the meeting, Chen Guojie as the head of the expert group, introduced the study and the conclusions of the CAS. The CAS had concluded differently from the MWR in that they found the negative environmental impact would outweigh the positive aspects, and construction of such a project in the near future was not feasible. Li Boning got very angry and instructed Chen to leave the meeting (*yinggai tuichu huichang*). However, Ma Shijun, a highly respected scientist, supported Chen by stating that if Chen had to leave, he would also leave. Thus, Qian Zhengying had no choice but to mediate between the two groups and stated that they should all remain at the meeting. This example from the debate about the environmental issues of the dam project illustrates that a scientific view was interpreted as a personal view. Chen did not represent his own opinion, as the views were the conclusion of the study. One may conclude from this that the highly political flavour of this project has had extensive influence on the environmental discussion.

Recapitulation

Summing up the above, a new era began in the 1980s when the first comprehensive study of the environmental impact of the project was begun by CAS in 1984. The study marks a shift in the approach to the environmental issue in the project history, as the research begun in 1984 was more independent (in the sense that CAS was not a project proponent), comprehensive and complete than previous studies. Extensive research was for the first time carried out on the environmental capacity in relation to the potential environmental impact of the large population to be resettled, as well as their potential for a sustainable living in the area. Previous research was often initiated by the MWR and carried out by subordinate institutes, one of the project proponents, and the research agenda was often fixed. CAS concluded that the Three Gorges project would have negative environmental impact in the area, and that further research needed to be carried out before the project could be launched. Based on the studies by CAS and MWR, they compiled a *Statement*/EIA in 1991. The EIA was approved in 1992 by NEPA, while the dam project had already been approved two weeks earlier by the State Council. This questions the importance of the EIA in the decision-making process.

1993-1997—The Period Following the NPC Approval of the Three Gorges Project

This section will look into the environmental developments following the approval to construct the project, until Zhu Rongji took over as Premier in March 1998. There are four issues of importance for the environmental developments that will be discussed: i) a shift in scientific focus; ii) the increased emphasis on environmental networking, iii) environmental supervision and management; and iv) NEPA's role in the environmental process.

A New Reality: Shift in Scientific Focus

In the period leading up to the approval of the dam, the debate and the research reports published were mainly focused on arguments for and against the construction of the dam. With regard to the environment and resettlement for the project, some interest groups may have portrayed the impacts more favourably than was realistic, such as the unproblematic view about resettling a large number of people in an area with limited environmental capacity. On the other hand, some scholars[590] seem to think that opponents during the discussion process may have painted a darker picture than necessary by exaggerating the number of relocatees and potential environmental impacts. According to these scholars, the truth would be somewhere in the middle between the views of proponents and opponents. Following the NPC approval to construct the dam, a shift in the focus of the discussion took place and the participants in the discussion no longer engaged in a debate about the pros and cons of constructing the dam. Now the purpose of the discussions and studies was to assess the environmental situation before, during and after construction. Moreover, the studies were carried out in order to minimise environmental degradation, in particular in relation to the relocatees' potential for sustaining livelihoods. The researchers who had rejected the project for scientific reasons, now continued to carry out research in order to reduce the environmental impact. The research carried out now would, for instance, involve proposing different scenarios in order to provide advice for decision-makers in the construction process of the project. Since the construction work on the project was initiated in 1994, monitoring stations have been established in the reservoir area and reporting on the environmental state in the area from all stations is carried out annually. The information is used by Chinese authorities in environmental policymaking. One could thus conclude that the focus of the present research is to come up with counter-measures. A large number of research institutions both under the ministries involved and under CAS were and still are involved in the environmental and resettlement research for the project, too numerous to be listed here.

Since the establishment of the Three Gorges Project Construction Committee (TGPCC) in 1993, most research for the dam project seems to go through this

---

590 Based on interviews with researchers during field work in 1999.

committee. A procedure exists where potential research institutes give their bids, and the most suitable research institutes receive the funding. The majority of the research funding for projects is channelled through the TGPCC, which controls what research is to be carried out. Which institutes that in the end receive the funding depends on several matters, such as experience from the research field in question as well as good contacts and relations in particular.[591] Typical projects involve a number of research institutes; they are called split-the-food-projects (*fenzhi chi fan*). One example of such a project is a RMB 3 million-project involving Qinghua, Chongqing Environmental Sciences Institute, Chengdu University, Research Institute for Protection of Yangtze Water Resources, and China Insitute of Water Resources and Hydropower research.[592] Co-operation between the institues is sometimes difficult, and the TGPCC must often spend much time co-ordinating and mediating between the research institutions involved.[593] In the project, different scenarios are developed for certain years and periods, where implications for, for instance water quality change are studied, and recommendations are given to the authorities. It is uncertain to what extent research results are modified, as some sources say that the TGPCC checks everything before publishing.[594] According to some interviewees, the TGPCC has a monopoly to decide which research projects should go ahead. CAS and other departments must all go through the committee to obtain funding for the Three Gorges research projects. It is possible to find funding in other ways (foreign funding for instance), but it is not very common for the Three Gorges Project. Due to the monopoly by the TGPCC, and the difficulty in doing independent research, some scholars stated during interviews that they were unwilling to participate in research funded by the committee, due to its partiality.

---

591 80 percent of the success can be attributed to contacts and good relations. In the soft science area this type of relation is even more important. Interview no. 8, Beijing, 1 September 1999.

592 Another example of complex organisation involving several institutions is the following: the Three Gorges Project Development Corporation is the entrusted unit, CAS is the responsible department, the organisational and co-ordinating unit is the office of the TGPCC, the Institute of Soil Sciences, CAS, Nanjing is the responsible institute; participating units are Institute of Geography, CAS; Changsha Institute of Modernisation of Agriculture, CAS; Shenyang Institute of Applied Ecological Research, Beijing Normal University and People's government of Zigui county in Hubei province. See the report from the project: Ma and Yang (1996).

593 The institutions that co-operate in a project write a contract. Sometimes if the contracts are not clear or do not explicitly include the exchange of certain materials, it will be very difficult to obtain these materials at a later stage. The different institutions do different types of research and the intention is to exchange information.

594 Interview no 8, Beijing, September 1999.

Environmental Network for the Three Gorges Project

NEPA established a Yangtze River Three Gorges ecology and environment monitoring network in 1993, *Changjiang ji Sanxia shengtai huanjing jiance wang* (Yangtze and the Three Gorges Ecology and Environment Monitoring Network). The monitoring network consisted of EPBs from Chongqing, Wuhan, Nanjing, and other relevant areas.[595] The Chinese government eventually realised the need for a comprehensive approach to the environmental challenges in the Three Gorges area. This is reflected in the establishment of the Co-ordination Small Group for the Ecology and the Environmental (*Shengtai yu huanjing baohu xietiao xiaozu*) under the TGPCC following the official initiation of the construction of the project in 1994. The purpose of this group was to co-ordinate the ecological and environmental monitoring network that began monitoring in the mid-1990s. A number of ministries and institutions are involved in this monitoring work, which makes the co-ordination of the project very difficult. The implementation plan (*Changjiang Sanxia gongcheng shengtai yu huanjing jiance xitong shishi guihua*, The Implementation plan for the ecological and environmental monitoring system for the Yangtze River Three Gorges project) was approved by the TGPCC and the Co-ordination Small Group for the Ecology and the Environment.[596] The majority of the members of the Co-ordination Small Group are involved in the monitoring network and their tasks are the following:[597]

- SEPA is involved in strengthening the management of pollution sources in the reservoir area, as well as monitoring the water quality in the different seasons (flood, normal and dry season) in the sections along towns and cities.
- The Ministry of Water Resources (MWR) is involved in the water and land conservation work in the area, as well as hydrology and sedimentation work. It is also involved in monitoring water quality as well as the protection of water resources.
- The Ministry of Agriculture and the State Forestry Administration are involved in constructing a national monitoring network for the ecology and the environment in agriculture and forestry. They carry out monitoring work in relation to their tasks in the area.

---

595 Zhongguo Huanjing nianjian bianji weiyuanhui (1995), p. 164.

596 Zhongguo Huanjing nianjian bianji weiyuanhui (1996), p. 103.

597 Zhongguo Sanxia jianshe nianjian bianjibu (1997b), p. 45. The committee is headed by the vice-director of the TGPCC office, Li Shichong. In addition, members are the Yangtze River Water Resources Commission, administrations for geology and mineral resources (since 1998 under the Ministry of Land and Natural Resources) and China Bureau of Seismology, and members from the TGPCC, the Three Gorges Project Resettlement Bureau and the TGPDC. For a comprehensive list of all members (including individual names) of the Co-ordinating Small Group, see Zhongguo Sanxia jianshe nianjian bianjibu (1998), p. 26.

- The Chinese Academy of Sciences (CAS) is involved in monitoring and protection of various species in the area. In addition CAS carries out research in a number of areas and comes up with suggestions and measures for the policy-makers. CAS also runs two research and monitoring stations in Wanxian in Chongqing municipality and Zigui in Hubei province.[598]
- The State Meteorological Administration is involved in weather and disaster forecasting and monitoring. One purpose is to look into how the expected change of climate can be utilised for agricultural and fruit production.
- The Health Ministry works on issues in relation to avoiding the spreading of diseases in the area.
- The Ministry of Construction, Chongqing Municipal government and Hubei provincial government are also involved in environmental monitoring work for the project.
- The TGPCC Resettlement bureaux of Chongqing and Hubei are also involved in the environmental planning.
- The Ministry of Communications was invited to join at a later stage; their area would be to monitor the mobile pollution sources (boats on the river).[599]

These ministries, administrations and research units hand in their results, which are edited and compiled into an annual report, *Changjiang Sanxia gongcheng shengtai yu huanjing jiance gongbao* (*The Yangtze River Three Gorges Project Ecological and Environmental Monitoring Bulletin*), by the China National Monitoring Station and published by SEPA.[600] The report gives a very simple overview of the situation in various areas, with a little information about each topic. Topics vary: resettlement, the resettled population and the potential economic development in the area, agriculture, the ecological situation in the area, terrestrial and aquatic animals, waste disposal, pesticides, and pollution. The annual report also gives descriptions of how the work is carried out in the reservoir area, and who is responsible for the tasks. Apart from a couple of pages at the beginning of the report which may point to both achievements and future challenges,[601] the report does not provide an overall analysis of the situation.

---

598 One example of the work carried out is the report from Wanxian and Zigui counties: Yang Linzhang, (1998) et al, *Sanxia gongcheng shengtai huanjing shiyan zhan jianshe 1998 nian niandu baogao* (The Annual Report of the Three Gorges Project Ecological and Environmental Experimental Station). This project is part of a larger project organised by the TGPCC, *Changjiang Sanxia gongcheng shengtai yu huanjing jiance xitong* (The Yangtze River Three Gorges Ecological and Environmental Monitoring System).

599 Zhongguo Sanxia jianshe nianjian bianjibu (1997a), p. 38. The ministry's participation in the group was suggested by other members in the co-ordination group.

600 There are 16 units altogether that participate in compiling the annual report.

601 For instance the annual report 2000 (SEPA 2000a), points to the potential water pollution problems due to the lack of processing plants, p. 3.

Environmental Supervision and Management

The Three Gorges project is expected to set off economic development activities in the area, and the challenges with regard to pollution control and the protection of natural resources in the Three Gorges area are numerous, in particular in the construction area. Therefore, *huanjing guanli* (environmental management) has become an important term for the Chinese authorities. Environmental management and supervision is an issue that has become more common in the development of the dam project, in particular in one part of the Three Gorges reservoir area that is called the construction area. The construction area is situated at Sandouping in Hubei, and is where the actual dam is being constructed. The China Yangtze River Three Gorges Project Development Corporation (*Zhongguo Changjiang Sanxia gongcheng kaifa zonggongsi*), the Three Gorges dam project proprietor, is responsible for environmental protection in the construction area. There is concern about the environment in the construction area, due to the great construction activity, which involves dust and noise. In addition, between 10,000-20,000 people will be living there for the duration of the construction. Thus, problems such as waste disposal and the potential for disease are focused on. Following the project approval in 1992, several studies were initiated in order to assess the environmental and ecological situation in the Three Gorges reservoir area.[602] Also, frequent monitoring of the water quality at the construction area takes place.[603] Following this, two important documents for environmental management in the area were issued. One was the *Sanxia gongcheng shigongqu huanjing baohu shishi banfa* (Implementation Measures for Environmental Protection of the Three Gorges Project Construction Area) that was issued by the TGPCC. In addition, the *Sanxia gongcheng shigongqu huanjing baohu shishi guihua* (Implementation Plan for Environmental Protection of the Three Gorges project construction area) was issued jointly by the SEPA and the TGPCC of the State Council in 1995. The Plan mainly concerns measures for the protection of the water quality, atmosphere, noise, greenification, health, garbage disposal, and monitoring of the environmental situation in the construction area.[604] The *Implementation Measures for Environmental Protection of the Three Gorges Project Construction Area* is according to Chinese scholars the first document in China that formally requires environmental management and supervision.[605]

602 Such as the Zhongnan (South China) Survey and Design institute (kance sheji yuan) of the Ministry of Power that carried out research, and the Environmental Monitoring Station in Yichang city carried out investigation on the level of TSP in the construction area, as well as noise level. Zhongguo Sanxia jianshe nianjian bianjibu (1996), p. 128.

603 Ibid., p. 128.

604 The Plan was inspected and approved jointly by NEPA and TGPCC in December 1994 in the Three Gorges construction area. Ibid., p. 129.

605 Pan and Chen (1998), p. 12.

In both the construction area and the rest of the reservoir and the resettlement area, the existing national environmental laws and regulations are relevant. For instance, one important policy instrument is the policy of 'Three Simultaneouses' (*san tongshi*, discussed in Chapter 6) which is regarded and stressed by leaders as a management tool for environmental protection in the Three Gorges area. The policy requires that design, construction and operation of pollution treatment facilities be co-ordinated with design, construction and operation of the overall project. This policy is also stressed for all projects in relation to the construction of the Three Gorges project. The Implementation Plan for Environmental Protection of the Three Gorges project construction area is an important policy document in this respect, as all these construction projects include environmental aspects. For the Three Gorges reservoir area in general, the relocation of enterprises and the construction of new enterprises may provide an opportunity to install new and environment-friendly equipment. However, as discussed in Chapter 5, the new resettlement regulations 2001 state the need for the specific areas or enterprises to carry the financial burden of expansion themselves, and it is therefore questionable whether the funding potential exists for new environment-friendly technology in all new enterprises.

To further integrate environmental management for the entire Three Gorges area, *the Sanxia diqu jingji fazhan guihua gangyao* (The outline for the economic development programme for the Three Gorges area) was put forward in 1996 by the State Planning Commission. As economic development is expected to occur in the Three Gorges area, this Outline reflects concern on the part of the Chinese authorities for a need to protect the environment while developing the economy.[606] The Outline stresses the necessity for an overall plan for the area, with overall focus on the development of agriculture and industry.

Despite measures and policies for environmental management, there are prevailing problems. Even though the environmental management system and responsibilities have been made clear, the implementation of the system is still problematic. One reason for this is that the environmental management responsibilities have been placed with the contractors and the supervisory units of the project, but they have not yet provided specialised personnel who could carry out the task of developing the environmental supervisory system. The suggestions for environmental supervision are therefore difficult to carry out. Furthermore, projects that have been inspected have lacked the environmental installations. Attention to environmental issues on the part of the participating construction parties is lacking. However, there is no form for punishment mechanism, therefore making environmental protection management in the construction area more difficult than for instance in projects funded by international loans.[607]

606 Zhongguo Huanjing nianjian bianji weiyuanhui (1997), p. 74.

607 Pan and Chen (1998), p. 13.

NEPA's Role in the Environmental Process

The Three Gorges project will have an impact on the environment in the area and one reason is the large population to be resettled. This is a fact that is acknowledged by both officials and researchers. As the environmental agency, NEPA has been responsible for all aspects of environmental policy development and implementation in China. For the Three Gorges project, as was described above, NEPA began to set up a network in the early 1990s to monitor the situation in the Three Gorges area by using the EPBs. Inspection of the Three Gorges area has also been among NEPA's annual tasks, and shortly after the official construction of the project began, NEPA inspected the area (in June 1994). In addition to officials from NEPA, the inspection group consisted of experts, scientists and officials as well as media. The purpose of the tour was to inspect the pollution situation and the environmental management and supervision system in the Three Gorges area.[608] NEPA had also compiled several reports on different aspects related to the pollution problem and environmental problems in the area.[609] The Chinese Research Academy for Environmental Sciences (CRAES), a subordinate unit of NEPA, has carried out research about the ecology and resettlement of the project.[610]

As the agency responsible for environmental matters, it would be natural for NEPA to be heavily involved in the environmental policymaking for the dam project. Nevertheless, the impression from this study and interviews is that NEPA's authority during this period appears modest. Apart from co-ordinating the monitoring work and responsibility for publishing the annual environmental report for the project, it seemed to play a minor role and held little authority in the development and implementation of the Three Gorges Project environmental policy. This may not be surprising, as NEPA, due to its status as agency at the time, had not been a very strong organisation among the bureaucratic actors in China. Its main involvement in the dam project seems to have been in the preparation of the annual report, as well as monitoring the water quality in the river sections along towns and cities. Furthermore, this is an impression given the author by the scholars interviewed who have been involved in environmental research projects for the dam, and others who somehow have been engaged in the project. Nevertheless, it may be possible that NEPA has played a greater role behind the

608 Zhongguo Huanjing nianjian bianji weiyuanhui (1995), p. 164.

609 An annual report that was first published in 1994, *Changjiang Sanxia kuqu ji zhong, xia you shengtai huanjing zhiliang baogaoshu* (The Report Regarding the Ecological and Environmental Quality of the Three Gorges Reservoir Area and the Middle and Lower Reaches of the Yangtze River), and *Changjiang Sanxia kuqu wuranyuan paiwu zhuangkuang baogao* (The report on the State of Affairs regarding the Yangtze River Three Gorges Reservoir Area Pollution Sources and Discharges), Zhongguo Huanjing nianjian bianji weiyuanhui (1995), p. 164.

610 Interview no. 14 with SEPA and CRAES, September 1999. At the time of the interview, CRAES was applying for funding from the TGPCC to carry out ecological and environmental monitoring.

scenes than has been apparent, such as airing views regarding the *Statement*/EIA, and that NEPA's attitudes have possibly had some influence on the developments that will be discussed in the next section.

In sum, there are four issues of importance for the environmental developments for the dam project following the approval of the project, that have been discussed here. Due to a change in reality, i.e. the approval of the dam, a shift took place in the scientific focus. From a main focus where the proponents attempted to justify the dam, the project approval resulted in scientists exploring ways to minimise the potential environmental damage created by the dam itself and the resettled population. The second area concerns the environmental network that has been established among the many ministries to monitor the situation before, during and after construction. A third area is the beginning importance placed on environmental supervision and management, in particular in the construction area. Nevertheless, implementing the policy is hampered by the lack of specialised personnel. The rest of the reservoir area relies mainly on the existent environmental laws and regulations. The fourth area concerns NEPA's role in the project. NEPA was mainly involved in monitoring of water quality in the river sections along cities and towns, as well as being responsible for compiling the annual report. NEPA has carried out inspection tours to the area, but does not have political leverage in the environmental decision-making process of the project, and has too little funding to carry out important environmental protection tasks in the area.

Environment and Resettlement—a New Awareness? 1998-2001

In 1998, a new stage in the environmental history of the dam project commences. There are two important events that are interrelated, that may be regarded as the cause for this shift. One is the leadership change that took place in March 1998 when Zhu Rongji became Premier and the head of the TGPCC. This change is assumed to have had a certain impact on the Three Gorges project environmental policymaking, as discussed in Chapter 8 about the leadership change, media and changes in Chinese society. The second factor is the devastating floods that took place in the spring and summer of 1998 on the Yangtze River. As mentioned in Chapter 5, the floods in 1998 made the authorities issue a prohibition regarding tree felling on the upper reaches of the Yangtze River. Moreover, the floods were an important eye-opener for the Chinese authorities with regard to the erosion problems in the Three Gorges area and the large population to be resettled. Even graver erosion problems could be anticipated after squeezing rural population into areas that are already over-populated. When the decision to relocate a large number of the relocatees out of the reservoir was announced by the Premier, Zhu Rongji, the stated reason was environmental and ecological concern. He warned in his speech that relocating the people into the mountains along the river, pushing them up the hills (*houkao*), might have grave environmental consequences, as farmers would cultivate steep land and worsen soil erosion. This is a concrete example of how environmental concern has influenced the policymaking for the dam project. In order to understand the developments that have taken place in this period, it is necessary to take a look at some events that may have influenced the environmental decision-making process for the dam project, and that have linked the resettlement and environment issue areas in a new way. There are a number of factors that may have been important in raising the environmental awareness in the environmental policymaking process for the Three Gorges project.

The first factor is Zhu's apparent focus on environmental protection. Already during his first visit as Premier to the Three Gorges area in December 1998, Zhu signalled increased environmental focus. In particular he stressed the need to protect the environment in relation to the large-scale resettlement, thus linking the two very clearly. Other topics that were emphasised were the need to avoid creating new pollution sources, the protection of the forests, avoiding erosion, and control of agricultural activity.[611] Later, Zhu again stressed the environmental aspect in relation to resettlement regarding the decision to move 125,000 people out of the reservoir area.

A second factor that has been important in the process that needs to be mentioned is a petition by 53 experts in March 2000 that was written to President Jiang Zemin, NPC Chairman Li Peng, Premier Zhu Rongji and CPPCC chairman Li Ruihuan. The letter was written and organised by Lu Qinkan, who was formerly a CPPCC member, and who had also participated in one of the large feasibility studies organised by the MWREP that was completed in 1988. He was one of 10 experts who had refused to sign the study.[612] According to the letter, this was the third time that experts had expressed their concern to the leaders. They claimed that they had received no replies earlier (in 1998 and 1999).[613] The contents of the letter was therefore made known to Probe International and published on their website in order to bring about international attention.[614] No attention has been given to the petition in the state media. The experts were concerned about the potential siltation problem and environmental pollution in relation to the drainage outlets that would be submerged by the reservoir in Chongqing. The method planned to get rid of the silt in the reservoir is *xuqing paihun*, i.e. retaining clear water in the dry season and flushing out the silt in the flood season.[615] However, there is discussion regarding the potential impact this may have on the back end of the reservoir, where Chongqing is situated, should this method be unsuccessful.[616] A second serious problem from sedimentation concerns the reservoir's ability to

611 *Renmin ribao* (31 December, 1998).

612 See Dai (ed. 1989), p. 100, and Heggelund (1993), pp. 37-8.

613 This was negated in the official response. See Probe International (17 April, 2000).

614 Probe International (3 March, 2000).

615 The discussion about the siltation problem in the reservoir concerns the potential silt problems for Chongqing. For the discussion about the siltation issue, see for instance Dai (ed. 1989), pp. 56-7.

616 For instance research carried out about sedimentation shows the silting up of one of the biggest wharfs in Chongqing, Jiulongpo. See Probe International (3 March, 2000).

retain the flood water, as the silting up of the reservoir reduces the storage space. Moreover, the resettlement issue was one of the issues that was discussed in the letter, in relation to both the environmental capacity, silt deposit and the water level in the reservoir. The experts stated that the number of people who would need to be relocated (in addition to the 1.13 million) due to siltation may reach 300,000 in 20 years. A reply from the Three Gorges Project Construction Committee[617] was eventually sent to the experts, but it concluded differently than the expert group. They stated that the silt problems in the Three Gorges project would be solved, but noted that the experts' concern would be taken into consideration. Nevertheless, the letter may have had a certain impact on the decision-makers, as the petition group of 53 experts consisted of water and hydro experts, ecologists, engineers, and others.[618] Many of these experts had participated in earlier studies and evaluations of studies, and had not signed their names to the final *Statement*/EIA.

A third significant factor which put increased emphasis on the potential environmental damage from the reservoir was internal correspondence in the spring and summer 2000 between important actors in the environmental policymaking process: the Qinghua University professor Zhang Guangdou, Guo Shuyan, the director of the Three Gorges project office[619] and the Environmental Protection Bureau (EPB) of Chongqing municipality. Zhang Guangdou was the head of the group of 55 experts that investigated and approved the final *Statement*/ EIA for the dam project in 1992,[620] and had been an eager supporter of the Three Gorges dam for decades. In February 2001, the contents of the correspondence was made known by Probe International, and revealed that from April through June 2000, a discussion had been going on between the Three Gorges project leadership and the professor about the environmental situation in Chongqing municipality. The leakage disclosed that the Qinghua professor was seriously concerned about a potential environmental catastrophe in the Three Gorges area and the resettlement activities, which he had written about in a letter to the TGPCC leadership. The background for the letter was an inspection trip to Chongqing on 16-18 March, 2000 made by professor Zhang and Qian Zhengying, head of the Quality Inspection Group that was established by Zhu Rongji in June, 1999. In his letter, Zhang explains that he is very much concerned about the protection of the ecology and the environment in the area, as well as the potential for bringing the existing and future problems under control. The reason for his concern was a meeting with the director of Chongqing EPB, which had revealed the following: i) the Chongqing EPB had not known enough about the situation with regard to waste

617 Probe International (17 April, 2000).

618 For the complete list of names and backgrounds see Probe International (3 March, 2000).

619 This information was provided by Probe International, which had obtained leaked correspondence between the professor and the project authorities. See Probe International (14 February, 2001).

620 For the list of participants in the evaluation of the EIA, see Zhongguo Sanxia jianshe nianjian bianjibu (1996), p. 132-33.

water discharge in the reservoir area; ii) resettlement was implemented without paying attention to the protection of the environment, and the situation had even deteriorated due to the resettlement activities; iii) EPB had seemed more interested in the situation close to Chongqing (and did not care that much about the areas further down in the reservoir), and iv) they had demanded more funding to do environmental surveys in the area without even knowing the amount needed. Zhang was particularly troubled by the possibility of negative forecasting by the foreign media about the environmental problems in the Three Gorges reservoir, and he thus wished to point out the problems in order to avoid disaster. Following this letter, of which a copy with comments had been sent to the Chongqing municipal government, the vice-director of the Chongqing EPB, Yu Dengrong, wrote a letter (9 May, 2000) to the project authorities in order to defend the EPB. His main message was that there was not enough funding for extensive environmental efforts in the reservoir area, and the Yangtze River Water Resources Commission and Chongqing municipal government should be blamed for the lack of work carried out. Professor Zhang then (17 May, 2000) gave a statement to Guo Shuyan in person about the importance of the environmental action in the area, and mentioned that foreigners were following the developments closely. Following this, according to Probe International, Guo Shuyan passed Zhang's statement to Wu Bangguo, the vice-Premier and the vice chairman of the TGPCC, who immediately informed Zhu Rongji. Zhu decided to pass on the information to Li Peng and Lu Youmei, the director of the Three Gorges Project Development Corporation. The discussions that followed this round of communication is not known, but a possible policy outcome is discussed later in this chapter.

The media did not report anything about these internal discussions. However, on 21 September 2000, the article 'Sanxia kuqu wuran yanzhong huanbao zhihou' (The environmental pollution in the Three Gorges reservoir area is serious and stagnant), Chongqing youguan lingdao tichu yanli piping (Relevant leaders in Chongqing raise severe criticism) appeared in the *People's Daily Overseas edition*, and in *Chongqing chenbao* (Chongqing Morning News).[621] The environmental pollution problem was criticised for the first time publicly by the vice-mayor of Chongqing, Chen Jiwa, and the leaders of the Chongqing EPB, who had travelled to Fuling in the reservoir area to have a meeting with the leaders of 11 districts and counties of Chongqing. The contents of their criticism concerned the following: more than half of the resettlement construction projects in the area did not carry out the policy of *san tongshi* ('Three Simultaneouses'), which instructs the enterprises to incorporate environmental measures during design, construction and operation.[622] Some of the problems stated concerned the fact that 240 enterprises did not satisfy the discharge standards, and hospitals discharged wastewater directly into the river without any cleansing. The article also points to the fact that the SEPA leadership had been on inspection tours four times that year and had

621 See Renmin ribao haiwai ban (21 September, 2000). It was also available on the *Zhongxin Chongqing wang* (18 September, 2000).

622 Renmin ribao hai waiban (21 September, 2000).

criticised the environmental protection efforts in the reservoir area. CCTV had also exposed the lack of environmental efforts in the reservoir area.

Only one month later, in October 2000, *Guangming ribao* published an article with the unusually strong title: '*Sanxia kuqu huanjing baohu pozai meijie*' ('The need for environmental protection in the Three Gorges reservoir area is extremely urgent').[623] The author expressed grave concern about the lack of environmental protection measures for the area in surprisingly strong wording. He stated several examples of how bad the situation was, such as: Chongqing waste water reaches 1.185 billion tonnes annually of which industrial waste water constitutes 940 million tonnes and city waste water amounts to 245 million tonnes. Neither the industrial sector nor the city sector satisfies the national standards. The entire situation in the Yangtze river was depressing according to the author, who stated that 22 cities along the river had 394 waste water outlets, of which 30 percent had not reached the national discharge standards for waste. The discharge in the outlets that still had not reached national standards amounted to 20 billion tonnes per year, and was increasing by 2-3 percent per year. At that time, in the Chongqing area, the density of pollutants in the water had increased by 34 percent, while in Wanzhou it had increased by 573 percent. According to the author, the Yangtze River was becoming a 'public sewage channel' (*gonggong wushui gou*). Furthermore, in addition to the pollution issue, the author stressed the resettlement issue in relation to the environmental situation, as the resettled population would aggravate the erosion problem in the area by cutting down trees and cultivating the steep slopes along the river. The article stated that eroded land amounts to 63 percent of the total area in Chongqing municipality (each year 200 million tonnes of soil is eroded, of which 140 million tonnes enters the Yangtze river).

The process described above including the petition, the leakage of information from professor Zhang's letter, the letter from Chongqing EPB and the following actions from China's top leaders, the critical articles in September and October 2000 by Li Shijie, may all have been important factors for a decision to increase environmental spending in the Three Gorges area. On 3 January 2001, it was announced in the media that the Chongqing municipal government had approved spending RMB 44.48 billion to protect the environment in the Three Gorges area in the next ten years.[624] Furthermore, the Chongqing government had approved a programme for Chongqing, *Chongqingshi Sanxia kuqu huanjing baohu he shengtai jianshe guihua* (Plan for environmental protection and ecological improvement for the Three Gorges reservoir area in Chongqing municipality). According to Chongqing EPB,[625] the funding for the environmental clean-up would come from the central government as well as from Chongqing municipality. Results from the increased funding would be 34 waste water treatment plants to be constructed in the Chongqing municipality by 2005.[626] Thus, one may conclude that the

623 Li (30 October, 2000). According to sources, the author is a scholar at CAS.

624 Cai (3 January, 2001).

625 Personal communication April 2001.

626 Interview with Chongqing EPB, April 2001.

environmental discussion about the environmental problems in the Chongqing part of the reservoir area has had a positive effect on the area, and the increased spending on the environment will be beneficial for the resettled population.[627]

Summing up

In the early discussions of the Three Gorges project, little attention was paid to the environmental issues related to the construction of the dam, or to the relationship between the environment and resettlement. In the 1980s, the debate about launching the Three Gorges project intensified, and numerous studies were carried out. 1984 was the year when the government really began to pay attention to the environmental issues of the Three Gorges project, when CAS initiated important studies about the project's environmental impacts. The study was significant due to its comprehensive character and thoroughness with regard to the environmental impact and resettlement. This was a new approach compared to research that had been carried out by other institutes, where the research agenda was more or less set. New issues appeared as a result of the research carried out that made clear the need for further studies. One might therefore conclude that researchers played an important role in setting the agenda for the environmental discussion for the project. For instance, the Three Gorges environment was included as a research topic in the Seventh Five-Year plan (1986-1990).

In 1991, CAS and MWR completed the compilation of the *Statement*/ EIA for the dam project. Despite the differing opinions in the conclusion by CAS and MWR, and the fact that several experts among the 55 experts who evaluated the *Statement*/EIA did not sign, NEPA approved the *Statement*/ EIA in 1992. There are several points to be noted from the research process leading up to the approval. Construction and design units under the MWR have carried out EIAs for the project, a fact which can lead to questioning the objectivity of the studies. Therefore, the NPC Environment and Resources Committee put pressure on the MWR to do further studies for neutrality and quality purposes. Institutes under CAS and MWR therefore compiled jointly the *Statement*/EIA for the dam project, which was based on earlier research carried out by these two. The State Council approved the project before the *Statement*/EIA was approved, which can lead to

627 Regarding the environmental spending and funding for the dam project, the author asked the various officials interviewed about the figure for environmental spending for the project. However, it was impossible to obtain an answer as to how much is being invested. One of the problems seemed to be that the spending for the environment is divided into a number of different budgets. Also, defining what is environmental spending/costs was also a problem. According to one official, the funds for dealing with erosion, for instance, is included in the investment for the project proper, not as an environmental issue. Part of the spending for the environment will come from the resettlement funding, as was indicated in the 2001 resettlement regulations.

questioning its significance and the seriousness of the Chinese government at the time with regard to the environment and resettlement problems. Knowing that objective reporting has been difficult in the process, and that scientists from CAS were criticised because their scientific view was interpreted as a personal view, it is significant that a few scientists did not sign the final evaluation of the *Statement*/EIA.

Succeeding the project approval by the NPC in 1992, a new era began. Four issues that have been important in the developments following approval of the dam are: i) A new reality resulted in a shift in scientific focus. The debate about the project leading up to the 1992 approval was oriented towards positive and negative aspects of the dam project, including the environment and resettlement. After the project had been approved, the purpose of discussions and research was to minimise environmental degradation and propose counter-measures for environmental damage. The establishment of the TGPCC in 1992 has resulted in most research applications going through the TGPCC, and it has to a certain extent become a control organ before publication. Numerous institutes and organisations are involved in research for the project, and co-operation is problematic at times. ii) An environmental network was set up with participation from a number of ministries such as NEPA, MWR, Ministry of Agriculture and others. NEPA initiated an annual report about the environmental situation in the Three Gorges area with input from the participants in the network. iii) Environmental supervision and management have become increasingly important with regard to environmental protection, in particular in the construction area. Environmental management regulations have been issued for this area. Despite these regulations, problems exist with regard to implementation and specialised personnel. iv) NEPA's role in the environmental process has been modest, and its main involvement appears to be through inspection tours to the area and compiling the annual environmental report as well as monitoring water quality for the project. This is related to NEPA's lack of strength in comparison with other ministries. Nevertheless, NEPA's perspectives towards the environmental issues of the Three Gorges project that have occasionally been reported, may have had some impact on the decision-making process.

In 1998, a new stage in the environmental developments of the dam project began, when Zhu Rongji became Premier and head of the TGPCC. When the resettlement policy change took place in 1999 due to the environmental capacity in the reservoir area, resettlement and the environment were linked together in a new way and the issue of mutual dependency was lifted up to the highest political level. Furthermore, several factors have been important in influencing environmental policy developments for the project: i) Devastating floods in 1998 made the government realise the need to curb erosion along the Yangtze River; ii) A petition written by 53 experts to Chinese leaders pointed to siltation problems as well as environmental and resettlement problems; iii) Internal communication between a Qinghua professor (head of the group of 55 experts that investigated and approved the final *Statement*/EIA for the dam project in 1992), Chinese leaders and the EPB of Chongqing municipality was leaked to Probe International regarding the potential catastrophe for the environment and resettlement activities in Chongqing municipality. Furthermore, significant environmental criticism was published in two major newspapers in China: environmental concern was illustrated in the *People's Daily Overseas Edition* when the Chongqing vice-mayor and the city's EPB criticised the environmental work being carried out, and a bleak picture of existing and future problems was drawn in the *Guangming ribao*. All of the above factors may have led to the increased environmental funding for Chongqing, which will be beneficial for both the environment and the population living in the area.

The process from the 1970s until 2001 illustrates that positive environmental developments have taken place for the Three Gorges project. It has revealed an increased emphasis on comprehensive environmental research for the project, and science has become increasingly important for the project's decision-making. Nevertheless, it must also be noted that where politics and science have contended about the environmental policymaking for the project, the political forces involved were stronger than science. Dissenting scientific views were interpreted as personal views, which indicates the dilemma in carrying out scientific research for the Three Gorges project.

Chapter 8

China's Changing Shape of Decision-making and the Three Gorges Project: Interaction between Leadership, Knowledge and Media

This chapter will discuss a factor that may have contributed to the resettlement policy change for the Three Gorges dam project: the changing shape of decision-making in China. Chinese society has been changing in the 1980s and 1990s. It has become an increasingly pluralistic society, where diversified opinions may be aired in scientific journals and to some extent in the media. Decision-making in China today differs greatly from the Mao period when communist ideology in principal was the foundation for decision-making. The present trend seems to be that decision-making increasingly is based upon information and knowledge. New institutional actors have thus appeared on the scene, as consultative organs are on a frequent basis included in the policymaking process.[628] How then, does this relate to the Three Gorges dam project? This chapter intends to illustrate that the developments in Chinese society portrayed above have been important for the issue being discussed in this book: the resettlement policy change. Factors to be discussed in the chapter that may have been important for the policy change development are: change of leadership (Premier) in China in 1998 and the subsequent change in the Three Gorges project leadership; the increase of information provided to the leadership; and the media coverage in the late 1990s following the leadership change. I presume in the chapter that the interaction between these factors, the leadership, knowledge-based decision-making and the media, has had an important impact on the dam project, and may have made the resettlement policy change possible. The issues above are all interlinked, and are all part of the developments in Chinese society. Thus, they will be discussed as components of a complex picture, where one is dependent upon the other, and vice-versa. This chapter, in discussing the above-mentioned developments in relation to the dam project, will present examples of how this change has manifested itself.

First, a general description will be given of the political changes that took place in Chinese society from the late 1970s until present. Two particular periods when discussions about political reform took place will be focused on, since these

---

628 See Shambaugh (2001), p. 103.

periods also had an impact on the developments of the Three Gorges project. Also, a description of the background for the recent developments of the Three Gorges project decision-making will be given. The purpose of the historical review is to understand the dynamics in Chinese society, as some of the policy decisions that were made in the 1980s have been important for the development of the Three Gorges project. Thus, it is necessary to study the developments over time in order to be able to place the dam project into a larger context.

Secondly, the role of scientific debate in relation to the Three Gorges project resettlement policy change will be looked into. Scientific research results and knowledge appear more important as a basis for decision-making in China today. This trend was initiated in the 1980s and blossomed under Zhao Ziyang. Since then, it has gradually developed. This development may have had a certain impact on the Three Gorges project policy change in 1999. The scientific debate about the project resettlement and environment is discussed in the chapter, and its potential impact on the resettlement decision-making. Examples from various journals are given in order to illustrate that the areas of concern among scholars in China may have been heeded.

Third, in addition to the knowledge-based decision-making, I assume that the change of leadership, i.e. Premier, has directly influenced the Three Gorges Project resettlement policy change. I assume that the new leadership, in particular Zhu Rongji, may be more willing to pay attention to advice, and base decisions on these. Construction began on the Three Gorges dam in 1994. The resettlement plan follows the construction schedule, and is divided into three phases, the first phase ending in 1997 with the closure of the river. Five hundred and fifty thousand people are to be resettled in the second phase, which ends in 2003. From 1994 to 1997, under the leadership of the then Premier Li Peng, the resettlement seemingly proceeded according to the original plan, without major changes and few problems reported. Nevertheless, after Zhu Rongji became Premier and responsible for the dam project, certain changes took place. This chapter will discuss the importance of the leadership change in relation to the resettlement policy change. The chapter also discusses the possibility of Zhu Rongji being more open for scientific advice than Li Peng.

Finally, the chapter will aslo reflect upon the possible effect of the new leadership on the information given about the resettlement issue, and the project itself, in the state media. I assume that one indication of leadership change may be the changed reporting in the media about a number of topics related to the dam project. Furthermore, it is assumed that the conclusions from scientific debates that have been largely restrained to scientific journals, have to a greater extent than earlier been reflected in the media. The media will therefore be used to illustrate the effect of the leadership change on the Three Gorges resettlement policy.

The Political Developments in Chinese Society from the 1970s to the Present

After Deng Xiaoping took over in the late 1970s, the political message of the time was to restore the country back to normal from the chaotic years of the Cultural Revolution (1966-76). One topic that Deng Xiaoping stressed following the experiences of the Cultural Revolution, was political reform. Furthermore, separation of responsibilities between the party and the government was part of this plan. His main rationale at the time was that the party should take care of political leadership, and 'should be aquainted with government work and check up on it, but should not attempt to take it over'.[629] In 1978, a reversal of the verdict of the 1976 demonstrators (to mourn Zhou Enlai) took place, and thousands of Chinese began to put their thoughts, poems and essays down on paper, which were posted on the Democracy Wall in Beijing.[630] One of the most prominent writers was Wei Jingsheng, who proposed that until China embraced a fifth modernisation, i.e. democracy, the other four would not succeed.[631] In addition to the posters on the wall, several small-scale demonstrations took place in Beijing that asked for democracy and human rights. Finally, the government decided to crack down on the movement, as it was seen to have been 'challenging the fundamental premises of the CCP itself'.[632] The most vocal participants of the democracy movement were arrested, among them Wei Jingsheng, who was arrested in March 1979 and sentenced to fifteen years in prison. Deng's talk of political reform eventually died down in 1981; the democracy wall movement was one factor that influenced such a development.[633]

China began a second wave of political reform in the mid-1980s when the request for reform was raised among intellectuals, students and local government leaders.[634] Intellectuals and students called for freedom of the press and freedom of speech. Some of the pressure came from local leders who believed that the lack of political reform hindered economic development. The leaders at the time, Deng Xiaoping, party general secretary Hu Yaobang and premier Zhao Ziyang also believed that without political reform and the separation of Party and government, China's economic reform would not proceed smoothly. Deng's wish for economic growth was the main purpose of his support for political reform. Even though he conceded to some political relaxation, Deng was not interested in full democratic reforms. Zhao Ziyang believed that a certain amount of political liberalisation and

---

629 Chen (1995), p. 134. Quoted from a speech by Deng Xiaoping at the opening ceremenoy of the National Science Conference, March 18, 1978.

630 To the west of the Forbidden city in Beijing.

631 The four modernisations in industry, agriculture, science and technology and national defense.

632 Spence (1999), p. 628.

633 Other factors were political difference within the Party regarding political reform, as well as a priority shift towards the reconstruction of the economy. Chen (1995), pp. 137-38.

634 Ibid., pp. 139.

transparency (*toumingdu*) was necessary to pave the way for economic reform. Deng and Zhao worked closely with the work for political reform, and Deng had intended to formally put political reform on the political agenda in 1986 (of the Central Committee meeting that year). A Political Reform Research Group was established (with a Political Reform Office under it) consisting of five top leaders including Zhao Ziyang. In 1987 the political reform process met with setbacks, as student demonstrations took place in 1986/87, and Hu Yaobang was forced to resign as party general secretary. A campaign to fight 'bourgeois liberalism' (*zichan jieji ziyouhua*) was initiated in early 1987 by conservative leaders such as Li Peng. Through Zhao's efforts, the campaign was confined to the party, and was ended in late 1987. 'The General Program for Political Reform' was approved by the CCP Central Committee in 1987 and incorporated in Zhao's report to the CCP National Congress in 1987, which made the political reform decision official. In his report, Zhao also highlighted three important principles for media coverage where the press should exercise supervision of the work and conduct of officials: supervision by public opinion (*yulun jiandu*), the press should inform the public about important events (*zhongda qingkuang rang renmin zhidao*) and also discuss important events with the people (*zhongda wenti jing renmin taolun*). Nevertheless, the implementation of the political reform ran into obstacles, as the reforms were perceived as a threat to leaders' power, and implementation was more or less stopped in 1988. The Party changed its policy on media and reform in late 1988, when a retrenchment of the economic reform programme occurred.[635] With the crackdown of the student demonstrations in June 1989, the conditions for media and reporting became even worse.

As is described in the history chapter, the late 1980s may be regarded as a relatively open period in China's modern history. At the time, Zhao Ziyang was General Secretary of the Communist Party, and he promoted political reform and supervision by the public. This was of importance in the debate about the Three Gorges project, which was particularly heated in 1988. It then reached a peak in the period leading up to the NPC in 1989, when the Three Gorges project was to be decided upon. One of the books against the project published at the time, *Changjiang Changjiang*, uses the principles of informing and discussing important matters with the public introduced by Zhao Ziyang to illustrate the need to discuss the Three Gorges project.[636] Negative aspects had been effectively silenced in the state media in the process leading up to the 1989 NPC. The announced postponement until 1995 of the construction of the dam illustrates temporary success for the individuals opposing the project. Nevertheless, following the 1989 crushing of the demonstrations in Beijing, the critics of the dam project became silent. Zhao Ziyang had to leave his post as CCP General Secretary because he failed to accept the view of the eight Party Elders,[637] and because he had appeared

635 See Lee (ed. 1990), pp. 42-3.

636 Dai (ed. 1989), on the third page of the February, 1989 version (first edition).

637 See Nathan and Link (2001), pp. 256-64, and pp. 268-72. A meeting was held on 21 May by Deng Xiaoping, Chen Yun, Li Xiannian, Peng Zhen, Deng Yingchao, Yang

sympathetic towards the demonstrations. Proponents of the dam project used the opportunity to launch the project again in 1991, by using the floods that year to put the flood control project on the agenda again. Li Peng, then Premier and head of the TGPCC was instrumental in pushing the project, which resulted in approval by the NPC in April 1992. When the official construction of the dam began in 1994, it was accompanied by positive media reports about developments, often with photographs of a smiling leader at the dam site. The closure of the river took place in November 1997, which ended phase one of the construction. During this period, one may say that the trend in the media reporting about the project was mainly positive and optimistic, praising the advantages of the project.

Since the crushing of student demonstrations in 1989 and the approval of the dam project in 1992, Chinese society has gradually been changing. The early 1990s were still affected by the happenings in 1989, and experienced a rigid political climate. Eventually, a process of gradual relaxation has occurred and the recent developments of the Three Gorges Project are taking place within a society that is becoming increasingly pluralistic and where the role of information is growing in importance. Despite control and a complicated legal framework for publications, a constant flow of new books, newspapers and magazines are appearing that are not mouthpieces of the CCP.[638] They report more freely on matters of importance than for instance newspapers such as the *People's Daily*. Thus, one may conclude that the Party's control of the media has declined in China, which has resulted in reduced control over newspapers and magazines. Even though the CCP propaganda apparatus still exercises considerable control over what is printed in the media, it is no longer entirely monopolistic.[639] One such example is the *Nanfang zhoumo* (Southern Week-end), a newspaper that often challenges the official version and is known for its candour in discussing negative trends in developments in the Chinese society. In addition, books are being published where crtitical topics on the Chinese society are discussed, such as the *Shendu youhuan* (Profound hardship) and *Xiandaihua de xianjing* (The Pitfalls of Modernisation) which are part of a series of books called 'China's Problems'.[640] The first book discusses China's existing environmental problems and the causes quite openly. The second book discusses topics such as the economic politics, 'underground economy' ('*dixia jingji*'), who are the winners and losers of economic reform, the

Shangkun, Bo Yibo and Wang Zhen where it was discussed whether Zhao Ziyang had committed a violation of Party discipline and the Elders decided that Zhao had to be replaced. Zhao Ziyang was relieved of his duty as General Secretary on 22 May, p. 264.

638 Three bureaucracies control publications: China's Press and Publications Administration, the State Council Office of Information, and the Ministry of Propaganda.

639 The apparatus consists mainly of CCP Propaganda Department, the Central Leading Group on Propaganda and Education, the Central Leading Group on Foreign Propaganda, and New China News Agency. Shambaugh (ed. 2000b), pp. 178-9.

640 Zheng and Qian (1998); and He (1998), both published by Jinri Zhongguo chubanshe (China Today Press).

Chinese countryside and the situation for the peasants, pluralism in Chinese society and the potential for an emerging civil society (*gongmin shehui*).[641] Many intellectuals engage actively in an intellectual discourse about China's reforms. Their views are independent, as many are no longer employed by the state, as well as diversified.[642] Moreover, TV stations also engage in critical reporting, for instance about the environment, and pollution incidents are reported on national TV, such as the *Zhonghua huanbao shijixing* (the China century information campaign on the environment) described in Chapter 6. There are channels for citizens to write complaints on environmental matters. SEPA for instance, receives a number of complaints each year. Hotlines have been established in many cities in China; residents can make complaints or discuss environmental problems.[643] Officials participate on radio programmes to directly answer questions from the public. Access to the Internet and to news from outside China provides the Chinese public with diversified information. The number of Internet users is growing rapidly and is expected to reach 200 million users by 2005.[644] Internet discussion sites are established, and even the *People's Daily* has an Internet discussion site, *Qiangguo luntan* (Powerful Nation Forum) where heated and open discussion takes place among anonymous participants about various topics.[645] The discussion site has become a 'platform for an unprecedented wave of criticism of officialdom in China'.[646] The newspaper also routinely provides links to the speeches and writings of China's ministerial and provincial leaders, thus increasing the transparency of views of individual leaders.[647] Social organisations and NGOs have been established in several areas, including the environment. The participants in these groups are few in comparison to the Chinese population, but their significance in a changing society should nevertheless not be overlooked.

---

641 See He (1998), p. 323 for a list of ways for people to enrich themselves, such as smuggling and narcotics. He points out the many ways by which people may enrich themselves illegally, such as 'insiders' that are able to obtain 'black income' (*heise shouru*) through stocks, etc.

642 The views are different and not dichotomous, for or against the reform; rather, the intellectuals discuss the priorities, purposes, promises and pitfalls of reform. See Li (2000a), pp. 124-5.

643 In Dalian, Liaoning province, a 24-hour telephone hotline was installed. Furthermore, the city has a radio talk show to discuss environmental concerns. Ma and Ortolano (2000), p. 71.

644 The Internet has become an important source of information for the Chinese and has grown quickly. In 1999 there were 8.9 million Internet users; by June 2001 the number had reached 26.5 million users. Zhou (7 November, 2001).

645 http://bbs.people.com.cn/cgi-bbs/ChangeBrd?to=14. Examples of some topics that are discussed are the Taiwan issue, how the WTO affects China, the Beijing Olympics in 2008 and the financial consequences. Users also attack corruption, bad governance and social inequality.

646 See Gilley (2001b).

647 Li (2001), p. 18.

The developments mentioned above are changing the Chinese society, and are fostering a society that differs much from the society under Mao Zedong for instance. Nevertheless, the positive trend is also countered by frequent setbacks, and China is still far from being a society where one is free to openly discuss and criticise all topics. Political reform, for instance, is not a priority—although it is being discussed.[648] As is commonly known, it is impossible to criticise the CCP, challenge its power, or participate in political activities other than those in the name of the CCP. The 16th Party Congress that convened in November 2002 made Chinese authorities wary of criticical voices, and they put restrictions on the media by issuing orders through a party document to tighten mainland censorship leading up to the congress.[649] Furthermore, a popular Internet bulletin board for journalists was shut down recently because it allegedly leaked secrets, slandered state leaders and attacked government bodies.[650] The flight by He Qinglian, the author of the *Pitfalls of Modernisation*, is another example of how difficult it is to write about critical issues in Chinese society.[651] Moreover, several editors from *Southern Weekend* (*Nanfang Zhoumo*) were dismissed in 2001; the newspaper had reported on sensitive issues such as the official abuses in China, the spreading of AIDS and organised crime as well as irregularities in relation to the Three Gorges project.[652] Criticism of the Three Gorges project is difficult, as has been described in several chapters. One may conclude that whatever criticism may exist in China, it is never aimed directly at the CCP. It is important to acknowledge that the positive developments in certain areas of Chinese society occur at the same time that serious tightening of control in other areas takes place. The positive trends are also a part of the picture and need to be mentioned when discussing developments in Chinese society. It becomes particularly important since the trends may have impact on the decisions being made by Chinese leaders.

Increased Importance of Information as Basis for Decision-making

One assumption in this book based on interviews and literature,[653] is that information and knowledge have become more important for the Three Gorges

---

648 See Gilley (2001c). One important issue being discussed is the expansion of the CCP to include different groups in society in order to become more representative of the country.

649 Chan (27 June, 2001).

650 Reuters (18 October, 2001).

651 He Qinglian fled to the United States on 14 June, 2001, following raids of her apartment by the police. Gilley (2001a), p. 15.

652 Probe International (11 June, 2001). In her book, He Qinglian refers for instance to many articles from the *Southern Weekend* about the Chinese countryside and peasants.

653 Interviews no. 24, Nanjing, 1999, and no. 16, Beijing 1999. Li (2001); Xiao, Zhengqin (1999).

decision-making process, as is the trend in Chinese politics in general. When Deng Xiaoping took over as leader in China in the late 1970s, science and technology were given prominent roles in the future development of the country as one of the four modernisations of the country: agriculture, industry, national defence, science and technology. Furthermore, science was given a key role in achieving the other three modernisations.[654] Thus, science was to become both a tool and a basis for decision-making in the years to come. In recent years, policymaking in China has become more consultative under Zhao Ziyang, Jiang Zemin and Zhu Rongji. This development is according to some scholars (such as Lieberthal, Oksenberg and Lampton) a result of the economic reform process that has made consensus building necessary, and has created a realisation of the need for more information to make qualified decisions. This has resulted in increased importance for and involvement of research and policy institutions, which have also become involved in the policymaking process. The background for this development is that in the 1980s, a gradual change was taking place in China. Professional groups such as economists and scientists were included at a greater rate in the decision-making process in the country in general. Moreover, these were young and professional people. There was a recognition in this period of the need to obtain information, advice and support from key sectors of the population.[655] Thus, an increase in the inclusion of social scientists in policymaking in China is taking place, and leaders establish advisory bodies in the areas where they are most interested in receiving advice.[656] During the 1980s a set of research groups was established under the Premier's office to provide policy advice on topics such as economics, technology and foreign affairs.[657] These centres again relied on information from relevant researchers and staff from outside the centres such as CAS, the ministries, CASS, the universities and professional associations. Halpern argues that professional associations that concern themselves with various aspects of economics, for instance, have multiplied greatly in post-Mao China. Also, interaction between communities exists through meetings and journals. The established research centres under the State Council have for instance brought economists into regular contact with economic bureaucrats.[658] Also, initiatives taken by researchers have also influenced policymaking, such as in the case of China's accession to the Montreal Protocol.[659]

654 See Miller (1996), p. 72.

655 O'Brien (1990), pp. 176-7.

656 Halpern (1988), p. 228.

657 Halpern (1992), pp. 130-32.

658 Halpern (1988), p. 230.

659 China's accession to the Montreal Protocol for instance, was initiated from the scientific community. Professor Tang Xiaoyan at Bejiing University wrote a paper on the depletion of the ozone layer that received leadership attention. Eventually, the interest of the SSTC was evoked, and China signed the agreement in 1991. See Oksenberg and Economy (1998), pp. 29-30; and Ma and Ortolano (2000), p. 18.

For instance, both Zhao Ziyang and Zhu Rongji set up and apply think tanks in order to receive advice in policymaking. One of the think-tanks established in 1982 by Zhao Ziyang, *Guojia jingji tizhi gaige weiyuanhui* (the State Committee for Restructuring the Economy) became one the most important think-tanks for Zhu Rongji, and has been responsible for planning China's economic reforms. It was demoted to an office at ministerial level (*zheng buji*) in 1998 by Zhu Rongji, and was named *Guojia jingji tizhi gaige bangongshi* (the Office for Restructuring the Economy).[660] Under Zhu, the Office became a consultative-type organ consisting of ministers, and an organ to co-ordinate the ministries.[661] The *Guowuyuan yanjiu shi* (State Council Research Office) is the most important think-tank providing research to the Premier, according to Xiao Zhengqin. The tasks of this office are in addition to research, to draft reports and speeches, propose research topics for the Premier's, vice-Premiers' and state councillors'consideration.[662] Both the Office for Restructuring the Economy and the State Council Research Office are working bodies under the State Council. The Chinese Academy of Sciences, with Lu Yongxiang as President, is one of the most important think-tanks for Zhu Rongji and the State Council, as well as the central authorities in general. CAS is the organisation that carries out scientific research on the highest level in China, and in this way differs from the ministries, which also provide advice to the leadership. As is discussed in Chapter 7 about the Three Gorges environmental developments, CAS has had the main responsibility for Three Gorges related environmental research for a number of years. After Zhu became Premier, CAS has increased its importance because Zhu apparently places more importance on science and technology than any other Premier.[663] Moreover, Jiang Zemin relies on think tanks in the decision-making process to a greater extent than did Deng Xiaoping.[664] The

660 During the 1990s, Zhu Rongji has established two offices in addition to the Office for restructuring the economy, such as the *Guowuyuan shengchan bangongshi* (the State Council Production Office) in 1991, and the *Guowuyuan jingji maoyi bangongshi* (the State Council Economic and Trade Office) in 1992, see Li and Xiao (eds. 1998), pp. 10-17.

661 The office was headed by Wang Qishan, who is presently acting mayor of Beijing. See Zheng (2000), p. 77; and Xiao, Zhengqin (1999), pp. 227-35, about Wang Qishan's background. For more information on the work of the office see www.chinaonline.com/refer/ministry_profiles/c00121168.asp (The site names the previous head, Liu Zhongli as head of the the Office; however, Liu became vice-chairperson of the SDPC in September 2000.)

662 The research office is divided into eight departments with responsibility for research of specific topics such as economy, agriculture, industry, education, science, etc. Xiao, Zhengqin (1999), pp. 301-2. Researchers must be party members, since they will write drafts for the Party.

663 Ibid., pp. 314-15. CAS is an institution on ministry level, directly under the State Council.

664 Li (2001), pp. 18 and 235; and Xiao, Chong (1999), p. 299. These books state that for instance Liu Ji, former vice-president of the Chinese Acadamy of Social Sciences and advisor to Jiang Zeming, praised and endorsed the publication of He Qinglian's book *The Pitfalls of Modernisation*. He also invited He Qinglian to Beijing to discuss the

Chinese Academy of Social Sciences (CASS) is an institution that gives important policy advice to China's leaders. The fact that former politburo member Li Tieying was the President of CASS may have contributed to greater importance being put on social sciences in China.[665] This development has had important implications for the decision-making process in China, as feasibility studies and advice are sought out. Thus, there is actually a possibility for research centres and institutes to have impact on the policy decisions that are made at the highest levels.

From the above, one may conclude that the present Chinese leadership seeks advice and information in ways that differ from the Mao period. This may have positive influence in the sense of improved government, albeit not necessarily increased democracy. One important factor for the increased focus on information may be the change in the educational background of leaders. Zhu Rongji, Li Peng and Jiang Zemin all belong to the third generation of leaders, and all have technical backgrounds (engineers).[666] A main difference between the revolutionary leaders of the first and second generations and the technocrats of the third generation is that they are problem-solvers by training; 'they focus on problems not "isms"'.[667] Moreover, the educational background of the fourth generation of leaders is an important change from the third generation of leaders. An increasing number of financial experts and lawyers who belong to the fourth generation are included in important posts in China.[668] Zhu Rongji's portfolio in Chinese politics has been economic and financial issues, and the majority of the members of Zhu's think-tanks are economists.[669] One important leader from the fourth generation is Hu Jintao, Politburo standing committee member, who is now the CCP Secretary General position after Jiang Zemin 'retired' in 2002. Others are NPC chairman Wu Bangguo, and Premier Wen Jiabao, both of them politburo members. Wu Bangguo

---

book with her. Liu Ji is according to Li a 'chief adviser' to the China Today press that published He's book.

665 Halpern (1988), pp. 228-9, states that among social scientists, economists and foreign affairs specialists have been given an institutionalised presence in China. This situation may have changed somewhat, since the increasing number of advisers participating in providing input to the policymaking process have backgrounds in law and political science, in addition to economy, foreign affairs and technology. Halpern writes that sociologists are less likely to have an impact on policymaking.

666 Li (2000b), pp. 4-5. A political generation is often defined as a group of birth cohorts whose combined length approximates 15-22 years, as well as participation in similar historical and social circumstances.

667 Fewsmith (2001b), p. 84. Li Cheng describes the Qinghua University as one of the most important educational and political establishments in post-Mao China. Li states that political networks have been established at Qinghua that are important for policymaking in China. See The 'Qinghua Clique', in Li (2001), pp. 87-126.

668 Several of Jiang Zemin's top aides are trained in law and social sciences, such as Wang Huning, political scientist and former dean of Fudan University law school and Cao Jianming, law professor and vice president of Supreme People's Court, Li (2001), p. 225.

669 Xiao, Zhengqin (1999), p.V. One example is Wang Qishan who heads the Office for Restructuring the Economy.

is also a Three Gorges Project Construction Committee (TGPCC) vice-chairperson, and thus works closely with Zhu on the dam project. According to some scholars, the rise of the fourth generation of leaders is even more evident at the ministry and provincial levels.[670] The fourth generation group is also more diversified in terms of experiences, politics, ideology, and occupational background than the third generation of leaders. An important change in post-Mao politics in general is the increase in political elites with higher education, in particular in engineering and natural sciences. Approximately 75 percent of the fourth generation of leaders have university education. In the Ministry of Finance for instance, the average age of division and bureau heads is 44.4.[671] Furthermore, some of the leaders have education from European or North-American universities, as opposed to the third generation of leaders who were educated in the former Soviet Union, and whose experiences are greatly related to their participation in the Long March or the Anti-Japanese War. In sum, the increased use of think-tanks and diversified backgrounds has changed the way decisions are made in China. This has also had impact on decision-making for the Three Gorges dam project, as will be discussed in the next section.

Basis for Decision-making—Three Gorges Project Scientific Research and Publications

In the preceding sections we have seen that knowledge and scientific research results have become increasingly important for the decision-making process in China. This has to do with developments in the Chinese society as well as certain leaders' emphasis on scientific research as a basis for decision-making, such as Zhu Rongji. The emerging trend in Chinese society seems to be that officials in general increasingly base political decisions on knowledge, instead of making decisions for the sake of ideology.[672] With regard to the Three Gorges project, interviews as well as written materials leave an impression that Zhu Rongji and his leadership to a greater extent than the former Premier Li Peng, did pay attention to information from research reports and reporting 'on the spot' by local officials, or to resettlement work meetings.[673] One may conclude that Zhu is perceived as a

670 Li (2000b), p. 2. Central authorities have reportedly issued a new rule stating that all provincial Party standing committees must have at least three members in their 50s or younger, as well as at least two people in their 50s or younger who are governors and vice-governors.

671 Ibid., p. 19.

672 The Maoist era was characterised by politics organised around ideological lines, while during the Deng era ideological polarisation was repudiated, and commitments were to 'bureaucratic rationalization and reform', see Dittmer (2001), pp. 57-8; See also Fewsmith (2001a), p. xv, that stresses the importance of ideology in the political system, and that 'changes in policy had to be justified by changes in ideology'.

673 I will not go into the reporting matters or the meetings in this book. However, it is clear that a combination of written and oral reporting from ongoing resettlement must

leader who is more receptive to results from research reports, which is relevant for the Three Gorges resettlement policy change. In order to set the discussion about the potential impact of research and information into a context, it is useful to give some examples of the types of articles by researchers that may have been instrumental in prompting the resettlement policy change.

Examples of Articles in Academic Journals by Researchers

In addition to the research reports described in Chapter 7, numerous articles about the Three Gorges project have been published both in the 1980s and 1990s in newspapers, books and magazines. As has been mentioned earlier, most of the media coverage leading up to the decision to launch the project in 1992 was supportive of the construction. This was criticised by the project opponents as one-sided and 'speaking with one voice'[674] in a number of books that were published in the fairly open period of the mid- and late 1980s. Examples are *Changjiang Changjiang* (Yangtze, Yangtze), *Lun Sanxia gongcheng* (On the Three Gorges project), *Lun Sanxia gongcheng de hongguan juece* (A Discussion of the Macroscopic Decision-making of the Three Gorges Project), *Zai lun Sanxia gongcheng de hongguan juece* (Another Discussion of the Macroscopic Decision-making of the Three Gorges Project).[675] These books with contributions from scholars, CPPCC representatives and officials are critical towards the decision-making process and towards the promises of the dam project. In addition, the dam project has been the research focus of many scientists throughout China for decades, and their articles have been published in scientific journals. These articles have been based upon scientific research reports in the Three Gorges project area both in the 1980s and 1990s. The scientific articles are often more varied in their description of the effects of the dam on the Three Gorges area than were the newspaper articles, both with regard to the environment and the resettled people. These articles often describe the existing or potential problems in the TGP area, and come up with potential solutions to the problems. In addition, in some cases the author's critical view is reflected clearly, and the articles are quite outspoken towards a number of issues of the project. Below is a brief introduction to a few publications that frequently write about the Three Gorges project. My purpose for giving an account of these is twofold:

- To illustrate that a debate among scholars in China about the resettlement and environmental issues of the project is taking place, and has been for years. It has been possible to engage in a debate about these issues in scholarly

---

have been important factors for the resettlement policy change. This is also confirmed by the interviewees.

674 For more details on the discussion about the dam project in the 1980s, see Heggelund (1993), chapter 6.

675 Dai (ed. 1989), Li (1985); Tian, Lin and Ling (eds. 1987); Tian and Lin (eds. 1988), respectively.

journals. However, their discussions have not usually been reflected in the media. It has been possible to be more critical in scientific journals, as these are often limited to groups, such as researchers, with specific interests.

- To say something about the potential impact such articles may have had on the TGPCC leadership, and Zhu Rongji in particular. I am aware that it may not be possible to verify the extent to which these particular articles had impact. Nevertheless, when taking the developments in Chinese politics into consideration, where information is more valuable for decision-making in general, and knowing that the institutions have become more important in the policymaking process, I assume that some impact is likely. Thus, I wish to illustrate the issues of a debate that may have influenced the decision to make the resettlement policy change. The examples below are chosen as they are representative for the articles that have been published in scientific magazines over the years, and the topics that have been discussed.

I have chosen to divide the publications into three different categories. The categories may not be completely accurate, as there is some overlap between them. Nevertheless, dividing the information described below into different categories will provide easier access for the readers.

Social science One category is the publications that can be classified as social science, and they focus on policy and politics in relation to both resettlement and the environment. Examples of such journals are *Gaige* (Reform), *Zhanlüe yu guanli* (Strategy and Management), *Renkou yanjiu* (Population Research), and *Renkou yu jingji* (Population and Economics). Articles about the Three Gorges focus on social aspects of the implementation of the resettlement policy. They may discuss advantages and disadvantages of certain resettlement methods, such as social impact from distant migration and cultural differences.

Natural science The second category belongs to natural science, and consists of scientific publications that focus on environmental and scientific aspects, such as the *Changjiang liuyu ziyuan yu huanjing* (Resources and Environment in the Yangtze Basin), *Keji daobao* (Science and Technology News), *Chongqing huanjing kexue* (Chongqing Environmental Sciences), *Dili kexue* (Scientia Geographica Sinica), *Dili xuebao* (Acta Geographica Sinica) and *Sichuan Sanxia xueyuan xuebao* (Journal of Sichuan Three Gorges University).[676] Articles in these publications are often written by scientists who are engaged in environmental and

676 *Sichuan Sanxia xueyuan xuebao* (Journal of Sichuan Three Gorges University) seems to be a more official journal, which may belong in the third category. Some of the articles (columns) are jointly published with the the Resettlement Bureau of the Wanzhou Resettlement Development area. The reason for including the *Journal* in the second category is that the articles seem to point to critical issues in the reservoir area, are realistic, and the tone and contents are similar to many of the journals mentioned in the second category.

resettlement research related to the project. The articles often depict the potential negative impact that the resettlement will have on the Three Gorges area, or the problems in the implementation process. To a certain extent they also reflect the authors' concern for social instability due to issues such as economic weakness among peasants and high expectations among relocatees. The majority of the articles used for this book come from this category.

Journals published by organisations or the bureaucracy These are a third category of journals that have been established in relation to the Three Gorges Project in the beginning of the 1990s, and that specifically discuss the Three Gorges Project from an 'official' point of view. Examples are *Zhongguo Sanxia jianshe* (China Three Gorges Construction)[677] published by the TGPCC and the newspaper *Zhongguo Sanxia gongcheng bao* (China Three Gorges Project News). These are official publications by the Three Gorges Project Construction Committee, where the TGPCC is the editor and publisher. Both publications are sponsored by the China Three Gorges Project Development Corporation (*Zhongguo Changjiang Sanxia gongcheng kaifa zonggongsi*). These journals also discuss potential problems for the project, and the articles appear to be similar to the ones in the second category listed here. However, as they often are written by employees of the TGPCC, heads of local resettlement bureaus, or other agencies, they tend to be less critical and more optimistic about finding solutions.[678] This literature is more descriptive of developments in the area, and one intention appears to be spreading of information regarding the progress made in resettlement.

I focus below on the two first categories, as these are fairly independent journals (i.e., not subordinate to the TGPCC) that discuss the resettlement and environment issues for the project. The active role of the TGPCC in these two publications may make the contents seem more or less biased and politically coloured. It is, however, necessary to mention them in the context of existing literature about the resettlement issues for the Three Gorges project.

Social Science Category

This section will mainly deal with an article in the journal *Zhanlüe yu guanli* (Strategy and Management) in January 1999, that caught the attention of people who have followed the developments of the Three Gorges project, including officials and researchers.[679] The reason for the attention is that *Zhanlüe yu guanli* is a conservative journal with some links to the government. The journal is published

---

677 *Zhongguo Sanxia jianshe* (China Three Gorges Construction) has two versions, one is the resettlement version (*yimin zonghe ban*), and the other concerns the project's scientific and technological issues (*gongcheng keji ban*).

678 'Countermeasures' seems to be a favourite word in this connection.

679 Wei (1999), pp. 12-20. Other articles in relation to the Three Gorges policymaking or resettlement policymaking in general in this category are Xu (1999); Gu and Zhong (1998); Wang, Huang, and Ding (1999).

by the *Zhonguo Zhanlüe yu guanli yanjiuhui* (The China Strategy and Management Research Association), which has attachments to the State Council's Office for Restructuring the Economy (*Guojia jingji tizhi gaige bangongshi*). The article discussed the failings of the resettlement efforts so far, and the difficulties facing the rural relocatees of the dam project. The article was very critical towards the resettlement policy of the dam project. The title is 'Sanxia yimin gongzuo zhong de zhongda wenti yu yinhuan' (The Significant Problems and Hidden Dangers in the Three Gorges Resettlement), written by Wei Yi, a pen name. The author criticises the dam building in China because the resettlement has been regarded as a subsidiary problem, as well as an economic problem. In reality, the author says, the resettlement problem is first of all a complex social problem. It also illustrates many important issues in China today such as equality, development, disparity between the regions, and social stability.

The author praises the developmental resettlement policy as an improvement, but points to new problems that have emerged during the years of trial resettlement since the mid-1980s and during six years of formal practice. The author warns further that if the problems are ignored, and one waits until the 'flood water arrives at our feet, and thousands upon thousand of relocatees arrive at the government's door', it will be too late.[680] The article is significant in several ways, as it criticises the government directly and it was definitely noticed by a number of people, including officials, and had some impact on the government.[681] Below are some of the main points made by the author that are worthwhile noticing:[682]

Criticism of authorities There is direct criticism of authorities about the way involuntary resettlement is treated. One point is aimed at respecting the rights of the involuntary migrants. A dialogue should take place between the migrants and the authorities in order for both sides to air views and come up with the best solutions. The author points to the importance of participation of migrants in the decision-making process. Presently, he states, migrants who express their wishes and viewpoints are labelled 'unruly people' (*diaomin*).[683]

Economic transition The author points out for instance that the government behaviour in the Three Gorges case is typical of the planned economy period; the government is in charge of everything from organising and supervision to construction of new towns. It is necessary to take into consideration the dynamics of the economy when resettling the large population as well as the relocation of enterprises. The author points to the big gap in the resettlement funding between the projected funding in the planning process to actual needs at the present. It is a

680 Wei (1999), p. 13.

681 Written communications with the author.

682 In addition to these, one issue is the shortsightedness on the part of the county and township level cadres which is related to their short terms in office and the difficulty involved in the resettlement work.

683 Wei (1999), p. 16.

critique of the government's lack of comprehension of the market forces, which will have negative effects on the resettled people.

Lack of legal protection The author points out that, apart from the Three Gorges project resettlement regulations from 1993, there exists no legal system to protect the relocatees nor to guide the resettlement process. There is no law for reservoir resettlement. Furthermore, it takes place in a market system which is difficult for the resettled people to enter. The resettlement is carried out by using administrative methods only.

It is interesting to note that the article appeared in January 1999, shortly after the inspection tour by Premier Zhu, and around the same time as the article 'Sanxia gongcheng women jixu guanzhu ni' (The Three Gorges Project, We Continue to Pay Attention to You) appeared in the *People's Daily* on 24 January, to be discussed later in this chapter. There may be two possible explanations to the appearance of the article. At the time of publishing a shift seemed to have taken place in the media reporting of the project, as mentioned earlier in the chapter. This may indicate that the article could have some support from above. The magazine is linked to the State Committee for Restructuring the Economy, which is considered to be one of Zhu's think-tanks. Still, this would be difficult to verify, and it may be unlikely that such a critical tone would be endorsed at higher levels. A second potential explanation is that the author had close connections to the magazine, which enabled him to publish the article. Academic journals frequently discuss critical issues, and the editors of the magazine may have agreed to publish a critical view of the resettlement process. This may be a more plausible explanation as it is known that the author has been criticised for his article.[684]

Natural Science Category

Articles in the second category may also point to critical issues. However, they differ from the *Strategy and Management* article as they do not appear to be as critical of the entire process. There are exceptions, which will be given below. However, the majority of the articles reveal concern for the environment as well as for the well-being of the relocatees. The articles may often be quite scientific, technical, and contain a great number of statistical figures. They also often come up with suggestions as to how the problems may be solved. As there are no articles in this category that stand out in the way that the *Strategy and Management* article does, I will briefly give an overview of some of the topics that are covered by some of these articles, and references will be made to examples of articles published in the 1990s.[685]

---

684 Written communications with the author.

685 I have also included a few articles from 2000 in order to illustrate one issue that was repeated by several scholars during interviews in 1999.

Conditions of the relocatees A number of the articles focus on the future conditions of the relocatees after having been resettled in their new homes and areas. The topics are similar to the ones mentioned in the chapter about resettlement, such as: the link between lack of farmland and erosion and environmental degradation; change of vocation for 40 percent of the resettled rural popoulation and the potential difficulty in finding work; the 'erci yimin' phenomenon (secondary resettlers, who leave their land for city development). These articles commonly suggest measures, and in this case it suggests moving from *yiminlou* (resettlement buildings) to *yimincun* (resettlement villages) or sometimes *yimincheng* (relocatee towns) where the relocatees can raise pigs, chickens etc. in order to ensure their income.[686]

Environmental impact and capacity Many articles discuss the potential impact of the project in the Three Gorges area on the environment, and some predict that the majority of impacts will be negative.[687] Some articles stress the environmental and ecological challenges in relation to resettlement due to the lack of available farmland; some are in favour of moving out peasants in order to save the environment.[688] Some articles focus on soil erosion of slopelands in relation to limitations on agriculture and improved living standards, as well as the need to focus on ecological agriculture.[689]

Legal framework and mismanagement of funding Strengthening of the legal framework in relation to the implementation of the resettlement policy is a recurring issue. Local officials on both high and low levels enrich themselves and their relatives through the resettlement funding, which is a cause for concern among many scholars.[690] Their message is that the legal framework must be strengthened in order to avoid social instability.

River basin management[691] In order to manage the natural resources and the environment of the Three Gorges reservoir area, some scholars state the necessity of establishing an organisation that cuts across the administrative borders.[692] One suggestion is to establish a Three Gorges reservoir area river basin management

686 By Gu and Huang (1999). See also Gao (2000).
687 Chen (1999c); and Chen (1999d).
688 Xia (1999), pp. 5-6. And Wu and Liao (1999).
689 Zheng and Shen (1998); and Deng (1997).
690 Huang (1998) pp. 5-9; and Cui (1999), p. 2.
691 The issue of regarding the Yangtze River as one integrated system and treating the Three Gorges issues as part of this system, was discussed by Chen Guojie in several articles in the 1980s and 1990s. One example is Chen (1999f), first published for *Kexuebao* (Scientific Daily), (23 January, 1987). The need to establish one river basin organisation with cross-provincial authority was stressed during interviews with Chen and other scholars in 1999.
692 He and Chen (2000).

commission directly under the State Council, with the authority to give orders to EPAs, the Land Management Bureaux, Water Resources Bureaux and the Urban Construction bureaux of provinces and cities.[693] At present, the responsibility is divided between several provinces, ministries and river basin commissions.

On the whole, the total number of articles regarding the Three Gorges resettlement and environmental issues may have had a certain impact over the years, and made the leadership aware of some of the problems related to the resettlement and environment in the Three Gorges area. The articles referred to above are only a few of the many articles published in the past few years about the project. One scholar, whose work stands out in the scientific debate about the project since the 1980s until present, is Chen Guojie at the CAS in Chengdu. He has written numerous articles about the potential impacts of the Three Gorges dam project on the environment, as well as on the decision-making process.[694] Chen has participated in a number of research projects about the dam project environmental issues, and was head of research groups for several environmental impact studies that were carried out by CAS in the 1990s. In 1999 he attempted to publish a volume with a collection of all his articles about the Three Gorges project (in addition to other environmental issues). However, no publisher has yet been willing to publish the collected works. For his scientific views he has received criticism, lost out on professional advancement and angered high level leaders in China.[695] Furthermore, as mentioned in Chapter 7, Chen and the research team reached scientific conclusions that were not 'acceptable' in 1992, when the environmental impact assessment was compiled. It is difficult to measure the influence of work by Chen and other scholars on the decision-making process for the dam project. However, their work has definitely gained the attention of leaders, and it is likely to conclude that their research work has had a certain influence on the resettlement policymaking for the dam project.

To sum up this section, extensive publishing about the Three Gorges resettlement and environmental issues have been published over the years. The *Zhanlüe yu guanli* article and the work of Chen Guojie stand out as exceptional in the context of the Three Gorges resettlement debate. Scientists write about the potential and actual problems with resettlement. Some are more outspoken than others, although their main focus is on solving the problems.

---

693 Ibid., p. 187. SEPA and the Ministry of Land and Natural Resources would have a guiding relationship with the proposed commission.

694 Chen (1988); first publishd in *Qun yan*, 1988, Vol.10. See also Heggelund (1993), pp. 109-13, for a discussion of the article.

695 Once by Zou Jiahua due to publishing the article 'Dui Changjiang Sanxia gongcheng shengtai yu huanjing wenti de tansuo' (Exploring the Three Gorges Project's Ecological and Environmental Problems) in *Keji daobao* (Science and Technology News) in 1987, 1, pp. 4-9.

A Change Takes Place in the Media Reporting

Compared to scientific and academic journals, the media has been more closed to diversified views on the dam project. Even in the 1980s, which is regarded as a fairly open period in China, it was difficult to post articles in the media that advocated negative sides of the dam project,[696] which is related to the controversy surrounding the Three Gorges project. It is still difficult to write critically about the project in the state media. However, as we shall see in this section, some changes took place in 1999. In China, despite the positive developments described earlier in this chapter, the state media is still tightly controlled, and some issues cannot even be discussed.[697] As mentioned above, reporting in the media on the Three Gorges project has been mainly positive since the construction of the project began in 1994. This section will look into the changes that followed when Zhu Rongji became Premier in March 1998. The assumption is that the increase of realistic and to some extent critical reporting in the media of the dam project has been allowed under Zhu. The articles have focused on the problems, and have not merely praised the project.

Zhu Rongji took charge of the Three Gorges project when he took over as Premier and became the TGPCC chairperson. In December (28-30) that year he made his first visit as Premier to the Three Gorges project dam site. During his visit there, the *People's Daily*[698] reported on issues of concern that were expressed by Zhu. The article quotes the Premier on a number of important issues, such as the quality of the construction, the resettlement problem, environment and corruption. In order to stress the importance of the quality issue, the Premier stated that this is the life or death for the project, and everyone involved in the construction should be aware of their historical role. Zhu does not explicitly criticise the lack of quality, but nevertheless, judging from the text it is clear that this is a concern. What further underlines the concern of the Premier, is his stated need for quality supervision of the project, and a suggestion to invite foreign companies with experience in dam construction to guarantee the quality of the dam. On the resettlement issue, the Premier stresses the need for successful resettlement in the second stage of the resettlement process, which he designated as the most challenging stage due to the large number of people to be resettled. One consequence would be delayed construction if the resettlement process slows down. Therefore, in order to speed up the resettlement the Premier suggests using several methods to resettle people, including moving some completely out of the reservoir area. At this stage however, the decision to move out a fixed number of

696 See for instance Heggelund (1994), pp. 113-14 regarding Dai Qing's description of the struggle to have articles published in newspapers and magazines. In the end the collection of articles were published as a book, Dai (ed. 1989).

697 Lists of topics that may not be discussed in the media are issued by the Ministry of Propaganda (*Xuanchuanbu*). These issues cannot be discussed without approval by the publication's editorial board. Yang (1998), p. 46.

698 Wang (31 December, 1998).

people had not yet been announced (the decision may have been made, but had not been made public yet). Relocation of enterprises and the environment are also concerns of the Premier, alhough in this newspaper report the quality issue receives the greatest attention. In sum, one could say that the entire article could be regarded as an expression of concern on the part of the Premier regarding several important issues related to the construction of the dam, as well as giving instructions on how to carry on the construction and resettlement of the dam project. Moreover, the article set the stage for the subsequent developments with regard to the media's reporting of the project.

Following the December 1998 article from Zhu's visit to the Three Gorges area, on 24 February 1999, four articles appeared in the *People's Daily* that emphasise the concern of the Chinese leadership for the Three Gorges project. Two concern the dam project directly, and the most significant of the articles in the 24 February edition of the *People's Daily* is the mentioned 'Sanxia gongcheng women jixu guanzhu ni' (The Three Gorges Project, We Continue to Pay Attention to You).[699] The significance of this article is that it marks the beginning of a period during 1999 and 2000 with a different and more critical reporting about the dam project in the official media. It is the first of several articles that raises concern over a number of aspects regarding the project, albeit it does not question the validity of the project itself. The second article concerns the lack of funding and time regarding the cultural relics that need to be excavated before the reservoir construction. The remaining two articles concern the project indirectly, but must also be seen in a context of leadership concern over the quality issue of the Three Gorges project. One is about regulations for project construction in general in China: specifically, a circular by the General Office of the State Council[700] demanding strengthening of project quality management, which included instructions such as the establishment of a responsibility system for project leaders. Furthermore, the circular requires lifelong responsibility for the quality of the project from the administrative leaders, legal representative and the legal representative of the participating construction units. The second article is a general commentary by the People's Daily commentator about the new circular[701] that praises the new responsibility systems. These two articles do not mention any project in particular, but it is likely that the Three Gorges project construction was one of the possible motives for issuing the circular, and could be seen in connection with the quality concern expressed by the Premier in December. Thus, one may say that 24 February 1999 marks the beginning of a new period with regard to the Three Gorges media reporting and indicates a different approach to the dam project developments.

Below I will go through some of the issues that have appeared in the media coverage of the dam project in more detail, beginning with 24 February, in order to illustrate how leadership change has manifested itself in the media.

699 Qian (24 February, 1999).

700 Guowuyuan bangongting (24 February, 1999).

701 Benbao pinglunyuan (24 February, 1999).

Let us take a closer look at the 24 February article 'Sanxia gongcheng women jixu guanzhu ni' (The Three Gorges Project, We Continue to Pay Attention to You).[702] It points to several factors that may be an indirect criticism of the project developments that may have led the way for more criticism. It raises, for instance, questions about the resettlement progress in the area. It begins by stating that during 1993-1997, the first period of the Three Gorges resettlement, altogether 103,400 people had been resettled successfully. It continues by stating that during 1998, the first year of the second resettlement period, Chongqing (that has the heaviest burden), only fulfilled 74 percent of the anticipated resettlement. In comparison, the moving of enterprises in the same period in the same area had reached 107.7 percent of anticipated resettlement. It states further that 'it is clear that the moving of enterprises is happening at a quicker pace than resettlement'. This could be interpreted as an indirect criticism of the priorities made in the construction process, where heavy focus is put on business instead of resettling the people on schedule. Furthermore, Chongqing resettlement officials are quoted regarding the great challenges of the second period of construction and resettlement and they state that the task appears even more complex and formidable. This could signify their wish for more assistance or funding. The article also refers to Chinese experts who have carried out research about the resettlement problem, who believe that it would be helpful for the economy and ecology of the reservoir area if more people are moved completely out. This may be regarded as a preparation for the policy change in May 99.

The quality issue is also discussed in the 24 February article, and the high-level Three Gorges Project officials who are quoted express some concern: a deputy director of the Three Gorges Project Development Corporation (TGPDC), senior engineer Wang Jiazhu, states that the Three Gorges Project construction quality is currently good, and satisfies design requirements. 'However, I cannot say that it has reached top international quality; the real test is in the future'. This could be interpreted as an indication of concern about the project quality. Yuan Guolin, another deputy director of the TGPDC expressed a wish for strengthening of public opinion and supervision of the dam project, since the project daily spends RMB 30 million. This was also raised by Zhu during his inspection tour of the Three Gorges area in December that year. The quality issue is also closely related to the corruption problem; it is thus possible to interpret the quality coverage as concern for the corruption and embezzlement of project funds that has grown gradually in the construction period of the dam project. The corruption issue, discussed in Chapter 5, has been given extensive coverage in the state media, beginning already in December 1998[703] when it was reported that resettlement funds had been embezzled. Thus, the quality issue discussed in the 24 February article is a continuation of the focus on quality and corruption in the project.

---

702 Qian (24 February, 1999).

703 Li (21 December, 1998).

The 24 May article in *People's Daily*[704] then announced the resettlement policy 'adjustment', which also signalled a new focus on and recognition of the environmental problems in the reservoir area. Zhu emphasised the need to think in alternative ways, as resettlement in the vicinity would lead to deteriorating ecology and environment, and stated that 'it seems now that we need to encourage even more people to move out'. The article also refers to Zhu's statement about the quality of the officials as an issue that will determine the success or failure of the dam project. The issue of mismanagement and corruption of resettlement funding is reflected in the statement that 'each penny must be spent on rebuilding homes for the relocatees'. Furthermore, the importance of quality in the construction of the project is emphasised. He pointed out hidden dangers in the resettlement and construction process such as the *doufu* 'scum' (or bad quality) projects. One of the reasons for talking about *doufu* 'scum' projects, was the collapse of a bridge (on 20 February 1998) in Badong county in Hubei, where 11 people died due to bad construction and bad quality materials.[705]

As discussed in Chapter 7, during autumn of 2000, the two articles regarding the environmental situation in the Three Gorges area further illustrate the change in reporting on the dam project. These two articles represent the first official criticism of the environmental protection work being carried out in the Three Gorges area. The article, 'Sanxia kuqu wuran yanzhong huanbao zhihou' (The Environmental Pollution in the Three Gorges Reservoir Area is Serious and Environmental Protection is Stagnant)[706] was also significant since it came from Chongqing's Vice-Mayor Chen Jiwa and the leaders of the city's EPB who are high-level decision-makers and officials. A few days later, on 26 September, Lu Youmei, the head of the Three Gorges Project Development Corporation, 'answered' the criticism at a special meeting about the Three Gorges project during the annual meeting of the World Commission on Dams, which was held in Beijing. Lu stated that 'the Three Gorges will definitely not become a pool of polluted water' (Sanxia jue bu hui biancheng yi tan wushui).[707] Lu acknowledged the fact that such problems exist, but insists that measures will solve the problems. In October, one month later, the second serious criticism of the environmental work in the Three Gorges area appeared in the *Guangming Daily* (Enlightenment Daily), which was even more condemning of the environmental situation in the area. Conclusions regarding the significance of the articles and the environmental criticism for the Three Gorges environmental decision-making process will be discussed in the conclusion (Chapter 9).

A conclusion to be drawn from the above is that critical views from the scientific debate have been 'allowed' to appear in the media. This is apparent

---

704 Lu and Jiang (24 May, 1999).

705 *Renmin ribao* (3 June, 1999). As a result the Quality Expert Group was established in 1999, described in chapter 7.

706 *Renmin ribao haiwai ban* (21 September, 2000).

707 Zhongguo xinwenshe (26 September 2000).

mainly in the articles that concern the environmental aspects of the dam,[708] where the environmental views portrayed are representative of the discussions taking place in academic journals. Moreover, the resettlement issue is most often discussed in the media in relation to the environmental capacity in the reservoir area, while the social problems associated with resettlement are seldom (or never) discussed. From this one may conclude that the resettlement issue appears to be a much more controversial issue than the environment issue; and that the social issue has not been lifted up to the same level of importance as the environment.

In sum one could say that under Li Peng, articles in the newspapers about the Three Gorges project have praised the greatness of the dam project, and did not focus on problems regarding the construction quality, environmental protection and resettlement. A number of issues have appeared in the media after Zhu became the chairman of the TGPCC that focus on existing and potential problems. Reporting in the media has, in the period studied for this book, therefore become more diversified. Some of the information being discussed in scholarly magazines, in particular about environmental problems, have increasingly been portrayed in the media.

The Significance of New Leadership on the Three Gorges Project

The preceding sections in this chapter have discussed the changes in Chinese society as one major factor that has influenced the resettlement policy change for this project. The growing importance of information in decision-making has made the Chinese leadership increasingly aware of the need to seek information from think tanks and research institutes. Relating this to the Three Gorges dam project, one may conclude that the scholarly discussion during the past two decades may have had considerable impact on the resettlement policymaking. The role of institutions such as CAS has been particularly important. In addition, one assumption of this book concerns the effect the change of leadership has had on the resettlement policy. From the discussion in the previous sections, one may conclude that the new leadership has had influence in two areas in relation to the Three Gorges project that are relevant for this book: one is the Three Gorges resettlement policy change, and the second is the change of tone in the articles about the dam project in the state media. Based on written sources about Zhu Rongji as well as interviews with scholars and officials, one may conclude that the resettlement policy change very likey is directly linked to the change of Premier. More importantly, the policy change may be linked to Zhu's character and style of leadership. One issue arises in this regard: the decision to move 125,000 of the rural population from the reservoir area to other provinces may not have been controversial, and may have received support from the leadership and the bureaucracy. On the other hand, the experience from other dam projects in China

---

708 The quality aspect of the dam may be a second area that reflects critical views. However, the scholarly debate about this issue is not within the scope of this book.

where distant removal has been unusscessful, may have made decision-makers hesitant about such a decision. The debate about the dam project has illustrated that even though the authorities did not publicly acknowledge that outmoving was contemplated, resettlers were moved to distant places in China for trial resettlement. These were not successful experiments, and the alternative to moving people far away has not been a favoured one. Thus, one may get the impression that outmoving may not have been an easy choice. Reflections regarding Zhu's role in the resettlement decison-making change, as well as the absence of policy change during Li Peng's tenure, are discussed below.

1. Zhu Rongji became Premier and the chairman of the Three Gorges Project Construction Committee (TGPCC) in 1998 when he replaced Li Peng. Li has been known as an ardent proponent and supporter of the dam project, and strived to have the project approved in the early 1990s. Under Li's leadership of the dam project from 1994 until 1998, little or no criticism appeared in the media about the Three Gorges Project. To many Chinese, Li is regarded as a mediocre leader, who is preoccupied with protecting himself and remaining on good terms with the elders.[709] Zhu Rongji has according to some never publicly praised the project, and it may be unlikely that he harbours the same feelings for the dam project as did Li Peng.[710] Only a few months after Zhu became Premier, changes in the media reporting took place, followed by the policy change in May 1999. One possible explanation for the attitude change may be Zhu's late involvement in the project, which may have given Zhu greater freedom to focus on the actual problems, as he is not as closely connected with the dam project as Li Peng was.
2. Zhu Rongji may want to distance himself from the policies of Li Peng, and criticising the project could be one way to indirectly criticise Li Peng. 'A change of person means a change in polic', said one of the interviewees who has worked on the project for many years. Ranked as number two in the politburo until 2002, Li is a politically strong person. However, he is not very popular with the Chinese people, partly due to his role in the crackdown of the student demonstrations in 1989. Zhu Rongji on the other hand is viewed as an efficient leader, albeit not always popular due to the downsizing of the bureaucracy and the reform of the State-Owned Enterprises (SOEs). Zhu is now in charge of Li's pet project and will be blamed if anything goes wrong. Therefore, self interest would be an important factor for him to wish for a successful project, and the resettlement policy change and increased focus on actual problems may be regarded as a way to obtain this.

---

709 See Xiao, Zhengqin (1999), pp. 242-5 for a discussion about Li Peng.

710 Interview with Dai Qing, Beijing, 1999 and 2000. According to scholars interviewed by the author, after Zhu became Premier many people both in and outside China wondered if the scale of the project would be reduced, or even abandoned, as Zhu was not known as an ardent supporter of the dam.

3. Zhu's pragmatism may be a third way to explain the diversified reporting in the media and the subsequent policy change. According to most interviewees, Zhu is viewed in general as more pragmatic than Li, and deals with concrete matters (*wushi, shi shenme jiu shi shenme*). His main area of concern is economic reform; political reform may not be on his agenda.[711] Zhu is perceived as a leader who does not desire power for the sake of power in order to exercise control.[712] He is regarded as a leader who understands how to grasp and to use power in order to reach certain goals. In short, Zhu is known as a leader who is efficient and goal-oriented. Some scholars interviewed by the author said: 'If it were not for Zhu Rongji, it would be impossible to know the state of affairs [for the project]' ('Zhu Rongji, mei you ta jiu bu zhidao you shenme yangzi').[713] Based on both written and oral material, one may conclude that Zhu's ability to be pragmatic and to deal with concrete matters may have been abilities that were important for the changes related to the dam project as well as the media.
4. Zhu's leadership style may have been an important explanatory factor. Zhu represents a different leadership style in China; he is open and at ease with the media and in public. For instance, Zhu held a press conference to introduce his new cabinet in 1998, which was unprecedented in China.[714] Zhu is responsible for restructuring the economy, and surrounds himself with economic planners. These people are often university educated, cosmopolitan and from coastal provinces.[715] Zhu in many ways is similar to Zhao Ziyang, the Secretary General who was ousted during the democracy movement in 1989;[716] Zhao also acted as a bridge between younger and older generations. 'Zhao was the first Chinese leader in his generation to sincerely embrace the younger generation and open up to fresh ideas'.[717] His economic think tanks were also filled with talented young people. Zhu Rongji also likes to employ

711 Some scholars state that Zhu's lack of political base, in addition to his focus on his portfolio of economic reform, would hinder Zhu in promoting political reform. Zhu being able to obtain Jiang Zemin's continued support has to do with his lack of ambition in political affairs. See Zheng (2000), p. 73.

712 Li and Xiao (ed. 1998), p. 10.

713 Interview no.29, Beijing May 2000.

714 Lecture by William C. McCahill, Jr., former Deputy Chief of Mission of the US Embassy in Beijing, 19 April, 2001 at NUPI, Oslo.

715 Shambaugh (ed. 2000a), p. 31.

716 Even though Zhao and Zhu share similar interests in using think tanks and young professionals, their work tasks differs in many ways. At the time when Zhao took over, China was at the beginning of economic reform. Zhao's main task was *fenquan* (division of power) or *po* (split up) which despite being necessary for the economic reform, resulted in a weakening of the authority of the central authorities in relation to the provinces. Zhu's main concern is to *jiquan* (to centralise state power) or *li* (erect) in order to establish a new economic system in China. See Zheng (2000), pp. 73-4, and Xiao, Zhengqin (1999), p. 409.

717 Chen (1995), p. 142.

young people, and differs from Jiang Zemin in that the members in his group of consultants are numerous, while Jiang depends on a fixed group of a few people.[718]

Some scholars state that Zhu lacks a power base (*zhengzhi genju*), and that he does not seem to have a patron-client relationship within the party apparatus, nor an elder patron to support him.[719] This could make Zhu vulnerable if his economic policy fails. His main task in government has been to take care of the economy, and his perceived lack of power base would also make it difficult for Zhu to propose controversial policy, such as political reform. His weak power base may be one reason to question his effectiveness in the Three Gorges resettlement policymaking. Nevertheless, although Zhu may have had turbulent periods during his Premiership such as his failure to secure China's entry into the WTO in 1999,[720] he nevertheless appeared to have the support of Jiang Zeming. During Zhu's period as Mayor in Shanghai in the 1980s he worked closely with Jiang Zemin who was Party Secretary there until the latter was called to Beijing in 1989; Zhu followed in 1991 to serve as vice-Premier of the State Council. Other factors that may suggest that Zhu has had impact on the resettlement policy change, are related to the changes regarding decision-making in China. Policy decisions now increasingly rest in the State Council and Leading Groups.[721] Moreover, the increasing influence of think tanks in making policy recommendations and evaluations is an important aspect in policymaking in China today. As Premier, Zhu was head of the State Council as well as several important leading groups and committees and would be able to influence policy.[722] In addition, the Qinghua network may be an important factor. Among the chairmen and the vice-chairmen of the TGPCC, Zhu Rongji, Wu Bangguo and Zeng Peiyan are all educated at Qinghua University, albeit graduated at different times, in 1951, 1962, and 1967 respectively.[723] According to

718 People such as Guo Shuqing, who has been the head of the department for national macro-economy of the SPC, and secretary-general of the State Council's Office for Restructuring the Economy. Others are Lou Jiwei, Zhou Xiaochuan and Li Jiange who all have doctoral degrees in economics. Xiao, Zhengqin (1999), p. 80; and Li (2001), p. 225.

719 The old revolutionaries strongly resisted promoting Zhu Rongji to leadership in the early 1990s, but with Deng Xiaoping's efforts Zhu was promoted. Zhu's initatives depend largely upon the support from Jiang. See Zheng (2000), p. 73; and Shambaugh (ed. 2000a), p. 29.

720 See Associated Press (11 November, 1999).

721 This has resulted in a continuing dispersion of power away from Party and the military. Shambaugh (ed. 2000a), p. 30.

722 State Leading Group for Scientific and Technological Education, Leading Group for the Development of the Western Regions, the Three Gorges Project Construction Committee.

723 Jiang Zhuping, the governor of Hubei province and TGPCC vice chairman also has indirect connections to Qinghua. He is the son of Jiang Nanxiang, former minister of education. Jiang Nanxiang initiated the 'double-load cadres' system at Qinghua,

Li Cheng, the Qinghua network has fostered a number of leaders in the third and fourth generations, and their technocratic identity and understanding is important for decision-making.[724] The significance of the Qinghua network may be the importance put on group identity (technocratic identity), closeness and loyalty that the network has fostered between members of this network, which may have had implications for the Three Gorges decision-making.[725] Political networks are important for political advancement, and as several scholars have pointed out,[726] informal networks and connections (*guanxi*) do actually count in China's policymaking. The extent to which it has been a factor in the Three Gorges resettlement policy change may be difficult to verify. However, based on the literature mentioned above, it is likely that the Qinghua network may have been a supportive factor in the resettlement policymaking for the project.

Summing up

This chapter discusses the possibility that the changing shape of decision-making in China may have indirectly prompted and enabled the resettlement policy change for the dam project. During the 1990s, Chinese society has been developing towards a more pluralistic society. It is becoming a society where information and knowledge are increasingly important for decision-making. The Chinese public and decision-makers have access to a wide range of information through books, newspapers, journals and the Internet. New institutional actors have appeared on the scene, as think tanks and consultative organs are frequently included in the policymaking process. One reason for the increased importance of research results and knowledge is that China's leadership has higher and more diversified education and backgounds. Zhu Rongji and other leaders rely on information from think tanks in their policymaking. Zhu also places great emphasis on science, which has resulted in increased importance for the role of CAS during his tenure. Although the positive trends exist, there are also setbacks. It is difficult to report criticially on certain developments in Chinese society as well as on controversial projects such as the Three Gorges, or to challenge the power of the Communist party.

---

which formed a basis for the present-day technocratic leadership. Jiang Zhuping is educated from another elite school, Harbin Military Institute of Engineering. Li (2001), p. 141 and p. 90.

724 See chapter 4, 'the Qinghua clique', which gives an interesting account of the importance of Qinghua as a base for developing technocratic leaders. Ibid., pp. 87-126.

725 The members of the 'Shanghai gang' and the 'Qinghua clique' often overlap. Jiang Zemin, Zhu Rongji and Wu Bangguo all have had important posts in Shanghai and worked together before they were promoted to high level posititons in Beijing. Zhu and Wu also belong to the Qinghua network.

726 Lucian Pye, Lowell Dittmer and Joseph Fewsmith.

A debate was onging regarding political reform during the 1980s that was ended in the late 1980s. This also influenced the Three Gorges project, which was openly dicussed in the 1980s. The debate was 'closed' again due to the rigid political environment following the crack-down on the student demonstrations in 1989, and this resulted in project approval in 1992. The project construction began in 1994, ending phase one of the construction. Under the leadership of Li Peng, the developments for the dam project had been given positive coverage in the state media, which constantly praised the project. Zhu Rongji took over as Premier and TGPCC chairman in 1998, when the second phase of the construction and resettlement process began. Important factors for the resettlement change that are discussed in the chapter are: change of leadership (Premier) in China in 1998 and the subsequent change in the Three Gorges project leadership; Zhu Rongji's leadership style that differs from previous leaders; the scientific advice provided to the leadership; and the media coverage in the late 1990s following the leadership change. I presume that the interaction between these factors, the leadership change, knowledge-based decision-making and the media, has had an important impact on the resettlement policy change.

A scholarly debate has been going on for many years about the resettlement and environmental issues. Their views, often critical, have been reflected in the many scientific journals described in this chapter. Moreover, the numerous articles in academic journals that point to a number of difficult factors in relation to the resettlement and environment may have had impact on the resettlement policy change. One way to study the leadership change and policy change is through the media coverage of the dam project. The chapter concludes that a shift took place after Zhu became Premier, and an increased number of articles were published in the media that point to problems in the process and portray more realistically the resettlement process and the environmental problems. Articles in the *People's Daily* and other state media have been more critical and focused on actual problems of the project such as quality, corruption, resettlement and the environment, which is a significant change from early reporting. One may conclude that critical views regarding the environmental issues that have been able to appear in scientific journals for years, have in the late 1990s appeared in the state media as well. However, critical views about the resettlement issue appear mainly in relation to environmental capacity; the social problems related to resettlement are rarely mentioned.

This chapter concludes that the change of leadership has had significant impact on two areas related to the Three Gorges project: the resettlement policy and the media reporting about the project. Some considerations that may be useful in explaining the change are: the project was not initated by Zhu, so he has greater freedom to focus on the actual problems; he is viewed as a pragmatic person who focuses on problems and how to solve them; Zhu represents a new leadership style, is more at ease with the media, and uses think-tanks to assist him in forming policy. However, Zhu is perceived as having a weak power base and does not appear to have any patron-client relationship. The effectiveness of Zhu's policymaking in the Three Gorges resettlement policy change may therefore be questioned. Nevertheless, Zhu's relationship with Jiang Zemin, increasing power of the State Council and Leading Groups in decision-making as well as the group identity of the Qinghua network may be factors that have enabled Zhu to exert critical influence on the policymaking for the Three Gorges resettlement.

Chapter 9

Conclusions

This concluding chapter is divided into two parts. The first part is a summary consisting of five sections, which consolidates the findings from the previous chapters in the book. The point of departure in the study is the resettlement policy change in the Three Gorges project that resulted in the decision to move out 125,000 people to other provinces, due to lack of environmental capacity. The main purpose of this book has been to explain the developments in relation to resettlement and the environment before and after the policy change, in order to understand how this change could take place. The first section begins with the long history of the dam project, which provides the necessary background on the controversy involved. Section two summarises the analytical approaches employed in studying the resettlement and environmental decision-making for the dam. Conclusions regarding the analytical approaches and the Three Gorges resettlement and environmental policy will be discussed in the second part of this chapter Section three describesthe international dam-building debate and outlines China's resettlement history and experiences. This is intended to form the background for the discussion of the resettlement policy change that took place in 1999, and the problems that have emerged in the process. Presumably, these problems and anticipated problems have been instrumental in bringing about the policy change. This section summarises the problems and links them to an international discussion by using the Impoverishment Risks and Reconstruction model by Michael M. Cernea. Section four summarizes China's environmental developments from 1972-2001. It is assumed that China's increased environmental focus has positively affected the environmental developments for the dam project, and the specific environmental developments for the Three Gorges project are discussed here. Section five discusses the resettlement policy-making change in relation to the changing shape of decision-making in China, the increased importance of science and information in the decision-making process, and the media.

In the second part of this chapter I want to highlight some key points related to the resettlement and environmental processes in the dam project. Although based on empirical findings, they may be somewhat speculative, not the least as a result of difficulties in verifying information about the intricate Chinese decision-making system. In doing so I introduce some new elements that I have not previously discussed in this book. I find it legitimate to speculate a bit more freely in this manner in a concluding chapter—after having summed up the more 'solid' findings. This may point towards important uncertainties and perhaps also gaps in our knowledge, and thereby indicate directions for future research on the Chinese society in general as well as to this dam project in particular. The two factors to be discussed are: i) the *analytical approach* is discussed in relation to the resettlement and environmental policymaking for the dam project, as well as additional approaches to explain policymaking; and ii) the *importance of resettlement and environment* in the structure of decision-making for the Three Gorges project. Comments will be given on the level of importance for these two areas in relation to the dam. Furthermore, some ideas regarding possible ways to increase understanding of the social aspect of resettlement by applying the IRR model will also be presented here.

Summary

Three Gorges History

Sun Yat-sen first proposed to construct a dam in the Three Gorges in 1919. Due to several reasons, such as civil war, the dam did not materialise during the Guomingdang rule. Following the establishment of the PRC in 1949, renewed discussion about the dam began in the 1950s. The debate continued until 1992, when the project was approved by the National People's Congress (NPC) despite widespread opposition; 1767 representatives voted for the project, while 177 voted against and 664 abstained (which is interpreted as rejecting the dam). As the NPC traditionally is known to be a rubber stamp organ, the number of representatives voting against the project signifies the great opposition to the dam project within China.

The purposes for the dam project are power generation, flood control and improvement of navigational facilities. Construction was begun in 1994 at Sandouping in Hubei province. The dam affects Hubei province and Chongqing municipality. 1.13 million people (official figures) will have to be resettled. Chongqing municipality is responsible for 85 percent, and Hubei province for 15 percent. Since the construction of the project began in 1994, approximately, 645,200 people had been resettled by the end of 2002.

The political significance of the Three Gorges project debate is that it was not only a debate about the technical issues for a dam project, but also a pretext to criticise the way decision-making in general is carried out in the PRC. This criticism comes mainly from parts of the bureaucracy itself, as well as from scientists and journalists. It was not a grassroots debate; most people are not concerned about the dam project. It would also have been difficult to take part in such a discussion, as a common belief is that scientific issues should be debated among scientific experts, and one does not question the leaders' decisions. The criticism concerns issues such as lack of democratic and scientific decision-making methods, one-sidedness on the part of the ministry responsible for the project (MWREP), manipulation of information, lack of press freedom and freedom of speech, etc. Nevertheless, these arguments were effectively silenced following the crackdown on the student demonstrations in 1989. This might have created a favourable condition for the approval of the project, as a harsher political climate had emerged with the changes at the central power structure. Furthermore, the floods during the spring and summer of 1991 on the Yangtze River had created increased awareness of the need for flood control in the area, with much positive coverage of the project's flood control ability in the media. These factors, in addition to the fact that top leaders were perceived as having given consent to the project, made the approval of the project possible.

In sum, the Three Gorges is a controversial project with a long history of debate and with many actors involved. The debate was also used as a pretext to criticise the decision-making process in general. In the next section we will turn to the analytical framework employed in the book, which sheds light on the resettlement and environmental decision-making for the project (discussed in the second part of this chapter).

Analytical Framework

Chapter 3 gives a brief introduction to three general approaches that have been employed in the studies of policymaking: the rational choice model, the organisational process model and the bureaucratic politics model. The chapter continues with a brief introduction to some selected central analytic approaches and theories that have been employed to discuss politics and decision-making in China by Western researchers. The totalitarianism and the two-line struggle models have more or less been abandoned by scholars as a method to explain Chinese politics, due to the changes on the political arena in China and opening up of the country to the outside world. This development has increased the availability of new research materials to Western scholars and access to Chinese people in general. This has contributed to a higher level of sophistication in the analysis of Chinese politics. The approaches discussed were factionalism and elite conflict, the clientelist model, the interest groups model, the culturalist approach, informal politics, the bureaucratic politics model and the fragmented authoritarianism. The conclusion from the brief overview is that the approaches described in the chapter must be regarded as mutually complementary in portraying the political reality in today's China, as they describe different components of the Chinese policymaking process.

As discussed in Chapter 3, the fragmented authoritarianism approach is the basis for the analytical framework in this book. The fragmented authoritarianism approach has been valuable in describing the decision-making process for the Three Gorges in general, as illustrated in the study by Lieberthal and Oksenberg. They conclude that the post-Mao administrative and economic reforms have contributed to a fragmented decision-making process, which has resulted in a fragmented structure of the central authorities. In their case study of the Three Gorges decision-making process that was concluded in the late 1980s (before the dam was approved), the complex bureaucracies model has shed light on the actors involved and their influence in a process where bargaining among equal bureaucratic units was common, and required co-operation from other ministries and departments. In addition, as the fragmented authoritarianism model was developed in the 1980s and changes have since taken place in Chinese society, it may be necessary to draw on theories that focus on additional aspects of policymaking, such as the importance of information in the decision-making process, as well as informal politics and personal relationships (*guanxi*).

Resettlement Experience and Resettlement Policy Change

Chapters 4 and 5 concern the resettlement issues. Chapter 4 looks into dam building and reservoir resettlement in general, which are often controversial issues due to the costs inflicted, and in particular due to the social, economic and environmental impact. The purposes for dam building are usually irrigation, electricity production, flood control, and improvement of navigational facilities on rivers involved. Dam construction is often part of a country's development scheme, and has an overall goal to benefit the entire nation. Dam building has shifted from the industrialised world to the developing countries. Reasons for this trend are that most suitable rivers in the industrialised world have been developed, and that more information is available revealing a series of problems in relation to dam building. Thus, in the developed parts of the world there is much resistance towards dam building, and instead of constructing new dams, dams are now being decommissioned. In the developing world, dam building is still perceived as an important part of national development and modernisation of the country. Moreover, opposition is often not possible due to the political system. A debate about the need for dam building has been going on in the international community for a number of years. The establishment of the Inspection Committee by the World Bank in 1994 and the World Commission on Dams in 1997 at the Gland meeting sponsored by the World Bank/IUCN's initiative, were attempts to clarify rights and responsibilities with regard to borrowers and to create understanding for the arguments of the stakeholders on both sides of the fence. WCD consisted of global expertise on dam building, where experts held diversified opinions on the subject. This was an attempt to create a dialogue between the opponents and proponents of dam-building in order to reach an agreement on the way ahead with regard to water access and energy services. In 2000, the WCD report *Dams and Development, a New Framework for Decision-making*, concluded that even if dams had been useful for human development, dams have also created misery and suffering. The report determined that a new approach is needed in the dam-building world. Despite negative attention being paid to dam projects in recent years, a number of countries in the third world continue to construct dams. Conclusions from this chapter are: i) Dams are symbols of national development as well as barometres of development progress. National interests are more important than local interests; and ii) The international debate about dam-building has had consequences for China, as international funding institutions such as the World Bank have avoided the project. Controversial projects are no longer easily funded, due to much information and lobbying by international NGOs.

In China, 10 million people have been resettled due to water conservancy and hydroelectric projects, of which one-third are regarded as successful, one-third as fairly successful (still many problems) and one-third are considered unsatisfactorily resettled. Therefore, resettlement due to dam building is regarded by Chinese authorities (and others) as unsuccessful until the beginning of the 1980s. There are several reasons for this: i) Emphasis was put on the construction of the project and little importance was placed on resettlement; ii) The lack of regulations at the time, and the lack of comprehensive planning; iii) The method employed: basically a 'lump sum' (*yicixing*) compensation was given to the resettled people, with no thought of how they were to sustain themselves in their new environments; and iv) A lack of sufficient funding. Gradually, the Chinese authorities have increased their understanding of the social needs and potential social instability if these are not met. In 1985, 1986 and 1991 important resettlement regulations in general were issued that intended to improve the situation for the relocated population. Due to the problematic resettlement in early dam projects, the developmental resettlement scheme (*kaifaxing yimin fangzhen*) was gradually developed. The scheme was formed during the planning process for the Three Gorges project, and it is employed in the Three Gorges project. It is a new resettlement method, which seeks to address the subsistence problems experienced in earlier dam projects. There are several points that make the new scheme different from earlier experience: part of the new scheme was the trial resettlement which took place in the 1980s, which is regarded by the authorities as an important experience in relation to the resettlement challenge of the project. Part of the plan is the economic development in the Three Gorges area, which is intended to provide work for peasants who must switch from agriculture to other professions. Specific regulations for the Three Gorges resettlement were issued in 1993. The regulations state that resettlement methods to be employed are moving the people within their original villages, townships, cities and counties. If these methods are not possible, other methods such as moving to other provinces must be considered. There are four ways to resettle the people: i) *Jiujinhoukao* (in the neighbourhood), literally pushed up the hills in the vicinity of their previous homes; ii) *Jinqian anzhi* (settling nearby) is a bit further away; iii) *yuanqian* or *waiqian,* basically out of localities or to other provinces; and iv) *touqin kaoyou*, to be resettled through the help of relatives and friends.

One of the assumptions of this book is that the problems which emerged in the resettlement process have prompted the leadership to introduce the resettlement policy change. The number of people to be moved in the Three Gorges project is unprecedented in reservoir resettlement in China. Chapter 5 discusses this policy change that was introduced in May 1999 by Premier Zhu Rongji, i.e. to move (*waiqian*) a large number (125,000) of the rural population out to other provinces. The stated reason for this policy change is the lack of environmental capacity in the Three Gorges area, where erosion and pollution problems are serious. The 1998 floods along the Yangtze River were an eye-opener for the Chinese authorities with regard to the environment and population pressure in the area. In addition to the stated environmental reason, it is assumed that concrete and potential problems in the resettlement implementation process have been important in bringing about the resettlement policy change. Despite the resettlement policy improvements that have taken place since the 1980s, a number of problems have appeared in the resettlement process. The problems, which are related to the environmental capacity of the Three Gorges area are: lack of available farmland on which to settle the rural population; lack of farmland has led to steep-hill farming (above 25 degrees), which has increased the erosion problem in the area; uncertainty regarding the exact number of people to be resettled; allocation of low-yield farmland to the peasants; problems of finding or switching vocation for former peasants; loss of economic strength which leads to loss of social status; broken networks; the phenomenon of secondary resettlers (*er ci yimin*) who have lost farmland to infrastructure construction; failed official policy; corruption of resettlement funding and loss of education. The issues are structured according to selected points in the Impoverishment Risks and Reconstruction model (IRR) by Michael M. Cernea. There are eight points in the IRR model: from landlessness to land-based resettlement; from joblessness to reemployment; from homelessness to house reconstruction; from marginalization to social inclusion; from increased morbidity to improved health care; from food insecurity to adequate nutrition; from loss of access to restoration of community assets and services; and from social disarticulation to networks and community rebuilding. The selected points discussed in this book are: landlessness, food insecurity, homelessness, joblessness, marginalisation and social disarticulation. Some of the categories are grouped together, as the issues discussed under the various points are interlinked to such an extent that merging them seemed natural and more practical. In addition, two topics are discussed that are not covered by the model: loss of education and corruption. Some issues are lack of funding for the construction of shools, as well as mismanagement and embezzlement of resettlement funding that results in a reduction in the amount of money available for this purpose.

The chapter also discusses the Chinese authorities' response to the problems, which has been twofold: to move people out of the reservoir area to other provinces in China, and the issuing of a new version of the resettlement regulations in February 2001. Moving people out of their villages to other provinces has been a controversial issue, since distant-moving often has been unsuccessful in early dam projects. Moving to a distant place may also create new problems for the relocatees, as family networks are broken. Emphasis has been put on moving the relocatees to provinces in the Jianghan plains, which are situated close to or along the Yangtze River. Farming methods would be similar, and the cultural differences not so great. Nevertheless, there are reports of unsuccesful resettlement from these areas. The new resettlement regulations are founded upon the many problems that have emerged in the resettlement process described above. The main changes from the 93-version are: the principle of agriculture as a basis (*yi nong wei jichu*) is omitted, thus diminishing the heavy emphasis on agricultural resettlement; there is increased emphasis on environmental protection and rational use of natural resources; increased emphasis on supervision and management of resettlement funding, as well as an expanded section on penalty measures aimed at both officials in charge of resettlement funding as well as peasants who refuse to move or move back home after having been resettled. Regarding the responsiveness of the Chinese authorities to the resettlement problems, there is evidently a capability and flexibility to adjust strategies. One alternative that would reduce the resettlement problems and reduce the pressure on the environment would be to lower the dam height. However, this may have a negative impact on the perceived benefits of the dam, such as flood control and power production. It is unknown whether or not this alternative has been discussed by the Chinese leadership. The social costs from resettling people still appear to be regarded as a lesser cost than the perceived benefits (flood protection, power production and navigational improvements).

One conclusion to be drawn regarding the relevance of the IRR model in China is that Chinese authorities have a perception of the need to reconstruct relocatees' livelihoods. This is reflected in the terminology used for describing resettlement in China, as well as in the resettlement policy for the project. This awareness is based upon many years of resettlement experience in China as well as on interaction with multinational agencies such as the World Bank. The World Bank praises China and its resettlement for thorough planning in resettlement projects, as well as for taking advantage of resettlement as an opportunity to develop economically. China already carries out a number of the suggested measures in the IRR model and the country's approach to resettlement is 'risk-conscious'. Nevertheless, the IRR model could be a useful planning tool for Chinese authorities even though awareness already exists, as the model covers areas that are important for reconstructing people's livelihoods, including the social aspect. The Chinese emphasis in resettlement is put on rebuilding relocatees' lives; it focuses less, or not at all, on the social aspects and the social trauma of broken networks.

In sum, the past resettlement experience in China has prompted Chinese authorities to develop new methods, hence the development resettlement scheme. Nevertheless, the environmental capacity will restrain the successful resettlement of the large rural population. The increased focus on the environment in China has led to increased awareness of the link between the peasants' livelihoods and the environment. Environmental developments for China in general as well as for the dam project are summarised below.

Environmental Developments for China and the Three Gorges Project

Chapter 6 is a brief and simplified introduction to China's environmental history from 1972-2001, which is intended to put the discussion regarding the specific developments for the Three Gorges project into context. The chapter illustrates that China's environmental development has moved gradually forward, in particular with regard to institution-building and legislation, although lack of enforcement remains an obstacle for effective implementation. China's participation in the Stockholm Conference in 1972 was an initial environmental awakening. National environmental pollution incidents that took place at the time also made China aware of the need for environmental protection. Consequently, environmental institutions were set up in the 1970s. A few developments in that first decade of environmental protection stand out as particularly important, such as the passing of the provisional environmental protection law in 1979, which was the turning point for environmental protection in China. A number of laws were passed during the 1980s, and it was a time of legislative strengthening. Perhaps the single most important event during that decade was when NEPA became an independent agency in 1988. The 1990s stand out as a time when China's leaders really began to understand the seriousness of the environmental situation in the country; that environmental pollution may seriously affect economic growth and people's health. This is illustrated by leaders taking part in national environmental conferences such as in 1996, which has lifted environmental protection up to a new level. Furthermore, the public's concern about these environmental issues has been manifested in the number of environmental organisations, including NGOs, that have emerged during this decade. The major event in the 1990s is the elevation to ministerial level of the environmental administration, SEPA, in 1998. Nevertheless, institution building and policy making is not identical to environmental action and enforcement of environmental legislation. The country still faces great challenges with regard to implementation of its environmental policy. As this goes beyond the intended scope of the chapter, it has not been discussed. A challenge ahead concerns the ability to implement central policy at the local level, which is related to behaviour challenge and increased awareness among lower level leaders. The extent to which SEPA will be able to carry out its role (since it is only on ministerial level, and not yet a full ministry), also remains to be seen.

For the Three Gorges project these environmental developments have been important. This is discussed in Chapter 7. In the early discussions of the Three Gorges project little attention was paid to the environmental issues related to the construction of the dam, as well as the relationship between the environment and resettlement. In the 1980s, the debate to launch the Three Gorges project intensified, and numerous studies were carried out. In 1984 the government really began to pay attention to the environmental issues of the Three Gorges project, when CAS initiated important studies about the project's environmental impacts. The study was significant due to its comprehensive character and thoroughness with regard to the environmental impact and resettlement. This was a new approach compared to research that had been carried out by other institutes, where the research agenda was more or less set. As a result of the research carried out, new issues appeared that made clear the need for further studies. One might therefore conclude that researchers played an important role in setting the agenda for the environmental discussion for the project. For instance, the Three Gorges environment was included as a research topic in the Seventh Five-Year plan (1986-1990).

In 1991, CAS and MWR completed the compilation of the *Statement*/ EIA for the dam project. Despite the different opinions in the conclusion by CAS and MWR, and the fact that several experts among the 55 experts who evaluated the *Statement*/EIA did not sign, NEPA approved the *Statement*/ EIA in 1992. There are several points to be noted from the research process leading up to the approval. Construction and design units under the MWR have carried out EIAs for the project, which can bring the objectivity of the studies into question. The NPC Environment and Resources Committee therefore put pressure on MWR to compile an EIA for neutrality and quality purposes. Consequently, institutes under CAS and MWR jointly compiled the *Statement*/EIA for the dam project, based on earlier research carried out by the two. The State Council approved the project before the *Statement*/ EIA was approved, which brings into question its significance and the seriousness of the Chinese government at the time regarding the environment and resettlement problems. Knowing that objective reporting has been difficult in the process, and that scientists from CAS were criticised because their scientific view was interpreted as a personal view, it is significant that a few scientists did not sign the final evaluation of the *Statement*/EIA.

Succeeding the project approval by the NPC in 1992, a new era began. Four developments during this period stand out as important: i) A new reality resulted in a shift in scientific focus. The debate about the project leading up to the 1992 approval was oriented towards the positive and negative aspects of the dam project, including for the environment and resettlement. After the project had been approved, the purpose of discussions and research was to minimise environmental degradation and propose counter-measures for environmental damage; ii) An environmental network was set up with participation from a number of ministries, such as NEPA, MWR, Ministry of Agriculture, etc. Each year a report is compiled by SEPA about the environmental situation in the Three Gorges area, with input from the participants in the network; iii) Environmental supervision and management have become increasingly important with regard to environmental protection, in particular in the construction area. Environmental management regulations have been issued for this area. Despite these regulations, problems exist regarding implementation and specialised personnel; iv) NEPA's role in the environmental process has been modest, and its main involvement has been inspection tours to the area and compiling the annual environmental report as well as monitoring water quality for the project. One important reason for this was NEPA's lack of strength compared to other ministries.

In 1998 yet another stage in the environmental developments of the dam project commenced when Zhu Rongji became Premier and head of the TGPCC. When the resettlement policy change took place in 1999 due to the environmental capacity in the reservoir area, resettlement and environment were linked together in a new way, and the issue of mutual dependency was lifted up to the highest political level. Furthermore, several factors have been important in influencing environmental policy developments for the project: devastating floods in 1998 made the government realise the need to curb erosion along the Yangtze River. A petition written by 53 experts to Chinese leaders pointed to siltation, environmental and resettlement problems. Internal communication between a Qinghua professor (head of the group of 55 experts that investigated and approved the final *Statement*/EIA for the dam project in 1992), Chinese leaders and the EPB of Chongqing municipality was leaked to Probe International regarding the potential catastrophe related to the environment and resettlement activities in Chongqing municipality. Furthermore, significant environmental criticism was published in two major newspapers in China: environmental concern was illustrated in the *People's Daily Overseas Edition* when Chongqing vice-mayor and the city's EPB criticised the environmental work being carried out, and a bleak picture of existing and future problems was drawn in the *Guangming ribao*. The process above may have led to increased environmental funding for Chongqing, which will be beneficial for both the environment and the population living in the area.

In sum, the process from the 1970s until 2001 illustrates that positive environmental developments have taken place for the Three Gorges project. It has revealed an increased emphasis on comprehensive environmental research for the project, and science has become increasingly important for the project's decision-making. Nevertheless, it must also be noted that where politics and science were in conflict regarding environmental policymaking for the project, the political forces involved were stronger than science. Dissenting scientific views were interpreted as personal views, which indicates the dilemma in carrying out scientific research for the Three Gorges project. The resettlement and environment developments are linked to the changing shape of decision-making in China that are summarised in the next section.

The Changing Shape of Decision-making, and the Changing Chinese Society

Chapter 8 discusses the possibility that the changing shape of decision-making in China may have indirectly prompted the resettlement policy change for the dam project. The chapter gives a brief review of political developments in Chinese society which is necessary to understand the recent changes in the Three Gorges project. A debate was ongoing regarding political reform in the 1980s, which also influenced the Three Gorges discussion. The project was discussed more openly in the 1980s, and was 'closed' again due to the rigid political environment following the crack-down of the student demonstrations in 1989. In addition, the beginning trend in the 1980s when decision-making increasingly became based upon information and knowledge, was important for the developments of the Three Gorges project. New institutional actors thus appeared on the scene, as consultative organs on a frequent basis were included in the policymaking process. Policy decisions now increasingly rest in the State Council and Leading Groups. Think tanks are more important now, as leaders such as Jiang Zemin and Zhu Rongji depend on these for information in the decision-making process to a greater extent than before. Some of the think-tanks used by Zhu are named *Guojia jingji tizhi gaige bangongshi* (the Office for Restructuring the Economy), The *Guowuyuan yanjiu shi* (State Council Research Office) and the Chinese Academy of Sciences which has increased its status under Zhu. One reason for the increased importance of research results and knowledge is that China's leadership now has higher and more diversified education and backgounds. An increasing number of financial experts and lawyers, who belong to the fourth generation of leaders, are included in important posts in China.

The above developments have been important for the resettlement policy change of the Three Gorges project as well. Specific factors discussed in this chapter which might have been important are: change of leadership (Premier) in China in 1998 and the subsequent change in the Three Gorges project leadership; the scientific advice provided to the leadership; and the media coverage in the late 1990s following the leadership change. In this chapter I assume that the interaction between these factors: the leadership, knowledge-based decision-making and the media, has had an important impact on the dam project, and may have made the resettlement policy change possible.

A scholarly debate has been going on for many years about the resettlement and environmental issues. This chapter introduces three categories of journals where scholars engage in discussions about the resettlement and environental issues; two of these categories are discussed in this study. Their views, often critical, have been reflected in the many scientific journals described in this chapter. One article in *Strategy and Management* (Zhanlüe yu guanli) in January 1999, stands out as particularly important in the scientific debate about the consequences of the Three Gorges dam resettlement. The author points to a number of difficult factors in the resettlement, and believes that the dam resettlement will fail. These critical points have seldom been reflected in the media. Nevertheless, the chapter concludes that a shift took place after Zhu became Premier. One way to study the leadership change is through the media coverage of the dam project. An increased number of articles were published in the media pointing to problems in the process and portraying more realistically the resettlement process and the environmental problems. Articles in the *People's Daily* focused on issues such as quality, corruption, resettlement and the environment. Criticism concerning the environment in the state media by high-level leaders in Chongqing is a significant change from earlier reporting. A second conclusion is that critical points and views that have been able to appear in scientific journals for years, have in the late 1990s been 'allowed' to appear in the state media. However, this concerns mainly the environmental aspects of the dam, and not the controversial issue of resettlement.

Thus, the chapter concludes that the change of leadership, i.e. shift of Premier and TGPCC chairman, has had significant impact on two areas related to the Three Gorges project: the resettlement and environmental policy, and the media reporting about the project. More importantly, the resettlement policy change may be linked to Zhu's character and style of leadership. One issue arises in this regard: the decision to move 125,000 of the rural population from the reservoir area to other provinces may not have been controversial, and may have received support from the leadership and the bureaucracy. On the other hand, it is known that past resettlement in China was unsuccessful, and part of the reason was that people were moved far away. Thus, one may get the impression that out-moving may not have been an easy choice. Moreover, Zhu is perceived as having a weak power base and does not appear to have any patron-client relationship. Therefore, taking into consideration the past resettlement experiences in China and that outmoving most likely was not a favoured alternative, the effectiveness of Zhu's policymaking in the Three Gorges resettlement policy change may be questioned. The chapter discusses Zhu's role in the resettlement decison-making change, as well as the absence of policy change during Li Peng's tenure. Some important factors behind these changes are the following: the project was not initated by Zhu, so he has greater freedom to focus on the actual problems; he is viewed as a pragmatic person who focuses on problems and solving them, as opposed to Li Peng and may explain the absence of policy change during Li's Premiership; Zhu represents a different leadership style—he is more at ease with the media, and relies on think-tanks to assist him in forming policy. The chapter concludes that Zhu's continued support from Jiang Zemin, the increasing power of the State Council and Leading Groups in decision-making, as well as the group identity of the Qinghua network may be factors that have enabled Zhu to exert critical influence on the policymaking for the Three Gorges resettlement.

In sum, the changing shape of decision-making in an increasingly pluralistic society has had an important impact on the Three Gorges resettlement and environmental policy. Zhu Rongji's direct and problem-solving leadership style differs from other leaders, and he has been instrumental in forging changes in the resettlement policy. After Zhu became Premier several articles appeared in the media that focused on problems with the project.

Concluding Remarks

As stated in the introduction to this chapter, the second part will concern comments on findings in the book that stand out as particularly important. This part discusses the decision-making in relation to the fragmented authoritarianism and other approaches as well as presenting personal observations from the course of the study.

Decision-making and Bargaining

According to the fragmented authoritarianism approach, decision-making in China involves building consensus among equal units and in the process of reaching consensus, bargaining takes place. The fragmented authoritarianism approach appeared to be a relevant way to explain decision-making for the dam project in the 1980s. Nevertheless, Chinese society has undergone many changes in the last decade. This book puts main emphasis on the developments in the 1990s and 2000/2001, and this model will be evaluated on the basis of its fruitfulness in relation to the topic discussed, i.e. how the analytical framework works in relation to the environmental and resettlement policymaking for the Three Gorges project. The issues to be looked into are: the role of bargaining in the process; in which areas did bargaining take place; and the interplay between the main actors and their role in relation to the decision-making structure identified by Lieberthal and Oksenberg. Furthermore, Lieberthal and Oksenberg stated five factors that can propel an issue onto the agenda of the top leaders (see Chapter 3). In this book, three points seem to have been particularly important in relation to the Three Gorges resettlement policy change: the particular interest of a top leader, the emergence of a critical problem, and bureaucrats (in the sense of centre versus locale) forcing an issue onto the agenda. It is important to stress that the analysis below is mainly based upon interviews and written sources, but that some parts may be called informed speculation, as it is not possible to confirm certain aspects of the analysis.

As we remember from Chapter 3, authority according to the fragmented authoritarianism model consists of four tiers. Among the core group of leaders that make decisions in China, several of them are in the TGPCC (see Table 2.3). This illustrates the importance paid to the project in China. Zhu Rongji, who is the leader of the TGPCC, ranked until the 16th Party Congress as number three in the party hierarchy (Li Peng still ranked as number two after Jiang Zemin). With the fragmented authoritarianism approach in mind, it is interesting to ask why and how the decision to move 125,000 people out to other provinces was made. First of all, the resettlement process has been problematic, as shown in Chapter 5. Zhu is percieved by scholars and officials as a person who acts and responds to problems, which according to intervieewes makes him different from many other high-level leaders of his generation. In relation to the Three Gorges project, leaders such as Li Peng are perceived as focusing on the positive aspects only. Zhu is not known to have shown a special interest in the dam (for or against), but as Premier and head of the TGPCC he has taken over the responsibility for the dam. Furthermore, Zhu is perceived as a leader who is interested in environmental protection within the framework of economic development. Thus, one may conclude that the particular interest of Zhu (his interest for the environment), as well as the emergence of a critical problem (the flood crisis in 1998) and the encountered environment related resettlement problems, have moved the issues on to his agenda. Zhu Rongji has thus been important for the resettlement policy change for the dam project, as has been illustrated in Chapter 8. Nonetheless, taking the fragmented authoritarianism approach as basis, where consensus is important, the decision to move out more than one-third of the rural population cannot have been made by Zhu alone. Furthermore, the issue may have been debated, due to the past experiences in out-moving in China. In the resettlement decision-making process, Chongqing municipality, formerly part of Sichuan province, stands out as a major stakeholder and bargainer for several reasons. First of all, it became a *zhixiashi,* an administrative city directly under the central government, in 1997 due to its role in the Three Gorges project. Without having made any specific study of this, one may imagine that Chongqing's becoming a *zhixiashi* also was a bargaining issue. When taking on the responsibility for resettling 85 percent of the people as a *zhixiashi*, Chongqing in return would receive increased funding and prestige, and would join the exclusive club of *zhixiashi* with Shanghai, Beijing and Tianjin. Secondly, Chongqing municipality has the largest resettlement burden. Following the increasing problems in the resettlement process, which were partly due to the environmental situation in the area, some sort of negotiation may have taken place between the TGPCC and Chongqing municipality. It is likely that pressure has come from leaders in Chongqing municipality due to the lack of available farmland and the existing erosion problems in the area. A third bargaining 'card' may have been the city's location at the back end of the reservoir. The potential silting up of the reservoir may have negative impact on ships reaching its port. Finally, the vice-mayor of Chongqing, Gan Yuping, and the former mayor, Pu Haiqing are both members of the TGPCC and would be sure to look after Chongqing's interests. Thus it may be possible to state that bureaucrats, in this case local interests, may have forced an issue onto the agenda of the TGPCC leadership and Zhu Rongji.

In the environmental funding process, the bargaining may appear more distinct. As discussed in Chapter 7, several critical articles were published by the state media. One was criticism by the vice-mayor of Chongqing municipality and the head of the municipality's EPB. The second was, according to sources, by a scholar from CAS. Both were very critical towards the environmental efforts in the reservoir area, or rather, the lack of attention paid to the environmental pollution problems. Moreover, leakage of information that was provided by Probe International suggested that some negotiations were taking place behind the scenes between Chongqing municipality, the EPB and the TGPCC. In their arguments, Chongqing EPB used Shanghai as an example and a means of bargaining, since Shanghai receives more funding than Chongqing for wastewater clean-ups. As both cities are directly administered by the central government, Chongqing, by mentioning Shanghai, demands equal treatment. Furthermore, as a poorer city with a large rural population situated in the less developed inland of China, Chongqing may feel inferior to Shanghai, which is a modern and richer city on the developed east coast. Thus, an element of competition for funding between these two cities, both situated along/by the Yangtze River, can be detected. Furthermore, the 1998 flood and its potential impact resulting in increased importance being put on the environment issue, may have been advantageous for Chongqing in the process regarding environmental protection funding. Following the internal negotiations between Chongqing and the TGPCC, it was then announced in the state media that Chongqing municipality would increase environmental spending and spend RMB 44.48 billion over the next ten years on environmental clean-ups. The article did not state specifically where the funding would be coming from. However, Chongqing EPB confirmed to this author that the central government would provide a large portion of the funding (in addition to some funding from Chongqing itself). This is a typical example of bargaining over financial resources, and illustrates that the environment is also an arena for bargaining.

Chongqing municipality and the Environmental Protection Bureau's (EPB) participation in the environmental criticism is also worth a comment, as the role of SEPA in general in this project has appeared rather modest. It is difficult to say to what extent SEPA was directly involved in the environmental criticism from Chongqing. EPBs, which are subordinate both to SEPA and the local governments, are often more loyal to the local governments who provide much of their funding. SEPA has carried out environmental inspection tours in the reservoir area over the years, and scholars and others have pointed out that the environmental administration is known to be critical of the lack of environmental protection in the reservoir area. However, as has been discussed in previous chapters, SEPA's authority in this process has been questioned. Scholars interviewed by the author have stated that SEPA does not dare to say anything against the project that will offend the leaders, or the Ministry of Water Resources in particular. Some scholars even stated that there is no use participating in research projects, since it would only be in the service of the proponents, the Ministry of Water Resources and other organs, and not to pursue the real issues. One put it this way in 1999: 'If there are problems, the only alternative is to find measures to solve them. You do not tell them [TGPCC] that they cannot be solved. Thus, SEPA's participation is only in the service of the proponents MWR and other organs, only in order for things to look better than they are (*tuzhi mofen*). As for the reservoir pollution, there are no concrete actions yet, including for discharge. SEPA should be in charge of this, but has no funding for it'. With regard to EPB's role in this, there may be a combination of several possibilities. SEPA's role in the system (*xitong*) and its bureaucratic rank in the vertical (*tiaotiao*) line implies that it has the authority to instruct the EPB environmental work in Chongqing. However, as pointed out above, this does not not always function, as the EPB is also 'subordinate' to the local government. This complicates matters further, as SEPA's bureaucratic rank, presently on a ministry level, implies that it cannot tell a provincial, or in this case a municipal government, what to do. Moreover, the local EPB and the local government would often have a conflict of interest regarding economic development and the environmental work, which most often results in the EPB conceding to the governemnt's wishes, being the weaker part. In the case of the environmental criticism for the Three Gorges project, it would be diffcult to verify what exactly took place. One option would be to assume that the criticism by the Chongqing EPB of the municipal government (referred to in the leaked correspondence to the TGPCC leadership) as well as the Yangtze River Water Resources Commission for not having carried out sufficient environmental work, would put the municipal government in an awkward position. This may have resulted in the government and the EBP finding common ground and ending up jointly criticising the lack of efforts in order to increase the possibility for funding, and perhaps to avoid being criticised by higher authorities. A second possibility is that the members of the TGPCC have intervened and 'suggested' the public criticism in order to increase the attention towards these problems. The publicity given to the environmental problem in the area both in the state newspapers and the CCTV implies that the reproach may have been backed by the TGPCC leadership. In concluding this section on the environmental criticism, SEPA may have played a more important role in putting the environment issue on the agenda than what first appears, by working behind the scenes together with the municipal EPB. Also, Xie Zhenhua, the minister of SEPA, is a graduate of Qinghua (in 1977), which could have opened the doors to the Qinghua network. Potential influence by SEPA may have been one of several actions that influenced the TGPCC to act, although it is impossible to know for sure what finally made it possible to highlight the criticism towards the lack of environmental efforts in several articles.

From the above, one may conclude that the fragmented authoritarianism approach still appears valuable in explaining developments in the Three Gorges project resettlement and environment policy. Leaders have become aware of the problems regarding both the environment and the resettlement by taking a personal interest in matters or by problems or disasters that have come forth. Bargaining has taken place, and Chongqing municipality stands out as a strong actor in the environmental and resettlement process. Nevertheless, as was emphasised in Chapter 3, the changes in Chinese society during the past decade may require additional approaches to the Three Gorges resettlement and environmental decision-making. With regard to environmental decision-making for the dam, I would say that in addition to the five points that were listed by Lieberthal and Oksenberg that determine how and what topics arise on the agenda of the leadership, a sixth point should be added, namely information and scientific research.

In the area of providing information, two important actors in the process stand out: the Ministry of Water Resources (MWR) and its subordinate institutions, and the Chinese Academy of Sciences (CAS) and subordinate research institutes. MWR has played an important role in the history of the project, as it has been involved in the dam as a proponent since the very beginning. During the discussions in the 1980s and 1990s, the MWR worked hard to have the project approved. With such a big project, a lot of money and prestige were involved. Furthermore, its subordinate institutes have carried out feasibility studies for the dam. For instance, the Yangtze River Water Resources Commission has been actively involved in many of the studies for the project. There is great prestige in solving the technological problems, and many of the staff would aspire to become academicians of the Academy of Engineering. Due to the substantial interests involved, research by the proponent's subordinate organs may have been biased to some extent. As discussed in Chapter 7, CAS initiated important studies beginning in the mid-1980s that were more comprehensive. The significance of this was that CAS was not a subordinate organ of the MWR, and was able to carry out more independent research, providing the TGPCC with diversified information. Yet, it should be acknowledged that researchers also may have a vested interest in the project, since it would provide research funding as well as opportunities for career enhancement. From the study, one may conlude that the MWR's role in providing information to the leadership diminished in the process, while CAS' participation received increased importance. This would be in line with the trends in Chinese society at large, where think tanks have become increasingly important and active in the decision-making process. It has been stated that it is difficult to verify the extent to which the scientists' opinions might be heeded in the shaping of policy. One scholar said 'CAS can influence the government; the opinions of the scientists are the actions of the politicians (*kexuejia de yijian, zhengzhijia de xingwei*)'. However, judging from what has been referred to earlier in the book, that science is a tool for politics, the exact opposite is implied. Yet technocratic problem solvers have increasingly turned to experts in various institutions in order to make decisions. It is plausible to think that with the ongoing generation change, from the third to the fourth generation of leaders, this trend will continue. Nevertheless, participation by intellectuals in providing information and thoughts to the leadership is not without risk. There are a number of people who have lost their jobs, missed out on promotion, or have been harassed due to being too outspoken about the negative impacts of the Three Gorges project. On the one hand, there is an ongoing positive trend where scientists, economists, and other experts are gradually becoming a greater part of the decision-making process. On the other hand, there is still a certain amount of control of the intellectual atmosphere despite occasional loosening. Finding the balance here may be a challenge which is illustrated by the recent escape by He Qinglian, and the fired editors in the *Southern Week-end* (Nanfang Zhoumo).

The fragmented authoritarianism approach has emphasised control over resources (particularly financial) in the decision-making process. With the changes in Chinese society, the importance of information must be given a greater role than is the case with this approach. Both Naughton and Halpern have stressed the importance of control of information in policymaking in China. We have seen in the book that information has gradually become more important for decision-making in general. Cheng Li has stressed the increased importance of think tanks in decision-making. This development has been particluarly important in relation to increasing competition for policy proposals and recommendations between the line minstries and the research institutions. In the Three Gorges project, the roles of the MWR and the CAS in the decision-making process are also related to the control of information. The MWR has attempted to control the information flows within the Three Gorges bureaucracy and had for a period monopoly on information. This had to do with the MWR's role as proponent of the project. However, in the mid-1980s, CAS entered the scene to a much greater extent and for two decades provided extensive studies about the Three Gorges resettlement and environmental impacts. The MWR was no longer able to control all input to the policy process, which eventually changed the policy agenda for the dam as new issues were added. On another level, control of information occurs between institutes that participate in research about the Three Gorges dam. As mentioned earlier in the study, partners in co-operation projects do not always exchange the necessary data, due to competition among the institutes in influencing the policymaking.

The increase of information and its importance is not only special for the Three Gorges project, but is also known from the Chinese bureaucracy in general, where it is difficult to share information across horizontal and vertical lines. This relates to what Joseph Fewsmith has pointed to: the importance of informal politics and the expanding role of 'intermediate' associations in Chinese society, while Dittmer and Pye emphasise the role of interpersonal relationships (*guanxi*) in decision-making. These ideas are also relevant in relation to the Three Gorges resettlement policy change, and should be viewed as supplementary to the fragmented authoritarianism approach. The fragmented model focuses mainly on the formal structure of decision-making and thus fails to shed light on the informal processes that most likely take place in decision-making in China. The increased importance of input from think-tanks and research institutions is partly the reason for this. Also, the pluralistic trend in Chinese society makes it possible for other groups to informally exert influence over the policymaking process. Such a development may have rendered personal relationships and common backgrounds even more meaningful, and leads to competition for the bureaucratic apparatus regarding the provision of advice and input. In the Three Gorges decision-making process, this aspect is important with regard to the common backgrounds of the TGPCC chairmen that belong to the Qinghua network. This may have been important in relation to bringing new issues on the agenda in a rapid manner. Fewsmith emphasises that informal politics has been critically important for introducing new policy recommendations. One example is the increased funding for environmental protection efforts which may not have occurred as rapidly through ordinary bureaucratic channels. Also, Li Peng, who has no longer an official role in the dam project, still appears to be consulted regarding issues relating to the dam, and is also frequently cited in the media. The information leaked from Probe International (discussed in Chapter 7) referred to Zhu informing Li Peng about the environmental problems in the Chongqing reservoir area, although he no longer has a formal position in the project bureaucracy. One recent example is the closure of the human-made diversion channel of the Yangtze River at the Three Gorges Dam site in Hubei province in November 2002. This event was carried live on national television and watched on site by Li Peng. Li's presence at the closure is interesting as it would seem natural for the present chairman of the Construction Committee, Premier Zhu Rongji, to be present at such an important event. Officially, Li Peng has no position in the project authorities, although being the chairman of the NPC that approved the dam in 1992 as well as having a background in hydro engineering, makes his continued interest legitimate. Nevertheless, it seems clear that Li still has an informal role in the project and closely follows the project developments. Li Peng as the main promoter of the dam still has great interest in seeing its successful completion. Thus, one may say that the informal aspect in the Three Gorges decision-making process definitely is important, and has become more important in the nearly 15 years that have passed since the fragmented authoritarianism model was published in *Policymaking in China*.

In sum, the fragmented authoritarianism approach has been a useful tool for analysing the decision-making process for the dam project. Bargaining has taken place in both the resettlement and environmental areas. In addition, information and informal politics are important aspects of the decision-making in China today.

Environment and Resettlement: Level of Importance at the National Level

The linkage between environment and resettlement has become apparent in this dam project, and culminated with the decision to move part of the rural population to other provinces. This indicates an increased awareness in both these areas. Nevertheless, the concern for the environment and for resettlement appear to be on different levels, and the potential explanations for this will be discussed below.

As discussed in Chapter 5, the environmental capacity in the Three Gorges area was the main reason for out-moving. The attention paid to the relationship between resettlement and the environment in this project, in particularly after 1998, may be unprecedented in a Chinese funded dam project. Increased funding for environmental clean-ups for the entire Three Gorges area in 2001, including the ten-year plan for the protection of the environment, illustrate the significant environmental developments for this project. The importance given to the environment is related to several factors. One important factor is that protection of the environment has been one of two national policies for several decades. This automatically raises the environmental issue to a national level of attention. Moreover, despite the fact that economic development has been given priority over environmental protection for a long time, the environment has reached a level of attention among China's leaders since the mid-1990s, as demonstrated in Chapter 6. The main reason for this development is the restraint that environmental problems and the depletion of natural resources put on economic growth. Thus, one may say that within the framework of economic development, the environment has received a higher status on the general agenda in China. This national focus on the environment has been positive for the Three Gorges environmental policymaking and has resulted in increased funding for environmental protection. The economic development aspect is important in this area as well, as it is expected that the project will induce economic growth, through for instance the counterpart support system (*duikou zhiyuan*) as well as the 'go-west' policy (*Xibu da kaifa*). Furthermore, the water level in the reservoir will rise in 2003 when power production begins, which puts pressure on the Chinese authorities to act swiftly. Without critical measures such as the planned construction of the water treatment plants in the area, serious water pollution in the reservoir and in the Yangtze River will be the result. One may therefore conclude that the protection of the Three Gorges environment has become a top priority for the Chinese government. Concern exists, nevertheless, regarding the efficiency of policy implementation.

It is nevertheless a paradox that this positive environmental development in the Three Gorges project will most likely also have negative human consequences. For environmental capacity reasons, one-third of the rural population is to be resettled out of the reservoir area. Scholars have stated (Chapter 5) that they expect the resettlement of this population to be mostly successful, but uncertainties with regard to both livelihood and their social inclusion in the host areas exist. There is no doubt that China has greatly developed and improved resettlement regulations and practices in the past decades. The reconstruction aspect is present, as well as a 'risk-conscious' approach where emphasis is put on restoring livelihoods. The great focus on the resettlement progress in the media illustrates the authorities' concern for and importance placed on the issue. However, in addition to the livelihood issue, the resettled population's satisfaction would depend on their ability to penetrate and become a part of the society in their new home areas. The side effects of moving into new environments, sometimes hostile as competition for resources increases, are not emphasised; nor are the human dimension and the issue of social reconstruction. What then is the reason for this? The central government sets the rules and regulations for resettlement, while the provincial and local authorities are in charge of the actual resettlement; it is their responsibility to solve the problems related to the resettlement. This has resulted in resettlement being regarded as a local issue, and to some extent a regional issue. The main focus in the Three Gorges resettlement process is on restoring people's livelihoods, such as compensating land and houses. Little emphasis is placed on restoring the social well-being of the relocatees. One may conclude that the *social* impacts from resettlement have not yet reached the highest level of attention. This may not come as a surprise in a country of nearly 1.3 billion, where dams are often perceived as symbols of national development (as in other developing countries), and one is obliged to sacrifice individual interests for the sake of the common good.

One conclusion to be drawn from the above is that Chinese authorities will have to respond to the problems that are intensifying as the resettlement process continues. One initial step in the right direction may be the planned study to be compiled by the Chinese Academy of Social Sciences (CASS) and the Resettlement Bureau under the TGPCC regarding the resettled population's conditions and human rights, as mentioned briefly in Chapter 5. This may be the beginning of increased focus on these issues in the Three Gorges project. Although, as has been indicated by Chinese scholars, the experience from the co-operation between CAS and MWR with the EIA was that the research results were heavily influenced by the proponent (MWR). Thus, it would be hard to predict the potential impact of CASS on this study. In addition to this recent inititative, I believe that carrying out research regarding the application of the Impoverishment Risks and Reconstruction (IRR) model in China could be useful. Chapter 5 concludes that the IRR model is important for China, and would be an important contribution to the country's resettlement practice. The model also contains a research part which could constitute the starting point for further systematic studies about resettlement in China in general, as well as for the Three Gorges project. Cernea (1997) states that the model can be used both for operational research, i.e. preparation for projects that involve resettlement, or for basic social research to formulate hypotheses and carry out testing of the model's variables under different circumstances. The most pressing aspects for China to focus on would be the points concerning 'from marginalisation to social inclusion, and social disarticulation to community construction'. As far as I know, this type of research has until present been very limited in China.[727] Some ideas for using the IRR model to carry out research, which subsequently may lead to policy, are as follows:

1. Analyse how structures could be established for including resettled people whose networks were dismantled into the host areas, identify the problems that occurred, and how they could be better solved in the host areas.
2. Look into differential impacts of resettlement in the project. One example from the study is children, who are particularly vulnerable as they may lose opportunity for schooling.

727 One example of ongoing research on these issues is a book that concerns various forms of insurance against the adversities/risks of displacement. Personal communication with Michael M. Cernea.

3. Investigate how channels could be established in order to obtain feedback from Three Gorges relocatees after they have been relocated (both within and outside the reservoir area).
4. Put forward ideas for establishing channels for communication between the relocatees and the authorities. Often when problems arise, the resettled people need to make contact with the authorities. However, the authorities in the old and the new host areas both claim that they are not responsible. The responsibility must be made clear.

The conclusion from the above is that the Chinese government needs to raise the social aspect of the resettlement up to a higher level than at present in the Three Gorges project, and needs to identify ways to do this. It needs to *acknowledge* that resettlement has social costs, and that it is problematic for the relocatees when families and friends are split up and when the ancestral land has to be abandoned. One important contribution in this regard would be the establishment of a law on the protection of people's rights and interests in reservoir-induced resettlement. It is too simple to state, as in the 2001 regulations, that the dam will create benefits for the 'country, the collective and individuals' (article 5). The Three Gorges area is already experiencing social disruption due to the problems from resettlement, and the response to the protests by local governments as reported by the Western press, has been to arrest the peasants who have dared to initiate protests. Arresting these individuals, who have voiced their concern about the resettlement process, will only increase the anger among fellow resettlers and influence social stability. Many of the problems are due to the rampant corruption which reduces the amount of money intended for resettlement. The challenges related to uprooting and moving to a new place come in addition to this. If left unresolved, the problems may only increase.

It should be emphasised that the case of the Three Gorges project is special even in China, due to the large size of the dam and reservoir, the potential environmental impacts, the long controversy surrounding the project, the many interest groups, the corruption problems and the number of people to be resettled. Thus, it may be difficult to make generalisations based on the developments for this dam project. It is important to acknowledge China's efforts in identifying resettlement practices where systematic measures are initiated for preventing impoverishment, which may not be matched in other developing countries. We have seen in the Three Gorges project that Chinese authorities have responded to the problems by moving people out and issuing new resettlement regulations. One may therefore conclude that these years have been a positive learning process for the Chinese authorities. However, the economic circumstances have changed for the dam project since it was first approved in 1994. An important issue is whether it has been possible for the Chinese authorities to adapt the project to the changes. The large size of the project would indicate the difficulty in doing this, as the project concerns such a vast number of areas in Chinese society. The resettlement funding for instance, has not been adjusted even though the outmoving may increase expenses for resettlement. Also, with regard to administration of the project at the local and regional levels, many scholars have suggested establishing one overarching organisation that can take care of the issues pertaining to the environment and natural resources. This would be an acknowledgement of the problems that exist in co-ordinating a number of ministries and river basins that are not eager to co-operate. The examples given here illustrate the need to question the feasibility of constructing such large and long term projects. This becomes particularly apparent in a society like the Chinese, which is experiencing such rapid changes both in the economy and in society at large.

The complexity of the issues in question for the Three Gorges project has resulted in an increased level of attention being paid to the dam project at large, including the resettlement issue. The long process and debate about the project has contributed positively to an increased awareness of the linkage between the resettlement and the environment. This may be important for other reservoir resettlement projects in China. Nevertheless, it is worthy to note that scholars interviewed by the author believe that the dam would never have been approved by the NPC today, due to the potential difficulties from resettlement and the possible environmental impact.

Bibliography

Books and Journals

Allison, Graham T. (1971), *Essence of Decision: Explaining the Cuban Missile Crisis*, Little, Brown and Company, Boston.

– (1969), 'Conceptual Models and the Cuban Missile Crisis', *The American Political Science Review*, Vol. LXIII(3), September 1969, Harvard University, Cambridge, MA, pp. 689-718.

Ash, Robert F. and Edmonds, Richard Louis (1998), 'China's Land Resources, Environment and Agricultural Production', *China Quarterly*, December, No. 156, pp. 836-79.

Barber, Margaret and Ryder, Grainne (eds. 1990), *Damming the Three Gorges, What Dam Builders' Don't Want You to Know*, Earthscan Publications Limited, London and Toronto.

Barnett, Doak A. (1985), *The Making of Foreign Policy in China: Structure and Process*, Westview Press, Boulder, Colorado.

– (1964), *Communist China: The Early Years 1949-55*, Pall Mall Press, London.

Bergesen, Helge Ole, Parmann, Georg and Thommesen, Øystein B. (eds. 1999/ 2000), *Yearbook of Interational Co-operation on Environment and Development*, The Fridtjof Nansen Insititute, Earthscan, London.

Bettelheim, Charles (1974), *Cultural Revolution and Industrial Organisation in China*, Monthly Review Press, New York.

Bissel, Richard E. (2001), 'A Participatory Approach to Strategic Planning, Dams and Development: A New Framework for Decision-making', *Environment*, Volume 43(7), pp. 37-40.

Brook, Timothy, and Frolic, B. Michael (eds. 1997), *Civil Society in China*, M.E. Sharpe, Armonk, NY.

Bruun, Ole and Kalland, Arne (eds. 1995), *Asian Perceptions of Nature: A Critical Approach*, Curzon Press, Richmond, Surrey, UK.

Brødsgaard, Kjeld Erik (1989), 'Kildegrunlag og indfaldsvinkler i studiet av kinesisk samtidshistorie', *Historisk Tidsskrift*, Vol. 89(2), Copenhagen (in Danish), pp. 300-321.

Brødsgaard, Kjeld Erik and Strand, David (eds. 1998), *Reconstructiong Twentieth-Century China. State Control, Civil Society and National Identity*, Clarendon Press, Oxford.

Buen, Jørund (2001), *Beyond Nuts and Bolts: How Organisational Factors Infuence the Implementation of Technology Projects in China*, Post graduate thesis, Norwegian University of Science and Technology, Trondheim.

Bureau of Resettlement and Development, State Council Three Gorges Project Construction Committee (1995), *An Introduction to the Construction of the Three Gorges Project and the Economic Development in the Area.*

Cannon, Terry (ed. 2000), *China's Economic Growth: The Impact on Regions, Migration and the Environment*, Macmillan Press Ltd., London.

Cannon, Terry and Jenkins, Alan (eds. 1990), *The Geography of Contemporary China, The Impact of Deng Xiaoping's Decade*, Routledge, London and New York.

Cernea, Michael M. (2000), 'Risks, Safeguards, and Reconstruction: a Model for Population Displacement and Resettlement', in Michael M. Cernea and Christopher McDowell

(eds.), *Risks and Reconstruction, Experiences of Resettlers and Refugees*, The World Bank, Washington D.C., pp. 11-55.
- (ed. 1999), *The Economics of Involuntary Resettlement, Questions and Challenges*, The World Bank, Washington D.C.
- (1998), *Yimin, Chongjian, Fazhan* (Resettlement, Rehabilitation, Development), *Shijie yinhang yimin zhengce yu jingyan yanjiu*, (II) (Studies on World Bank Resettlement Policies and Experiences (II), translated and edited by the National Research Centre for Resettlement (Shuiku yimin jingji yanjiu zhongxin), Hehai daxue chubanshe (Hohai University Press), Nanjing.
- (1997), 'The Risks and Reconstruction Model for Resettling Displaced Populations', *World Development*, Vol. 25(10), pp. 1569-87, The World Bank, Washington, D.C.
- (1996), *Yimin yu fazhan* (Resettlement and Development), *Shijie yinhang yimin zhengce yu jingyan yanjiu* (Studies on World Bank Resettlement Policies and Experiences), translated and edited by the National Research Centre for Resettlement (Shuiku yimin jingji yanjiu zhongxin), Hehai daxue chubanshe (Hohai University Press), Nanjing.
- (1990), *Poverty Risks from Population Displacement in Water Resources Development*, Harvard Institute for International Development, Harvard University, Cambridge, MA.
- (ed. 1985, 1991), *Putting People First. Sociological Variables in Rural Development*, Second edition, The World Bank, Washington, D.C.

Cernea, Michael M. and Guggenheim, Scott (eds. 1993), *Anthropological Approaches to Resettlement; Policy, Practice and Theory*, Westview Press, Boulder and Oxford.

Cernea, Michael M. and McDowell, Christopher (eds. 2000), *Risks and Reconstruction, Experiences of Resettlers and Refugees*, The World Bank, Washington D.C.

Changjiang shuili weiyuanhui (ed. The Yangtze River Water Conservancy Commission), (main editor Fu Xiutang, 1997), *Sanxia gongcheng yimin yanjiu* (Resettlement research for the Three Gorges project), Changjiang Sanxia gongcheng jishu congshu (The Yangtze River Three Gorges Project Science and Technology Collection) Hubei kexue jishu chubanshe (Hubei Science and Technology Press).

Chayes, Abram and Kim, Charlotte (1998), 'China and the United Nations Framework Convention on Climate Change', in Michael B. McElroy, Chris P. Nielsen, and Peter Lydon, (eds.), *Energizing China, Reconciling Environmental Protection and Economic Growth*, Harvard University Committee on Environment, Harvard University Press, Cambridge, MA and London, pp. 503-40.

Chen, Guojie (1999a), *Huanjing yingxiang pingjia yu zhanlüe* (Assessment and Strategy for Environmental Impact), Chengdu. Unpublished document.
- (1999b), 'Sanxia renkou yi chaoyue, jiudi houkao yimin nandu da' (Three Gorges population is already surpassed [capacity], on-the-spot resettlement is difficult) in *Huanjing yingxiang pingjia yu zhanlüe* (Assessment and Strategy for Environmental Impact), pp. 54-7; first published (1987), in *Shuitu baochi tongbao* (Water and Soil Conservation Journal), Vol. 7(5), pp. 42-6.
- (1999c), 'Sanxia gongcheng dui shengtai yu huanjing de yingxiang he duice' (The ecological and environmental impacts and counter measuers of the Three Gorges project) in *Huanjing yingxiang pingjia yu zhanlüe* (Assessment and Strategy for Environmental Impact), pp. 35-41; first published (1991), in Zhongguo kexueyuan yuankan (the CAS periodical), Vol. 4, pp. 297-304.
- (1999d), 'Sanxia kuqu shengtai yu huanjing wenti' (The ecological and environmental problems of the Three Gorges reservoir area) in *Huanjing yingxiang pingjia yu zhanlüe* (Assessment and Strategy for Environmental Impact), pp. 42-7; first published (1997), in *Keji daobao* (Science and Technology News), No. 2, pp. 49-52.

– (1999e), 'Dui Changjiang Sanxia gongcheng shengtai yu huanjing wenti de tansuo' (Exploring the Three Gorges project's ecological and environmental problems) in *Huanjing yingxiang pingjia yu zhanlüe* (Assessment and Strategy for Environmental Impact), pp. 5-12; first published (1987), in *Keji daobao* (Science and Technology News), Vol.1, pp. 4-9.

– (1999f), 'Changjiang shi yi ge wanzheng de da xitong' (The Yangtze River is an integrated big system), in *Huanjing yingxiang pingjia yu zhanlüe* (Assessment and Strategy for Environmental Impact) pp. 3-4; first published (23 January 1987), in *Kexuebao* (Scientific Daily).

– (1999g), 'Changjiang Shangyou Hongshui Dui Zhongxiayou De Yinxiang Yu Duice' (the Impact and Countermeasures of the Floods of the Upper Reach on the Middle and Down Reaches of Yangtze Valley), in Houze Xu and Qiguo Zhao (eds.), *Changjiang Luiyu Honglao Zaihai Yu Keji Duice* (The Yangtze River Valley Floods and Water Logging Disasters and Scientific and Technological measures), Kexue Chubanshe (Science Press), Beijing, pp. 10-13.

– (1998) 'Changjiang hongzai yinqi de sikao' (Reflections regarding the causes of the Yangtze River flood disaster, *Zhongguo Keji Yuebao* (China Science and Technology Monthly Magazine), No.7, pp. 20-21.

– (1988), 'Juece minzhuhua yu zhengdun lingdao zuofeng' (Democratise decision-making and rectify the workstyle of the leadership), in Tian Fang and Lin Fatang (eds.) *Zai lun Sanxia gongcheng de hongguan juece* (Another Discussion of the macroscopic decision-making of the Three Gorges project), pp. 89-93; first published (1988), in *Qun yan*, Vol. 10.

Chen Guojie, Xu, Qi and Du, Ronghuan (1995), *Sanxia gongcheng dui shengtai yu huanjing de yingxiang ji duice yanjiu* (Research on the Three Gorges project impacts on the ecology and the environment and counter measures), Kexue chubanshe (Science Press), Beijing.

Chen, Guojie and Chen, Zhijian (1993), *Sanxia Gongcheng Dui Shengtai Yu Huanjing Yingxiang de Zonghe Pingjia* (Comprehensive evaluation of the Three Gorges project's impact on the ecology and the environment), Kexue Chubanshe (Science Press), Beijing.

Chen, Jingqiu (ed. 1992), *Sanxia Meng Cheng Zhen* (*The Three Gorges Dream Becomes True*), Xinhua Publishing House.

Chen, Yizi, (1995), 'The Decision Process Behind the 1986-1989 Political Reforms' in Carol Lee Hamrin and Suisheng Zhao (eds.), *Decision-making in Deng's China, Perspectives from Insiders,* M.E. Sharpe, Armonk, NY and London, pp. 133-52.

Chen, Yongbai (1997), 'Sanxia gongcheng shigongqu huanjing baohu yu guanli' (Environmental Protection and Management in the Three Gorges construction area), *Huanjing baohu* (Environmental protection), pp. 2-4.

China Statistics Press (Zhongguo tongji chubanshe) (2001), *China Statistical Yearbook* (Zhongguo tongji nianjian).

Christiansen, Flemming and Shirin Rai (1996), *Chinese Politics and Society, An Introduction,* Prentice Hall, Harvester Wheatsheaf, London.

Cui, Guangping (1999), Sanxia yimin bu wending shijian tanjiu (Probing into the causes for [social] instablity incidents of the Three Gorges resettlement), *Sichuan Sanxia xueyuan xuebao* (Journal of Sichuan Three-Gorges University), Vol. 15(5), pp. 1-4.

Dai, Qing (1998a), *The River Dragon Has Come!: The Three Gorges Dam and the Fate of China' Yangtze River and It's People*, compiled by Dai Qing, edited by John G. Thibodeau and Philip B. Williams, M.E. Sharpe, Armonk, N.Y.

– (1998b), 'Three Gorges Project Threatens Soul of China', *Forum for Applied Research and Public Policy*, Spring, Vol. 13(1).

– (1994), *Yangtze! Yangtze!* edited by Patricia Adams and John Thibodeau, Probe International, Earthscan Publications Limited, London and Toronto.
– (ed. 1989), *Changjiang, Changjiang—Sanxia gongcheng lunzheng* (Yangtze, Yangtze—The debate about the Three Gorges project), Guizhou renmin chubanshe (Guizhou People's Press).
Deng, Ning (1997), 'Gaishan Sanxia kuqu shengtai huanjing guanjian – fazhan shengtai nongye' (Developing ecologic agriculture is the key to improving ecologic environment around the Three Gorges), *Keji daobao* (Science and Technology News), No. 4.
Ding, Qigang (1998), 'What are the Three Gorges Resettlers Thinking?', in Dai, Qing, *The River Dragon Has Come!*, pp. 70-89.
Ding, Qigang and Jiaqin Zheng (1998), 'A Survey of Resettlement in Badong County, Hubei Province', in Dai, Qing, *The River Dragon has Come!*
Dittmer, Lowell (2001), 'The Changing Shape of Elite Power Politics', *The China Journal*, January, Issue 45, pp. 53-67.
– (1995), 'Chinese Informal Politics', *The China Journal*, July, Issue 34, pp. 1-34.
Dittmer, Lowell; Fukuii; Haruhiro and Lee, Peter N.S. (eds. 2000), *Informal Politics in East Asia*, Cambridge University Press, Cambridge.
Economy, Elizabeth and Oksenberg, Michel (eds. 1999), *China Joins the World, Progress and Prospects*, A Council on Foreign Relations Press, New York.
Edmonds, Richard Louis (2000a), 'Recent Developments and Prospects for the Sanxia (Three Gorges) Dam' in Terry Cannon (ed.), *China's Economic Growth: The Impact on Regions, Migration and the Environment*, Macmillan Press Ltd, London, pp.161-83.
– (ed. 2000b), *Managing the Chinese Environment*, Oxford University Press, New York.
– (1994), *Patterns of China's Lost Harmony, A Survey of the Country's Environmental Degradation and Protection*, Routledge, London and New York.
EIAD/CAS (Environmental Impact Assessment Department of the Chinese Academy of Sciences) and the RIPYWR (Research Institute for Protection of Yangtze Water Resources) (1995), *Environmental Impact Statement for the Yangtze Three Gorges project*, China Yangtze Three Gorges Development Corporation, Science Press, Beijing.
Elvin, Mark and Ts'ui-jung Liu (eds. 1998), *Sediments of Time*; *Evironment and Society in Chinese History*, Cambridge University Press, Cambridge.
Ertan Hydroelectric Development Corporation (1994), *Ertan Hydroelectric Project, Environmental Assessment and Resettlement, Final Report*, Chengdu, Sichuan.
Falkenheim, Victor C. (1987), *Citizens and Groups in Contemporary China*, The University of Michigan, Ann Arbor.
Fewsmith, Joseph (2001a), *Elite Politics in Contemporary China*, M.E. Sharpe, Armonk, NY and London.
– (2001b), 'The New Shape of Elite Politics', *The China Journal*, January, Issue 45, pp. 83-93.
– (2000), 'Formal Structures, Informal Politics, and Political Change in China', in Dittmer et al (eds.), *Informal Politics in East Asia*, pp. 141-64.
– (1994), *Dilemmas of Reform in China, Political Conflict and Economic Debate*, M.E. Sharpe, Armonk, NY.
Fuggle, R. and Smith, W.T. (2000), *Experience with Dams in Water and Energy Resource Development in the People's Republic of China*, Prepared for the World Commission on Dams, Cape Town, South Africa.
Gao, Qi (2000), 'Sanxia kuqu nongcun yimin anzhi xinmoshi chutan, Guanyu xingjian "yimincheng" anzhi nongcun yimin de sikao' (Initial explorations about the settlement of the rural relocatees of the Three Gorges area; Thoughts regarding the construction of

'relocatee towns') *Sichuan Sanxia xueyuan xuebao* (Journal of Sichuan Three-Gorges University), Vol. 16(2), pp. 14-16.

Gilley, Bruce (2001a), 'Critic's Flight', *Far Eastern Economic Review,* July 12, p. 15.

– (2001b), 'Trojan Horse, a Chat Site Run by the Communist Party is Delivering Stinging Criticisms—of the Party', *Far Eastern Economic Review*, 24 May, p. 64.

– (2001c), 'The Right Stuff: Neo-authoritarians Come to the Fore as the Communist Party Grapples with Political Reform', *Far Eastern Economic Review*, 17 May, p. 26.

Goldsmith, Edward and Hildyard, Nicholas (1984), *The Social and Environmental Effects of Large Dams*, Wadebridge Ecological Centre, Camelford, Cornwall, UK.

Goodman, David S.G. (ed. 1984), *Groups and Politics in the People's Republic of China*, M.E. Sharpe, New York.

Gu, Chaolin and Chunxiao Huang (1999), 'Sanxia kuqu chengzhen yimin qianjian de wenti yu duice' (Study on the Resettlement and reconstruction of cities in the Three Gorges Reservoir Area), *Changjiang liuyu ziyuan yu huanjing*, (Resources and Environment in the Yangtze Basin) Vol.8(4), November, pp. 353-9.

Gu, Shengzu and Shuiying Zhong, (1998), Lun gongchengxing yimin de kechixu anzhi he fazhan duice (Sustainable Resettlement and Development of Project-induced Migrants), *Renkou yanjiu* (Population Research), Vol. 22(3).

Guojia tongjiju (National Bureau of Statistics, 1999), *Zhongguo tongji nianjian* (China Statistical Yearbook), No. 18, Zhongguo tongji chubanshe (China Statistics Press), Beijing.

Guo, Xuezhi (2001), 'Dimensions of Guanxi in Chinese Elite Politics', *The China Journal*, July, Issue 46, pp. 69-90.

Guowuyuan bangongting mishuju (State Council General Office Secretariat) (ed. 1998), *Zhongyang zhengfu zuzhi jigou* 1998 (Central Government Organsiation), Gaige chubanshe (Reform press), Beijing.

Halpern, Nina P. (1992), 'Information Flows and Policy Co-ordination in the Chinese Bureaucracy', in Lieberthal and Lampton (eds.), *Bureaucracy, Politics, and Decision-Making in Post-Mao China*, pp. 125-48.

– (1988), 'Social Scientists as Policy Advisers in Post-Mao China: Explaining the Pattern of Advice', *The Australian Journal of Chinese Affairs*, Issue 19/20, p. 215-40.

Hamrin, Carol Lee and Zhao, Suisheng (eds. 1995), *Decision-making in Deng's China, Perspectives from Insiders*, M.E. Sharpe, Armonk, NY and London.

Han, Guanghui, Wu, Yuezhao, Yan, Tingzhen and Chen, Xibo (2001), 'Wo guo kuqu yimin fangshi ji qi qishi' (The Migration pattern of our country's reservoir area and its instructive experiences), *Renkou yu jingji* (Population & Economics), No. 2, Tot. No. 125, pp. 44-7, and 70.

He, Dawei and Jingshen Chen (May 2000), 'Sanxia kuqu ziyuan yu huanjing yitihua guanli de jigou, falü, zhidu chutan' (Initial explorations regarding organisational, legal and system integrated management of resources and environment of Three Gorges reservoir area), *Changjiang liuyu ziyuan yu huanjing* (Resources and Environment in the Yangzte Basin), Vol. 9(2).

He, Qinglian (1998), *Xiandaihua de xianjing* (The Pitfalls of Modernisation), Jinri Zhongguo chubanshe (China Today Press), Beijing.

Heggelund, Gørild (1994), *Moving a Million, The Challenges of the Sanxia Resettlement*, report No. 181, October, Norwegian Institute of International Affairs, Oslo.

– (1993), *China's Environmental Crisis: The Battle of Sanxia*, report No. 170, August, Norwegian Institute of Inetranational Affairs, Oslo.

Hirsch, Philip and Warren, Carol (eds. 1999), *The Politics of Environment in Southeast Asia, Resources and Resistance*, Routledge, London and New York.

Ho, Peter (2001a), 'Greening Without Conflict? Environmentalism, NGOs and Civil Society in China', *Development and Change*, Vol. 32, pp. 893-921.

– (2001b), 'Who Owns China's Land? Policies, Property Rights and Deliberate Institutional Ambiguity', *China Quarterly*, June, No. 166, pp. 394-421.

Huang, Shunxiang (1998), 'Sanxia yimin zhong bu wending yinsu de falü sikao' (Legal reflections on unstable factors in the Three Gorges resettlement), *Sichuan Sanxia xueyuan xuebao* (Journal of Sichuan Three Gorges University), Vol. 14(4), pp. 4-9.

Huang, Zhenli, Fu, Bojie and Yang, Zhifeng (eds. 1998), *21 Shiji Changjiang daxing shuili gongcheng zhong de shengtai yu huanjing baohu* (Ecological and environmental protection in the Yangtze River large-scale water conservancy project of the 21st century), Zhongguo huanjing kexue chubanshe (China Environmental Sciences Press), Beijing.

Hubei sheng tongjiju (Statistical Bureau of Hubei province) (ed. 1999), *Hubei tongji nianjian* (Hubei Statistical Yearbook), Zhongguo tongji chubanshe (China Statistical Press), Beijing.

Human Rights Watch/Asia (1995), *The Three Gorges Dam in China: Forced Resettlement, Suppression of Dissent and Labour Rights Concerns*, February, Vol. 7(2).

Hutchings, Graham (2000), *Modern China. A Companion to a Rising Power*, Penguin Books, London.

Institut für Asienkunde Hamburg (1999), *China Aktuell Monatszeitschrift*, December.

Jahiel, Abigail R. (1998), 'The Organization of Environmental Protection in China', *The China Quarterly*, December, No. 156, pp. 757-87.

– (1994), Policy Implementation through Organizational Learning: the Case of Water Pollution Control in China's Reforming Socialist System, UMI Dissertation Services, the University of Michigan, Ann Arbor.

Jarman, Heather and Scrivener, Brian (eds. 1992), *Sardar Sarovar*, The Report of the Independent Review, for The Independent Review by Resource Futures International, Inc.,Vancouver, Canada.

Jiang, Di (1992), *Sanxia baiwan yimin chulu hezai?* (Is there a way out for the one million people to be resettled in the Three Gorges project?) Chongqing: Chongqing daxue chubanshe (Chongqing University press).

Jing, Jun (2000), 'Environmental Protests in Rural China' in Elizabeth J. Perry, and Mark Selden (eds.), *Chinese Society: Change, Conflict and Resistance*, Routledge, London and New York, pp. 143-60.

– (1997), 'Rural Resettlement: Past Lessons for the Three Gorges project', *The China Journal*, July, Issue 38, pp. 65-92.

Kalland, Arne, and Persoon, Gerard (eds. 1998), *Environmental Movements in Asia*, Curzon Press, Surrey, UK.

Knup, Elizabeth (1997), 'Environmental NGOs in China: An Overview', *China Environment Series*, Issue 1, The Woodrow Wilson Center, The Environmental Change and Security Project, Washington, D.C.

Lampton, David M. (1987), 'Chinese Politics: The Bargaining Treadmill', *Issues & Studies*, March, Vol. 23(3), pp. 11-41.

Lee, Chin-Chuan (ed. 1990), *Voices of China. The Interplay of Politics and Journalism*, The Guildford Press, New York and London.

Leng, Meng (1996), 'Huanghe da yimin (The Massive Resettlement on the Yellow River), Sanmenxia yimin shiwei (The whole story of the Sanmenxia resettlement)', *Zhongguo zuojia* (Chinese Writers), pp. 60-92.

Li, Boning (1992a), *Kuqu yimin anzhi* (Reservoir resettlement), Sanxia gongcheng xiaocongshu (the Three Gorges Project Collection), Shuili dianli chubanshe (Water Conservation and Electric Power Publishing House), Beijing.

– (1992b), *Lun Sanxia gongcheng yu kaifaxing yimin* (On the Three Gorges project and development resettlement scheme), Shuili dianli chubanshe (Water Conservation and Electric Power Publishing House), Beijing.

Li, Cheng (2001), *China's Leaders, the New Generation*, Rowman & Littlefield Publishers, Inc., Lanham, Boulder, New York and Oxford.

– (2000a), 'Promises and Pitfalls of Reform: New Thinking in Post-Deng China', in Tyrene White (ed.), *China Briefing 2000, The Continuing Transformation*, M.E. Sharpe, Armonk, NY, published in cooperation with the Asia Society, pp. 123-57.

– (2000b), 'Jiang Zemin's Successors: the Rise of the Fourth Generation of Leaders in the PRC', *The China Quarterly*, March, No. 161, pp. 1-40.

Li, Heming (2000), *Population Displacement and Resettlement in the Three Gorges Reservoir Area of the Yangtze River Central China*, Ph.D. dissertation, The University of Leeds, School of Geography.

Li, Rui (1985), *Lun Sanxia gongcheng* (*On the Three Gorges project*), Hunan kexue jishu chubanshe (Hunan Science and Technology Press, Hunan).

Li, Xiaozhuang and Xiao, Chong (eds. 1998), *Zhu Rongji renma* (Zhu Rongji's forces), Xiafei'er guoji chuban gongsi (Xiafei'er Press), Hong Kong.

Lieberthal, Kenneth G. (1999), 'China's Governing System and its Impact on Environmental Policy Implementation', *China Environment Series*, The Woodrow Wilson Center, The Environmental Change and Security Project, Washington, D.C.

– (1995), *Governing China. From Revolution Through Reform*, W.W. Norton & Company, Inc., New York and London.

– (1978), *Central Documents and Politburo Politics in China*, Michigan Monographs in Chinese Studies, No. 33, Center for Chinese Studies, Ann Arbor.

Lieberthal, Kenneth G. and Lampton, David M. (eds. 1992), *Bureaucracy, Politics, and Decision Making in Post-Mao China*, University of California Press, Berkeley and Los Angeles, CA, and Oxford, UK.

Lieberthal, Kenneth G. and Jackson, Bruce D. (1989), *A Research Guide to Central Party and Government Meetings in China 1949-1986*, M.E. Sharpe, Inc., Armonk, NY and London.

Lieberthal, Kenneth G. and Oksenberg, Michel (1988), *Policymaking in China: Leaders, Structures, and Processes,* Princeton University Press, Princeton, NJ.

Ling, Zhijun (1998), *Zhongguo jingji gaige beiwanglu (1989-1997)* (A Memorandum Book of China's Economic Reform), Dongfang chuban zhongxin (The East Publishing Centre), Shanghai.

Luk, Shiu-Hung and Whitney, Joseph (eds. 1993), *Megaproject: A Case Study of the China's Three Gorges Project*, M.E. Sharpe, Armonk NY.

Ma, Jun (1999), *Zhongguo shui weiji* (China's Water Crisis), Zhongguo huanjing kexue chubanshe (China Environmental Science Press), Beijing.

Ma, Xiaoying and Ortolano, Leonard (2000), *Environmental Regulation in China. Institutions, Enforcement and Compliance*, Lanham, Rowman & Littlefield Publishers, Inc., Boulder, New York and Oxford.

Ma, Yijie and Yang, Linzhang (1996), *Sanxia dianxing diqu fuhe nongye shengtai xitong jianshe yu gengzuo zhidu youhua yanjiu* (Research regarding establishing a complex agricultural system and improvement of the cropping system in a representative district of the Three Gorges), Chinese Academy of Sciences, Nanjing.

Maa, Shaw-chang (1993), *A Comparative Study of Provincial Policy in China: the Political Economy of Pollution Control Policy*, Ph.D. dissertation, Ohio State University, UMI Dissertation Services.

Mahapatra, Lakshman K. (1999), 'Testing the Risks and Reconstruction Model on India's Resettlement Experiences', in Michael. M. Cernea (ed.), *The Economics of Involuntary Resettlement, Questions and Challenges*, The World Bank, Washington D.C., pp. 189-230.

McCormack, Gavan (2001), 'Water Margins. Competing Paradigms in China', *Critical Asian Studies*, Vol. 33(1), pp. 5-30.

McCully, Patrick (1996), *Silenced Rivers, The Ecology and Politics of Large Dams*, Zed Books, London, UK.

McElroy, Michael B., Nielsen, Chris P. and Lydon, Peter (eds. 1998), *Energizing China, Reconciling Environmental Protection and Economic Growth*, Harvard University Committee on Environment, Harvard University Press, Cambridge, MA.

Meikle, Sheila and Zhu, Youxuan (2000), 'Employment for Displacees in the Socialist Market Economy of China', in Cernea and McDowell (eds. 2000), *Risks and Reconstruction, Experiences of Resettlers and Refugees*, pp. 127-43.

Merle, Goldman and MacFarquhar, Roderick (eds. 1999), *The Paradox of China's Post-Mao Reforms*, Harvard Contemporary China Series 12, Harvard University Press, Cambridge MA.

Miller, H. Lyman (1996), *Science and Dissent in Post-Mao China: The Politics of Knowledge*, University of Washington Press, Seattle and London.

Nathan, Andrew J. and Link, Perry (2001), *The Tiananmen papers: The Chinese Leadership's Decision to Use Force against Their Own People—in Their Own Words*, Little, Brown and Company, London.

Nathan, Andrew J. (1973), 'A Factionalism Model for CCP Politics', China Quarterly 53, January/March, pp. 34-66.

NEPA (National Environmental Protection Agency, Guojia Huanjing Baohuju, 1996), *The National Ninth Five-Year Plan for Environmental Protecion and the Long-Term Targets for the year 2010*, Beijing.

– (May 1997), *Changjiang Sanxia Gongcheng Shengtai Yu Huanjing Jiance Gongbao 1997* (Bulletin on Ecological and Environmental Monitoring of Three Gorges Project on the Yangtze River).

– (1993), *Zhonghua renmin gongheguo huanjing baohu fagui xuanbian* (A selection of environmental protection laws and regulations of the PRC).

– (1989), Zhonghua renmin gongheguo huanjing baohufa (The Environmental Protection Law of the PRC) in NEPA (National Environmental Protection Agency, Guojia Huanjing Baohuju, 1993), *Zhonghua renmin gongheguo huanjing baohu fagui xuanbian* (A selection of environmental protection laws and regulations of the PRC).

Nickum, James E. (1995), *Dam Lies and Other Statistics: Taking the Measure of Irrigation in China, 1931-91*, East-west Center Occasional Papers. Environment Series, No. 18, East-West Center, Honolulu, Hawaii.

O'Brien, Kevin J. (1990), *Reform Without Liberalization; China's National People's Congress and the Politics of Institutional Change*, Cambridge University Press, New York and Cambridge.

Oi, Jean (1989), *State and Peasant in Contemporary China. The Political Economy of Village Government*, University of California Press, Berkeley, CA.

Oksenberg, Michel (2001), 'China's Political System: Challenges of the Twenty-first Century' *The China Journal*, January, Issue 45, pp. 21-35.

– (1968), 'Occupational Groups in Chinese Society and Cultural Revolution', in Michel Oksenberg, Carl Riskin, Robert Scalapino and Ezra F. Vogel, *The Cultural Revolution: 1967 in Review*, Michagan Monographs in Chinese Studies, No. 2, Center for Chinese Studies, Ann Arbor, pp. 1-44.

Oksenberg, Michel and Economy, Elizabeth (1998), 'China's Accession to and Implementation of International Accords 1978-95', Reprinted from Edith Brown Weiss, and Harold K. Jacobson (eds.), *Engaging Countries: Strengthening Compliance with International Environmental Accords*, Chapter 11, The MIT Press, Cambridge, MA.

Pan, Dazhong and Chen, Yongbai (1998), 'Shilun Sanxia gongcheng jianshe de huanjing jianlizhi' (Discussion on the Environmental Supervision System in the Three Gorges Project Construction), *Huanjing Baohu* (Environmental Protection), No. 3, pp. 11-13.

Pearce, David W. (1999), 'Methodological Issues in the Economic Analysis for Involuntary Resettlement Operations' in Cernea (ed.), *The Economics of Involuntary Resettlement, Questions and Challenges*, pp. 50-82.

Perry, Elizabeth J. and Selden, Mark (eds. 2000), *Chinese Society: Change, Conflict and Resistance,* Routledge, London and New York.

Pye, Lucian (1985), *Asian Power and Politics: the Cultural Dimensions of Authority*, Harvard University Press, Cambridge, MA.

– (1968), *The Spirit of Chinese Politics*, MIT Press, Cambridge, MA.

Qi, Ren (1998), 'Is Developmental Resettlement Possible?' in Dai, Qing, *The River Dragon Has Come!*, pp. 50-62.

Qiu, Zhengguang, Wu, Lizhi and Du, Jinping (2000), 'Sanxia kuqu nongcun yimin anzhi moshi tantao' (Exploring the Three Gorges rural resettlement pattern), *Renmin Changjiang* ([People's] Yangtze River), March, Vol. 31(3), pp. 1-3.

Ross, Lester (1999), 'China and Environmental Protection', in Economy and Oksenberg (eds.), *China Joins the World, Progress and Prospects*.

– (1998), 'China: Environmental Protection, Domestic Policy Trends, Patterns of Participation in Regimes and Compliance with International Norms', *The China Quarterly*, December, No. 156, pp. 809-35.

– (1992), 'The Politics of Environmental Policy in the People's Republic of China', *Policy Studies Journal—Urbana*, Vol. 20(4), pp. 628-42.

– (1988), *Environmental Policy in China*, Indiana University Press, Bloomington and Indianapolis.

Saich, Tony (2000), 'Negotiating the State: The Development of Social Organizations in China', *The China Quarterly*, March, No. 161, pp. 124-41.

Sanxia gongcheng lunzheng lingdao xiaozu bangongshi (The Office of the Leadership Group of the Three Gorges Project Verification, 1988), *Sanxia gongcheng zhuanti lunzheng baogao huibian* (A Compilation by Topic of Verifying Reports on the Three Gorges Project), Beijing.

Schell, Orville (1994), *Mandate of Heaven, The Legacy of Tian'anmen Square and the Next Generation of China's Leaders*, Touchstone, Simon & Schuster, New York.

Scudder, T. (1985, 1991) 'A Sociological Framework for the Analysis of New Land Settlements' in Cernea (ed.), *Putting People First. Sociological Variables in Rural Development*.

SEPA (State Environmental Protection Administration, Guojia Huanjing Baohu Zongju, May 2000a), *Changjiang Sanxia gongcheng shengtai yu huanjing jiance gongbao 2000*. (Bulletin on Ecological and Environmental Monitoring of the Three Gorges Project on the Yangtze River).

– (June 2000b), *Zhongguo huanjing tongji nianbao 1999*, (The 1999 China Environment Statististical Annual Report).

– (May 1999), *Changjiang Sanxia gongcheng shengtai yu huanjing jiance gongbao 1999*. (Bulletin on Ecological and Environmental Monitoring of the Three Gorges Project on the Yangtze River).

– (May 1998), *Changjiang Sanxia Gongcheng Shengtai Yu Huanjing Jiance Gongbao* (Bulletin on Ecological and Environmental Monitoring of Three Gorges Project on the Yangtze River).

Shambaugh, David (2001), 'The Dynamics of Elite Politics during the Jiang Era', *The China Journal*, January, Issue 45, pp. 101-11.

– (ed. 2000a), *Is China Unstable? Assessing the Factors,* M.E. Sharpe, Armonk, NY and London.

– (ed. 2000b), *The Modern Chinese State*, Cambridge University Press, Cambridge.

Shapiro, Judith (2001), *Mao's War Against Nature, Politics and the Environment in Revolutionary China*, Cambridge University Press, Cambridge.

Shirk, Susan (1993), *The Political Logic of Economic Reform in China*, University of California Press, Berkeley.

Shuilibu Changjiang shuili weiyuanhui (1992) (The Yangtze River Water Conservancy Commission, Ministry of Water Resources) (neibu ziliao, restricted material), *Changjiang Sanxia shuili shuniu chubu sheji baogao (shuniu gongcheng)*, (The Preliminary Construction Report for the Yangtze River Three Gorges Key Water Control Project), Vol. 11, Environmental Protection, December.

Sinkule, Barbara J. (1993), *Implementation of Industrial Water Pollution Control Policies in the Pearl River Delta Region of China*, Ph.D. dissertation, Stanford University, UMI Dissertation Services, Ann Arbor.

Sinkule, Barbara J. and Leonard Ortolano (1995), *Implementing Environmental Policy in China*, Praeger, Westport, CT and London.

Skilling, H. Gordon (1966), 'Interest Groups and Communist Politics', *World Politics*, Vol.18(3), pp. 435-51.

Smil, Vaclav and Mao, Yushi (co-ordinators,1998), *The Economic Costs of China's Environmental Degradation*, Project on Environmental Scarcities, State Capacity and Civil Violence, University of Toronto and the American Academy of Arts and Sciences, Published by the Committee on International Security Studies, American Academy of Arts and Sciences, Cambridge, MA.

Sowell, T. (1996), *Migration and Cultures: A World View*, Basic Books, New York.

SPC (State Planning Commission) and SSTC (State Science and Technology Commission) (1994), *Priority Programme for China's Agenda 21*, Beijing.

Spence, Jonathan D. (1999), *The Search for Modern China*, W.W. Norton & Company, New York and London.

Stevenson-Yang, Anne (1998), 'Word Games', *The China Business Review*, May/June, pp. 42-8.

Tangen, Kristian, Heggelund, Gørild and Hu, Tao (2000), *Climate Policies in China: Institutional Setting and the Potential for the CDM*, FNI Report 12/2000, The Fridtjof Nansen Institute, Lysaker.

Tangen, Kristian, Heggelund, Gørild and Buen, Jørund (2001), 'China's Climate Change Positions: at a Turning Point', *Energy & Environment*, Vol. 12(2&3).

Tao, Jingliang (ed. 1996), *Changjiang Sanxia Gongcheng, 66 Wen*, Zhongguo Sanxia Chubanshe (China Three Gorges Press), Beijing.

Tian, Fang and Lin, Fatang (eds. 1988) *Zai lun Sanxia gongcheng de hongguan juece* (Another Discussion of the macroscopic decision-making of the Three Gorges project), Hunan Kexue chubanshe (Hunan Science and Technology Press), Changsha.

Tian, Fang, Lin, Fatang and Ling, Chunxi (eds. 1987) *Lun Sanxia gongcheng de hongguan juece (*A discussion of the macroscopic decision-making of the Three Gorges project), Hunan Kexue chubanshe (Hunan Science and Technology Press), Changsha.

Tien, Hung-Mao and Chu, Yun-han (eds. 2000), *China under Jiang Zemin*, Lynne Rienner Publishers, Boulder, Colorado.

Tsou, Tang (1995), 'Chinese Politics at the Top: Factionalism or Informal Politics? Balance-of-power Politics or a Game to Win it All?', *The China Journal*, July, Issue 34, pp. 95-156.

– (1986), *The Cultural Revolution and Post-Mao Reforms*, University of Chicago Press, Chicago.

– (1976), 'Prolegomenon to the Study of Informal Groups in CCP Politics', *The China Quarterly*, No. 65, March, pp. 98-114, reprinted in Tsou (1986), *The Cultural Revolution and Post-Mao Reforms*, pp. 95-111.

Van Slyke, Lyman P. (1988) *Yangtze—Nature, History and the River*, Addison-Wesley Publishing Company, Inc., Reading, MA.

Vermeer, Eduard B. (2000), 'Determinants of Agricultural Productivity in China: a Comparison of the New Provincial Census Data with Official Figures, and Some Implications' FAO Seminar on Agricultural Census Results, September, Beijing.

– (1998), 'Industrial Pollution in China and Remedial Policies', *The China Quarterly*, December, No. 156, pp. 952-85.

– (1995), 'An Inventory of Losses Due to Environmental Pollution: Problems in the Sustainability of China's Economic Growth', *China Information*, Vol. X(1).

Walder, Andrew (1986), *Communist Neo-Traditionalism: Work and Authority in Chinese Industry*, University of California Press, Berkeley, CA.

Wang, Rushu (2000), 'Sanxia gongcheng de huanjing yu yimin wenti' (The Environment and Resettlement of the Three Gorges Project), *Changjiang liuyu ziyuan yu huaning* (Resources and Environment in the Yangtze Basin), Vol. 9(1), February (in English).

Wang, Maofu, Huang, Qin and Ding, Lixian (1999), 'Quyu wenhua chayi dui yuanqian shuiku yimin fanqian de yingxiang' (The impact of regional cultural differences on the return of distant relocatees in reservoir resettlement), *Renkou yu jingji* (Population & Economics), No. 1, Tot. No. 112, pp. 43-6.

Wang, Shaoguang and Hu, Angang (1999), *The Political Economy of Uneven Development, The Case of China*, Asia and the Pacific Series, M.E. Sharpe, Armonk, NY.

Wang, Xi (2000), 'Implementing the Rio Declaration and Agenda 21 in China', *Asia Pacific Journal of Environmental Law*, Vol. 5, Issue 1, Kluwer Law International, pp. 9-32.

Wanxianshi Sanxia gongcheng yimin bangongshi (Wanxian Three Gorges Project Resettlement Office, 1996), *Wanxianshi yimin gongzuo qingkuang jianjie* (Brief introduction to the situation of Wanxian municipality resettlement work), October.

Wei, Yi (1999), 'Sanxia yimin gongzuo zhong de zhongda wenti yu yinhuan, Yi Chongqingshi Yunyangxian wei li' (The significant problems and hidden dangers in the Three Gorges Resettlement, Taking Yunyang county in Chongqing municipality as an example), *Zhanlüe yu guanli* (Strategy and Management), January, pp. 12-20.

White, Tyrene (ed. 2000), *China Briefing 2000: The Continuing Transformation*, published in co-operation with the Asia Society, M.E. Sharpe, Armonk, NY.

Wittfogel, Karl A. (1957), *Oriental Despotism, A Comparative Study of Total Power*, Yale University Press, New Haven.

World Bank (2001a), *China: Air, Land, and Water. Environmental Priorities for a New Millenium*, The World Bank, Washington D.C.

– (2001 b,c,d and e), see Internet listings.

– (1998), *Recent Experience With Involuntary Resettlement. China—Shuikou (and Yantan)*, Operations Evaluation Department, The World Bank, Washington, D.C.
– (1997), *China 2020; Clear Water, Blue Skies: China's Environment in the New Century*, The World Bank, Washington D.C.
– (1996), *Resettlement and Development, The Bankwide Review of Projects Involving Involuntary Resettlement 1986-1993*, The World Bank, Washington, D.C.
– (1994a), *Staff Appraisal Report, China, Xiaolangdi Resettlement Project*, March 25, Report No. 12527-CHA, The World Bank, Agriculture Operations Division, China and Mongolia Department, East Asia and Pacific Regional Office, Washington D.C.
– (1994b), *Staff Appraisal Report, China, Xiaolangdi Multipurpose Project*, March 25, Report No. 12329-CHA, The World Bank, Agriculture Operations Division, China and Mongolia Department, East Asia and Pacific Regional Office, Washington D.C.
– (1994c), *Resettlement and Development, The Bankwide Review of Projects Involving Involuntary Resettlement 1986-1993*, The World Bank, Washington, D.C.
– (1993), *China Involuntary Resettlement*, June 8, Report No. 11641-CHA, The World Bank, Office of the Director, China and the Mongolia Department, East Asia and Pacific Regional Office, Washington D.C.
– (1990), *Operational Directive 4:30 Involuntary Resettlement*, June, The World Bank, Washington, D.C.

World Commission on Dams (2000), *Dams and Development, A New Framework for Decision-making*, The Report of the World Commission on Dams, Earthscan Publications Ltd., London and Sterling, VA.

Wu, Lizhi (2001), 'Sanxia gongcheng nongcun yimin anzhi de tujing yu zhengce jianyi' (Approaches and Policy Suggestions for the Resettlement of Rural Migrants in the Three Gorges Project), *Renkou yu jingji* (Population & Economics), pp. 48-51.

Wu, Lizhi and Liao, Qinlan (1999), 'Cong Sanxia kuqu tudi rongliang lun yimin waiqian de biyaoxing (Discussing the necessity of moving out the relocatees from the point of view of land capacity in the Three Gorges area), Yi Chongqing shi Yunyangxian wei lie (Taking Yunyang county in Chongqing municipality as example)', *Changjiang liuyu ziyuan yu huanjing* (Resources and Environment in the Yangtze basin), Vol. 8(3), August, pp. 243-303.

Wu, Xiping, Chen, Jiong and Yu, Ke (2000), 'Sanxia kuqu ziyuan huanjing baohu yu shuishou zhengce yanjiu' (Research regarding the Three Gorges reservoir area environmental protection and tax revenue policy), *Gaige* (Reform), March, pp. 66-72.

Xia, Hongyuan (1999), 'Sanxia kuqu nongcun yimin waiqian anzhi zongheng tan' (Freely on the moving out of relocatees from the Three Gorges Reservoir), *Sichuan Sanxia xueyuanbao* (Journal of Sichuan Three Gorges University), Vol. 15(2).

Xiao, Chong (1999), *Beijing zhinangqun* (Beijing's Think-tank): *Zhongnaihai xin zhinang* (The new think tank of Zhongnanhai), Xiafei'er guoji chuban gongsi, Hong Kong.

Xiao, Zhengqin (1999), *Zhu zongli zhinang qunying* (The Think Tank of Premier Zhu Rongji), Taipingyang Shiji chubanshe youxian gongsi (The Pacific Century Press Limited), Hong Kong.

Xibu da kaifa zhinan bianji weiyuanhui (The editing committee for the Guide to the strategy of developing the western region) (2000), *Xibu da kaifa zhinan. Tongji xinxi zhuanbian* (Guide to the strategy of developing the western region. Special volume of statistical information), published jointly by Guowuyuan fazhan yanjiu zhongxin (The State Council Centre for Development Research), Zhongguo qiye pingjia xiehui (China Enterprise Evaluation Association) and Guojia tongji ju zonghesi (The Comprhensive Department of the Statistical Bureau), Zhongguo shehui chubanshe (China Society press), Beijing.

Xu, Qi and Liu, Yinong (1993), *Sanxia kuqu yimin huanjing rongliang yanjiu* (Research on the Resettlement and Environmental Capacity of the Three Gorges Reservoir Area), Kexue chubanshe (Science Press), Beijing.

Xu, Tangling (1999), 'Sanxia gongcheng yu kexue juece' (The Three Gorges project and Scientific Policymaking), *Zhanlüe yu guanli* (Strategy and Management), March, pp. 103-5.

Yang, Anne Stevenson, (1998), 'Word Games', *The China Business Review*, Volume 25(3), May-June, US-China Business Council, pp. 42-8.

Yang, Linzhang et al (1998), *Sanxia gongcheng shengtai huanjing shiyan zhan jianshe 1998 nian niandu baogao* (The Annual Report of the Three Gorges Projet Ecological and Environmental Experimental Station), Institute of Soil Sciences, the Chinese Academy of Sciences, Nanjing.

Yangtze Valley Water Resources Protection Bureau (YVWRPB), Ministry of Water Resources (MWR) and National Environmental Protection Agency (NEPA) (eds. 1999), *Questions and Answers on Environmental Issues for the Three Gorges Project*, Science Press, Beijing and New York.

Yin, Liangwu (1996), *The Long Quest for Greatness: China's Decision to Launch the Three Gorges project*, PhD dissertation, Department of History, Washington University.

Zhang, Zhongxiang (2000), 'Decoupling China's Carbon Emissions Increase from Economic Growth: An Economic Analysis and Policy Implications', *World Development*, Vol. 28(4), pp. 739-52.

Zheng, Du and Shen, Yuancun (1998), 'Podi guocheng ji tuihua podi huifu zhengzhi yanjiu (Studies on the process, restoration and management of degrading slopelands.), Yi Sanxia kuqu zise tupodi wei li (A case study of purple soil slopelands in the Three Gorges areas)', *Dili xuebao* (Acta Geographica Sinica), Vol. 53(2), March.

Zheng, Shiping (1997), *Party vs. State in Post-1949 China, The Institutional Dilemma*, Cambridge University Press, Cambridge.

Zheng, Yisheng and Wang, Shiwen (eds., 2001), *Zhongguo huanjing yu fazhan pinglun* (China Environment and Development Review), Zhongguo shehui kexueyuan huanjing yu fazhan yanjiu zhongxin, Shehui kexue wenxian chubanshe (Social Sciences Documentation Publishing House), Beijing.

Zheng, Yisheng and Qian, Yihong (1998), *Shendu youhuan, Dangdai Zhongguo de kechixu fazhan wenti* (Profound Hardship, Sustainable Development Problems in Modern China), Jinri Zhongguo chubanshe (China Today Press), Beijing.

Zheng, Yongnian (2000), *Jiang Zhu zhi xia de Zhongguo:gaige, zhuanxing he tiaozhan* (China under Jiang and Zhu: reform, change of pattern and challenges), Taipingyang Shiji chubanshe youxian gongsi (The Pacific Century Press Limited), Hong Kong.

Zhongguo 21 shiji yicheng bianzhi lingdao xiaozu (The A 21 compilation leadership group, 1994), *Zhongguo 21 Shiji yicheng* (China's Agenda 21).

Zhongguo Huanjing Nianjian Bianji weiyuanhui (ed. 1999, China Environmental Yearbook Editing Committee), *Zhongguo Huanjing Nianjian 1999* (China Environmental Yearbook), Zhongguo huanjing nianjian she (China Environmental Yearbook Press), Beijing.

– (ed. 1998), *Zhongguo Huanjing Nianjian 1998* (China Environmental Yearbook), Zhongguo huanjing nianjian she (China Environmental Yearbook Press), Beijing.

– (ed. 1997), *Zhongguo Huanjing Nianjian 1997* (China Environmental Yearbook), Zhongguo huanjing nianjian she (China Environmental Yearbook Press), Beijing.

– (ed. 1996), *Zhongguo Huanjing Nianjian 1996* (China Environmental Yearbook), Zhongguo huanjing nianjian she (China Environmental Yearbook Press), Beijing.

– (ed. 1995), *Zhongguo Huanjing Nianjian 1995* (China Environmental Yearbook), Zhongguo huanjing nianjian she (China Environmental Yearbook Press), Beijing.
– (ed. 1992), *Zhongguo Huanjing Nianjian 1992* (China Environmental Yearbook), Zhongguo huanjing kexue chubanshe (China Environmental Sciences Press), Beijing.
– (ed. 1990), *Zhongguo Huanjing Nianjian 1990* (China Environmental Yearbook), Zhongguo huanjing kexue chubanshe (China Environmental Sciences Press), Beijing.
Zhongguo Kexueyuan (1990), Sanxia gongcheng shengtai yu huanjing keyan xiangmuzu, *Changjiang Sanxia Shengtai yuhuanjing dituji*, Kexue chubanshe, Beijing.
Zhongguo Sanxia jianshe nianjian bianjibu (China Three Gorges Construction Yearbook Editing Department, 2000), *Zhongguo Sanxia Jianshe Nianjian 2000* (the China Three Gorges Construction Yearbook 2000), China Three Gorges Construction Yearbook Press, Yichang.
– (1999), *Zhongguo Sanxia Jianshe Nianjian 1999* (the China Three Gorges Construction Yearbook 1999), China Three Gorges Construction Yearbook Press, Yichang.
– (1998), *Zhongguo Sanxia Jianshe Nianjian 1998* (the China Three Gorges Construction Yearbook 1998), China Three Gorges Construction Yearbook Press, Yichang.
– (1997a), *Zhongguo Sanxia Jianshe Nianjian 1997,* (the China Three Gorges Construction Yearbook 1997), China Three Gorges Construction Yearbook Press, Yichang.
– (1997b), *Zhongguo Sanxia Jianshe Nianjian 1996* (the China Three Gorges Construction Yearbook 1996), China Three Gorges Construction Yearbook Press, Yichang.
– (1996), *Zhongguo Sanxia Jianshe Nianjian 1995* (the China Three Gorges Construction Yearbook 1995), China Three Gorges Press, Beijing.
Zhongguo Sanxia Jianshe Zazhishe (1996) *Zhongguo Sanxia Jianshe, Hedingben, Di 3 Juan, Zong 8-17 Qi.*
Zhonghua renmin gongheguo fensheng dituji (Collection of provincial maps of the Peoples' Republic of China, 1990), Zhongguo ditu chubanshe (China Map Publishing House).
Zhonghua renmin gongheguo guowuyuan ling (Decree of the PRC State Council), (25 February, 2001), *Changjiang Sanxia gongcheng jianshe yimin tiaoli* (The Resettlement Regulations of the Three Gorges Project), Di 299 hao (No. 299), Beijing.
– (19 August, 1993), *Changjiang Sanxia gongcheng jianshe yimin tiaoli* (The Resettlement Regulations of the Three Gorges Project), Di 126 hao (No. 126), Beijing.
Zhonghua renmin gongheguo tudi guanlifa (the Land Administration Law of the People's Republic of China) (1998), Falü chubanshe (The Law Press).
Zhu, Nong and Zhao, Shihua (eds. 1996), *Sanxia gongcheng yimin yu kuqu fazhan yanjiu* (Research on the Three Gorges resettlement and reservoir development), Wuhan daxue chubanshe (Wuhan University Publishing House), Wuhan.

Newspaper Articles

Benbao pinglunyuan (Commentator of the paper) (24 February, 1999), 'Zhuahao zhiliang, zhong zai luoshi' (Focus on quality, put emphasis on implementation), *Renmin ribao* (People's Daily).
Chan, Vivien Pik-Kwan (27 June, 2001), 'Beijing orders purge of media', *South China Morning Post.*
Chen, Jie (15 September, 2000), 'Love Nature and Create a Green Culture', *China Daily.*

China Daily (17 March, 2001), 'Law on resettlement' under the column 'What they are saying', (excerpts from speeches by NPC deputies).

– (22 November, 1999), 'Hotline for pollution problems'.

Cong, Yaping, (24 January, 2000), 'Sanxia yimin zijin cheng shenji zhongdian' (The Three Gorges resettlement funding becomes the auditing focus), *Beijing Qingnianbao* (Beijing Youth Daily).

Dong, Jianqin (18 November, 1999), 'Sanxia yimin waiqian panzi qiaoding, qi wan ren jiang qianru Chuan Hu deng shi yi shengshi anzhi' (The Three Gorges population to be moved out is fixed, 70,000 people will be resettled to Sichuan, Shanghai, altogether 11 provinces and cities), *Renminribao* (People's Daily).

Fu, Jing (8 February, 2003), '480,000 to make way for dam', *China Daily*.

Guowuyuan bangongting (the State Council General Office) (24 February, 1999), 'Jiaqiang jichu sheshi gongcheng zhiliang guanli, Guoban fachu tongzhi yaoqiu' (Strengthen quality management of basic installation projects; The General Office of the State Council issues Circular and demands), *Renmin ribao* (People's Daily).

Li, Jianxing (21 December, 1998), 'Quan guo shenji gongzuo huiyi zhuanchi xinxi: Shenji wei guojia zengshou jiezhi er bai duo yi yuan (The National Auditing Working Conference distribute information: Auditing increases income and saves expenses amounting to more than 20 billion yuan for the country)', *Renmin ribao* (People's Daily).

Li, Peng (11 January, 2000), 'Jianshe hao Sanxia erqi gongcheng, zuohao yimin gongzuo' (Do the construction of the Three Gorges project's second phase well, do the resettlement work well), *Zhongguo huanjing bao* (China Environmental News).

Liang, Chao (12 March, 1999), 'Environmental awareness high for urbanites', *China Daily*.

Lu, Yongjian and Jiang, Fu (24 May, 1999), 'Zhu Rongji zai Sanxia gongcheng yimin gongzuo huiyi shang qiangdiao tuokuan yimin anzhi menlu, jiada qiye tiaozheng lidu, quebao Sanxia gongcheng erqi yimin renwu yuanman wancheng' (Zhu Rongji emphasises the expansion of resettlement ways, increase of dynamics in the adjustment of the enterprises and to ensure the success in the second phase of the Three Gorges project resettlement), *Renmin ribao* (People's Daily).

Macleod, Calum and Macleod, Lijia (1 July, 2001), 'Flooded dreams in China', *The Washington Times*.

Qian, Jiang (24 February, 1999), 'Sanxia gongcheng women jixu guanzhu ni' (Three Gorges project, we are continuing to pay attention to you), *Renmin ribao* (People's Daily).

Renmin ribao (31 December, 1998, People's Daily), 'Zhu Rongji kaocha Sanxia gongcheng he kuqu shi yaoqiu: Quebao gongcheng zhiliang, tuoshan anzhi yimin jianding buyi ba Sanxia gongcheng jianshe hao' (Zhu Rongji visits the Three Gorges project; Ensure the quality of the project, resettle the people in an appropriate way and resolutely construct the Three Gorges project).

Reuters (19 December, 1992), 'No plans to move people along the Yangtze to remote border regions', *China News*.

Wang, Cujin (17 February, 2000), Sanxia kuqu kechixu fazhan, Guowuyuan wanshan Sanxia yimin zhengce (Promote sustainable development in the Thre Gorges reservoir area; The State Council improves the Three Gorges resettlement policy), *Renmin ribao haiwai ban* (People's Daily Overseas Edition).

Wang, Jinfu (31 December, 1998), 'Quebao gongcheng zhiliang tuoshan anzhi yimin jianding bu yi ba Sanxia gongcheng jianshe hao, Zhu Rongji kaocha Sanxia gongcheng he kuqu shi yaoqiu' (Assure the quality of the project, resettle the relocatees in a proper way, firmly complete the construction of the Three Gorges project), *Renmin ribao* (People's Daily).

Xinhua News Agency (8 January, 1992), 'Vice-governors of affected provinces comment on the Three Gorges project'.

Yu, Yilei (4 September, 2000), 'Greener Beijing is to greet great Olympics', *China Daily.*

Zhang, Shou (2 November, 2000), 'Sanxia wuran zhide zhongzhi' (The pollution of the Three Gorges deserves to be taken seriously), *Renmin ribao haiwai ban* (People's Daily Overseas Edition).

Zhang, Zhikui (31 October 2000), 'Zou Jiahua yaoqiu jiaqiang Sanxia kuqu shengtai huanjing baohu' (Zou Jiahua demands strengthening of ecological and environmental protection in the Three Gorges reservoir area), *Zhongguo Huanjing Bao* (China Environmental News).

Zhongguo Sanxia Gongchengbao (9 November 1997), 'Zai Sanxia Gongcheng Jieliu Yishi shang de Jianghua', No. 324 (afternoon edition), p. 1.

– (13 November 1997), 'Zai Jie Zai Li Yingjie Tiaozhan; Fang Guowuyuan Sanxia JianWeiYiminkaifaju Juzhang Qi Lin', No 325.

Articles and Reports from Internet and Subscription Lists

Asian Development Bank (ADB, 2001a), 'Workshop: Dams and Development', 19-20 February, www.adb.org/Documents/Events/2001/Dams_Devt/dams_development.asp.

– (ADB, 2001b), 'ADB's Ongoing and Planned Responses to the World Commission on Dams' Strategic Priorities, Best Practicres, and Institutional Responses', ADB Draft, July, www.adb.org/Documents/Events/2001/Dams_Devt/adb_response.pdf.

Associated Press (11 November, 1999), 'Fresh Hope as WTO Talks Open', *South China Morning Post*, www.scmp.com/News/Front/Ar...p_ArticleID-19991111135254773.asp.

Becker, Jasper (21 March, 2001), Three Gorges petitioners 'held by police', *South China Morning Post,* http://china.scmp.com/today/ZZZE4DOJDKC.html.

Behn, Sharon (6 April, 2000), 'Environmental Human Rights Activists Target Banks', Three Gorges Campaign highlights Discover card, Agence France Presse, http://irn.org/programs/threeg/000406.discover.html.

Bejing Qingnianbao (17 November, 2000, Beijing Youth Daily via Xinhua), Sanxia gongcheng xushui hou hui bu hui youfa dizhen? Zhuanjia renwei zhenji bu hui hen da, shang xian yinggai zai li shi 5.5 ji zuoyou (Will the Three Gorges dam project induce earth quakes? The experts belive that the magnitude will not be very big, the upper limit will be appr. 5.5 degrees), www.bjyouth.com.cn/Bqb/20001117/GB/4435^D1117B0615.htm.

Beijing Wanbao (6 December, 2000) 'Qintun nuoyong de Sanxia yimin zijin yi you 95 % bei shouhui' (95 percent of the embezzled and diverted Three Gorges resettlement funds have been recovered), www.sina.com.cn; http://dailynews.sina.com.cn/c/155149.html.

Cai, Min (3 January, 2001), 'Chongqing jiang zai 10 nian nei tou 400 yi ju zi baohu Sanxia kuqu huanjing' (Chongqing will make a huge investment of more than 40 billion to protect the environment in the Three Gorges area in the next ten years), http://dailynews.sina.com.cn/c/165039.html; www.sina.com.cn, Xinhuawang.

China Daily, (22 November, 2001) 'Eco-balance bottom line to develop west', www1.chinadaily.com.cn/cndy/2001-11-22/44756.html.

– (29 May, 2001), 'Environment plan outlined' www1.chinadaily.com.cn/news/cb/2001-05-29/9985.html.

– (16 July, 2000), 'Massive 15-year hydropower programme planned', http://search.chinadaily.com.cn/isearch/i_textinfo.exe?dbname=cndy_printedition&listid=12962&selectword=STATE%20POWER%20CORPORATION.

China Daily Hong Kong Edition, (13 November, 2001), 'State Environmental Protection Administration', http://www1.chinadaily.com.cn/hk/2001-11-13/43415.html.

Conachy, James, (1 February, 2000), 'Thousands of officials punished in China's anti-corruption purge', www.wsws.org/articles/2000/feb2000/chin-f01.shtml.

Dorn, James (21 September, 2000), 'State-Owned Enterprises Continue to Hinder Chinese Growth', www.cato.org/dailys/09-21-00.html.

Economic Restructuring Office of the State Council (SERO), www.chinaonline.com/refer/ministry_profiles/c00121168.asp.

Fang, Ning (13 September, 2001), *Sanxia yimin renquan baozhang yanjiu zai Jing qidong* (A study to ensure the human rights in the Three Gorges resettlement was initated in Beijing), http://202.84.17..../Detail.wct?RecID=61&SelectID=1&ChannelID=4255&Page=, www.xinhuanet.com, *Xinhuawang* (Xinhua net, China News Agency), Beijing.

Fei, Weiwei (11 November, 2001), 'Sanxia gongcheng jinzhan ruhe; Fang guowuyuan Sanxia jianshe weiyuanhui fuzhuren Guo Shuyan' (What is the progress of the Three Gorges project? Interview with TGPCC vice-chairman Guo Shuyan), *Renmin ribao* (People's Daily), www.peopledaily.com.cn/GB/jinji/31/179/20010611/485816.html.

Guangzhou ribao, (1 February, 2000, Guangzhou Daily), Sanxia yimin qianxiao hai que 11 yi (RMB 1.1 billion still lacking to resettle schools), from the *Zhongguo qingnianbao* (China Youth Daily). www.gzdaily.com/Class../....0000201/GB/dyw^58^1^zj4000237.htm.

He, Ping (14 January, 2001), 'Sanxia kuqu yimin zongtishang dadao guihua jindu yaoqiu', *Zhong Xin She* (China News Service), www.sina.com.cn.

Kamarck, Martin A. (Thursday, 30 May, 1996), Statement of the Board of Directors of the Export-Import Bank of the United States, by Martin A. Kamarck, President and Chairman, at a Three Gorges Press Briefing, www.exim.gov/press/may29.html and www.exim.gov/press/press2.html.

Knight, Danielle (31 May, 2001), 'Activists warn investors about banks of the Yangtze', www.probeinternational.org/pi/3g/index.cfm?DSP=content&ContentID=2147.

Li, Shijie (30 October, 2000), Sanxia kuqu huanjing baohu pozai meijie (The need for environmental protection in the Three Gorges area is extremely urgent), *Guangming ribao* (Enlightenment Daily), www.gmw.com.cn/0_gm/2000/10/20001030/GB/10^18589^0^GMB3-313.htm.

Liu, Xiaoqing (9 March, 2000), 'Gan Yuping: Chongqing jiang yancheng pinwu nuoyong yimin zijinzhe' (Gan Yuping: Chongqing will punish severly people who engage in corruption and diversion of resettlement funds), Zhongguo Xinwen she (*China News Agency*), http://dailynews.sina.com.cn/china/2000-3-9/70116.html.

Lu, Youmei, www.china3gorges.com/HTML/Information/hottalks/inht0708.htm.

Luoyang xinxi gang (26 October, 2000, Luoyang News), Sanxia gongcheng jianshe ji kuqu yimin gongzuo qingkuang xinwen fabuhui juxing (Press conference on the state of affairs regarding the Three Gorges project construction and reservoir resettelement), www.lyinfo.ha.cn/lynews/00-10-26/China/13.htm.

People's Daily (4 August, 1999), Commentator, 'Continued Efforts for Reforms by State-Owned Enterprises Necessary', http://english.peopledaily.com.cn/199908/04/enc_19990804001011_TopNews.html.

Probe International (5 November, 2001), 'Behind the dark curtains: Exclusive report on the Three Gorges resettlement', *Three Gorges probe* www.probeinternational.org/pi/3g/index.cfm?DSP=content&ContentID=2797.

– (11 June, 2001), 'Top editors dismissed from daring newspaper', *Three Gorges Probe*, www.probeinternational.org/pi/index.cfm?DSP=content&ContentID=2175.
– (20 April, 2001), 'Three Gorges migrants protest detentions, unfair treatment', *Three Gorges Probe*, www.probeinternational.org/pi/index.cfm?DSP=content&ContentID=2005.
– (29 March, 2001), 'Jianping Sanxia yimin tiaoli de xiuding' or 'Comments on the revised Resettlement rules and regulations of the Three Gorges Dam' (both in Chinese and English), by Wei, Yi, *Three Gorges Probe*, www.probeinternational.org/pi/3g/index.cfm?DSP=content&ContentID=1914.
– (23 March, 2001), 'Three Gorges Dam Petitioners abducted' A Three Gorges Probe exclusive, by Wang Yusheng, *Three Gorges Probe*, www.probeinternational.org/pi/index.cfm?DSP=content&ContentID=1894.
– (14 February, 2001), Probe International's Three Gorges Dam Campaign, Confidential documents: Correspondence between Three Gorges officials and Chinese government, *Three Gorges Probe*, www.probeinternational.org/pi/3g/index.cfm?DSP=content&ContentID=1716.
– (15 January, 2001), 'Three Gorges Dam Protesters Beaten, Town Held Under Guard', *Three Gorges Probe*, www.probeinternational.org/pi/print.cfm?ContentID=1606.
– (14 November, 2000), 'The global demise of dams', www.probeinternational.org/pi/index.cfm?DSP=content&ContentID=1394. www.probeinternational.org/pi/index.cfm?DSP=content&ContentID=1396.
– (17 April, 2000), *Official Response to Experts' Three Gorges Dam Petition*, www.probeinternational.org/probeint/ThreeGorges/reply.html.
– (3 March, 2000), Urgent appeal that the Three Gorges project should be operated at the initial retained water level of 156 metres in line with the National People's Congress's resolution in order to evaluate silt deposit and to reduce resettlement pressure, *Three Gorges Petition*. www.probeinternational.org/probeint/ThreeGorges/petition.html.
Qu, Guanjie (4 January, 2001), 'Xia jiang yong chun chao—Sanxia kuqu nongcun yimin jianwen' (A spring tide rises in the Gorges and river—information on the rural relocatees in the Three Gogres reservoir area), *Guangming ribao* (Enlightenment Daily), http://dailynews.sin.com.cn/c/165556.html.
Renmin ribao (4 June, 2001, People's Daily), 'Zhu Rongji zongli: Sanxia gongcheng chengbai de guanjian zaiyu zhiliang' (Premier Zhu Rongji: the quality of the Three Gorges project is the prerequisite for success), www.peopledaily.com.cn/GB/jinji/32/180/20010604/481254.html.
– (3 June, 1999), 'Sanxia yimin "doufu zha" gongcheng zeren ren bei pan xing' (The responsible people for the Three Gorges resettlement 'doufu scum' project have been sentenced). Wysiwyg://main.2/http:www.peopl…/newfiles/wzb_990603001021_2.html.
Renmin ribao haiwai ban (21 September, 2000, People's Daily Overseas Edition), 'Sanxia kuqu wuran yanzhong huanbao zhihou' (The environmental pollution in the Three Gorges reservoir area is serious and environmental protection is stagnant), 'Chongqing youguan lingdao tichu yanli piping' (Relevant leaders in Chongqing raise severe criticism).
– (31 January, 2000), 'Sanxia kuqu yimin zijin bei nuoyong jin wu yi yuan' (Close to RMB 500 million have been diverted from the resettlement funds of the Three Gorges project), www.people.daily.com.cn/haiwai/200001/31/newfiles/D102.html.
Reuters (18 October, 2001), 'Bejiing shuts popular media Net forum', *South China Morning Post*, Thursday. http://china.scmp.com/technology/ZZZIYMYXSSC.html.
– (2 March, 2001) 'Seven sentenced to die in China corruption case', www.cnn.com/2001/WORLD/asiapcf/east/03/02/China.execute/.

Shenzhen tequ bao (12 January, 2000, Shenzhen Special Zone Daily), 'Sanxia kuqu yimin jindu jiakuai' (The Three Gorges reservoir resettlement has quickened its pace), www.szszd.com.cn/khcd/2000/0112/newsfile/n7-9.htm.

Sullivan, Lawrence (summer 1996), 'Upheaval on the Yangzi: Population relocation & the controversy over the Three Gorges Dam', *China Rights Forum*, www.igc.apc.org/hric/crf/english/96summer/e7.html.

Turner, Jennifer L. (27 January, 2003), Clearing the Air: Human Rights and the Legal Dimensions of China's Environmental Dilemma, Congressional/Executive Commission on China Issues Roundtable: The Growing Role of Chinese Green NGOs and Environmental Journalists in China, Woodrow Wilson Center's Environmental Change and Security Project, China Environment Forum, http://wwics.si.edu/index.cfm?topic_id =1421&fuseaction=topics.item&news_id=20437.

UNEP (United Nations Environment Programme, 12 October, 2001), 'One of the worlds' most famous rivers, the Yangtze, set for pioneering flood reduction plan', Relief web. www.reliefweb.int/w/rwb.nsf/s/8430673EE852620185256AE3005F9D58.

U.S. Embassy (2001a), Environment, Science and Technology Section, Ninth Five-Year Plan Environmental Report Card, A March Report, www.usembassy-china.org.cn/english/sandt/9th5yearplan.htm.

– (2001b), Environment, Science and Technology Section, 'China's Year 2000 "State Of The Environment" Report', A June Report, www.usembassy-china.org.cn/english/sandt/SOTE4web.htm.

– (2001c), Environment, Science and Technology Section, Beijing Environment, Science and Technology Update, 'State of the Environment' Report, 15 June, www.usembassy-china.org.cn/english/sandt/estnews0615.htm.

– (2001d), Environment, Science and Technology Section, Beijing Environment, Science and Technology Update, 'Environment a Top Public Priority', 1 June, www.usembassy-china.org.cn/english/sandt/estnews0601.htm.

– (2001e), Environment, Science and Technology Section, Beijing Environment, Science and Technology Update, 'In Brief', 16 November, www.usembassy-china.org.cn/english/sandt/estnews111601.htm.

– (2000a), Environment, Science and Technology Section, 'China Revises its Air Pollution Law', A June Report, www.usembassy-china.org.cn/english/sandt/Clean airlaw.htm.

– (2000b), Environment, Science and Technology Section, 'Birth of an NGO? Development of Grassroots Organisations in the Land of Big Brother', A March 2000 Report, www.usembassy-china.org.cn/english/sandt/ngobirthweb.htm.

– (2000c), Environment, Science and Technology Section, 'Environmental Objectives and Investment Requirements for China's 10th Five-Year Plan' A November Report, www.usembassy-china.org.cn/english/sandt/10FYP.htm.

– (2000d), Environment, Science and Technology Section, Environmental Objectives and Investment Requirements for China's 10th Five-Year Plan, A November Report, www.usembassy-china.org.cn/english/sandt/10FYP.htm.

Wang, Diao (10 March, 2000), 'Youguan renshi chengjin ba cheng bei nuoyong Sanxia yimin zijin zhenggai daowei' (Regarding the embezzlements of the Three Gorges resettlement funding – rectification and reform has retrieved funding in 8 towns), *Nanfang Dushibao* (Southern Metropolis Daily) http://dailynews.sina.com.cn/china/2000-3-10/70325.html.

Wang, Xiangwei (23 February, 1999), 'Dam company in float plan', *South China Morning Post*, www.scmp.com/news/template/...vTemp=Default.htx//&PrevMFS=1289//.

Wenhuibao (12 February, 2001) Nisha yu ji yu yanzhong dangxin Changjiang cheng 'dishang he'(The more silt that amasses, the more serious raises concern that the Yangtze River will become 'an elevated river'), http://202.84.17..../Detail.wct?RecID=32&SelectID=4&ChannelID=4255&Page=.

World Bank (2001a) See main bibliography.

– (2001b), Operational Policies 4.12, Involuntary Resettlement, *The World Bank Operational Manual*, December. http://wbln0018.worldbank.org/Institutional/Manuals/OpManual.nsf/0/CA2D01A4D1BDF58085256B19008197F6?OpenDocument.

– (2001c), Operational Policies 4.12 Annex A, Involuntary Resettlement Instruments, *The World Bank Operational Manual*, December. http://wbln0018.worldbank.org/Institutional/Manuals/OpManual.nsf/0/46FC304892280AB785256B19008197F8?OpenDocument.

– (2001d), Bank Procedures 4.12, Involuntary Resettlement, *The World Bank Operational Manual*, December. http://wbln0018.worldbank.org/Institutional/Manuals/OpManual.nsf/0/19036F316CAFA52685256B190080B90A?OpenDocument.

– (2001e), 'World Bank's Revised Policy on Involuntary Resettlement (OP/BP 4.12)'. http://lnweb18.worldbank.org/ESSD/sdvext.nsf/65ByDocName/Policy.

World Commission on Dams, (26 January, 2001), 'Development Banks Announce Initiatives in Response to WCD Report', Press Release www.dams.org/press/pressrelease_84.htm.

Wu, Ming (1998), *Resettlement Problems of the Three Gorges Dam, A Field Report by Wu Ming,*a joint report by International Rivers Network and Human Rights in China, www.igc.apc.org/hric/reports/3gorges.html or http://irn.org/programs/threeg/resettle.html.

Xiamen Daily (13 March, 2000), 'Jianshe Changjiang shangyou jingji zhongxin, baozheng Sanxia yimin shunli wancheng' (Establish an economic centre in the upper reaches of the Yangtze River; Guarantee successful completion of the Three Gorges resettlement), http://dailynews.sina.com.cn/china/2000-3-13/71261.html.

Xinhua she (9 March, 2001, Xinhua News Agency), 'Qu Geping shuo, Sanxia gongcheng yijing kaolü huanbao wenti' (Qu Geping says environmental protection problems have already been considered for the Three Gorges project), www.xinhuanet.com. http://202.84.17.7.../Detail.wct?RecID=4&ChannelID=4255&Page=.

– (7 March, 2001), Zhongguo chenggong pojie Sanxia yimin gongzuo zhong de 'nanzhong zhi nan' (China has successfully analysed and explained the most difficult part of the resettlement work), www.xinhuanet.com, http://202.84.17.7.../Detail.wct?RecID=5&SelectID=1&ChannelID=4255&Page=.

– (22 February, 2000), 'Chongqing shi renzhen xiqu jiaoxun jiaqang Sanxia yimin zijin guanli' (Chongqing municipality draws a lesson in earnest in order to strengthen the management of the Three Gorges resettlement funding), http://202.84.17.11/chinese/2221020292.htm.

Xinhua wang (25 February, 2001, Xinhua net), 'Guowuyuan tongguo Changjiang Sanxia gongcheng jianshe yimin tiaoli (fu quanwen)' ('The State Council adopts the Yangzte River Three Gorges project construction and resettlement regulations', full text attached), http://dailynews.sina.com.cn/c/193764.html.

Xu, Yongheng (10 December, 2000), 'Sanxia gongcheng jianshe weiyuanhui yiminju juzhang cheng Sanxia yimin zijin zhong you "san wei"xianxiang', (The Director of the Resettlement Bureau under the Three Gorges Project Construction Committee states that the phenomenon of "three violations" exists in the Three Gorges resettlement funding) *Zhongguo qingnian bao* (ChinaYouth Daily) www.peopledaily.com.cn/GB/channel4/27/20001210/343763.html.

Yangcheng Wanbao (31 October, 2000, *Yangcheng Evening News*), 'Huanjing chengwei Zhongguo gongzhong zui guanzhu de shehui wenti' (Environment has become the social issue that receives most concern by the public), URL: url=www.ycwb.com www.e23.com.cn/asp/third.a...5.a&ondate=00-10-31.

Zhang, Guodong and Dehui Luo (23 January, 2001), 'Sanxia gongcheng jinzhan shunli; jin nian jiang quanmian zhankai gongjianzhan' (The Three Gorges project progresses smoothly, this year the overall assualt will be launched), http://dailynews.sina.com.cn /c/174794.html.

Zhongguo xinwenshe (25 November, 1999, China News Agency), Zhou Tienong shuai Quanguo zhengxie weiyuan shicha sanxia yimin gongzuo (Zhou Tienong leads CPPCC group to inspect the Three Gorges resettlement work, http://dailynews.sina.com.cn/ china/1999-11-25/35295.html.

Zhongguo xinwenshe (16 February, 2001, China News Agency), Guowuyuan changwu huiyi tongguo Sanxia gongcheng yimin tiaoli caoan (State Council's General office passes/adopts draft for Three Gorges project resettlement regulation).

– (26 September, 2000), 'Sanxia gongcheng zonggongsi zong jingli: Sanxia juebuhui biancheng yi tan wushui'(The General Manager of the The Three Gorges projcet Development Corporation: the Three Gorges will definitely not become a pool of polluted water), http://dailynews.sina.com.cn/china/2000-09-26/130598.html.

– (17 September, 2000), 'Sanxia kuqu yi chenggong banqian anzhi yimin ershi si wan duo ren' (More than 240.000 people have been successfully resettled in the reservoir area), www.chinanews.com.cn/2000-09-17/26/46851.html.

– (23 February, 2000), 'Sanxia kuqu Zigui xian yimin tupo liuwan ren' (Zigui county in the Three Gorges reservoir area reaches a breakthrough with 60,000 resettled), http://dailynews.sina.com.cn/china/2000-2-23/64367.html.

Zhonghua renmin gongheguo guowuyuan ling (15 February, 2001, Decree of the PRC State Council) No. 299, *Changjiang Sanxia gongcheng jianshe yimin tiaoli* (The Resettlement Regulations of the Three Gorges Project), Beijing. http://dailynews.sina.com.cn/c/ 193764.html.

Zhongxin Chongqing wang (18 September, 2000, China News Chongqing site), 'Sanxia kuqu huanbao shoudao yanli piping' (Environmental protection in Three Gorges reservoir is seriously criticised). www.cq.chinanews.com.cn, wysiwyg://36/www.cq. chinanews.com.cn/2000-09-18/1/3486.html.

Zhongxin Sichuanwang (24 August, 2001, China News Agency, Sichuan net), 'Ertan fadian liang nian kuisun 12 yi' (Ertan has experienced a loss of 1.2 billion [yuan] after two years of power production). www.peopledaily.com.cn/GB/channel4/988/20000824/ 200915.html.

Zhou, Jamila (7 November, 2001), 'Mainland looks at ways of controlling Internet', *South China Morning Post*, http://china.scmp.com/technology/ZZZOR4XX2TC.html.

'Zhuanjia huyu ba huanjing naru guomin jingji hesuan tixi' (31 October 2000) (Experts appeal to include environemnt into calculation system of the national economy), http//:www.yestock.com, wysiwyg://main.9/http://news.yestock.com//display.asp?WDL SH=329018.

Index

For Product Safety Concerns and Information please contact our
EU representative GPSR@taylorandfrancis.com Taylor & Francis
Verlag GmbH, Kaufingerstraße 24, 80331 München, Germany